Lecture Notes in Statistics 190

Edited by P. Bickel, P. Diggle, S. Fienberg, U. Gather,
I. Olkin, S. Zeger

Jérôme Dedecker
Paul Doukhan
Gabriel Lang
José Rafael León R.
Sana Louhichi
Clémentine Prieur

Weak Dependence: With Examples and Applications

Springer

Jérôme Dedecker
Laboratoire de Statistique, Théorique et Appliquée
Université Paris 6
175 rue de Chevaleret
75013 Paris
France
dedecker@ccr.jussieu.fr

Gabriel Lang
Equipe MORSE UMR MIA518 INRA
Agro Paris Tech, F-75005 Paris
France
lang@engref.fr

Sana Louhichi
Laboratoire de Mathématiques
Université Paris-Sud–Bât 425
914058 Orsay Cedex
France
sana.louhichi@math.u-psud.fr

Clémentine Prieur
Génie Mathématique et Modélisation
Laboratoire de Statistique et Probabilités
INSA Toulouse
135 avenue de Rangueil
31077 Toulouse Cedex 4
France
prieur@cict.fr

Paul Doukhan
ENSAE, Laboratoire de Statistique du CREST (Paris Tech)
3 avenue Pierre Larousse
92240 Malakoff
France
and
SAMOS-MATISSE
(Statistique Appliquée et Modélisation Stochastique)
Centre d'Economie de la Sorbonne
Université Paris 1-Panthéon-Sorbonne
CNRS
90, rue de Tolbiac
75634-Paris Cedex 13
France
paul.doukhan@ensae.fr

José Rafael León R.
Escuela de Matematica
Universidad Central de Venezuela
Ap. 47197 Los Chaguaramos
Caracas 1041-A
Venezuela
jleon@euler.ciens.ucv.ve

ISBN 978-0-387-69951-6 ISBN 978-0-387-69952-3 (eBook)
DOI 10.1007/978-0-387-69952-3

Library of Congress Control Number: 2007929938

Printed on acid-free paper.

9 8 7 6 5 4 3 2 1

springer.com

Preface

Time series and random fields are main topics in modern statistical techniques. They are essential for applications where randomness plays an important role. Indeed, physical constraints mean that serious modelling cannot be done using only independent sequences. This is a real problem because asymptotic properties are not always known in this case.

The present work is devoted to providing a framework for the commonly used time series. In order to validate the main statistics, one needs rigorous limit theorems. In the field of probability theory, asymptotic behavior of sums may or may not be analogous to those of independent sequences. We are involved with this first case in this book.

Very sharp results have been proved for mixing processes, a notion introduced by Murray Rosenblatt [166]. Extensive discussions of this topic may be found in his *Dependence in Probability and Statistics* (a monograph published by Birkhaüser in 1986) and in Doukhan (1994) [61], and the sharpest results may be found in Rio (2000) [161]. However, a counterexample of a really simple non-mixing process was exhibited by Andrews (1984) [2]. The notion of weak dependence discussed here takes real account of the available models, which are discussed extensively. Our idea is that robustness of the limit theorems with respect to the model should be taken into account. In real applications, nobody may assert, for example, the existence of a density for the inputs in a certain model, while such assumptions are always needed when dealing with mixing concepts. Our main task here is not only to provide the reader with the sharpest possible results, but, as statisticians, we need the largest possible framework. Doukhan and Louhichi (1999) [67] introduced a wide dependence framework that turns out to apply to the models used most often. Their simple way of weakening the independence property is mainly adapted to work with stationary sequences.

We thus discuss examples of weakly dependent models, limit theory for such sequences, and applications. The notions are mainly divided into the two following classes:

- The first class is that of "Causal" dependence. In this case, the conditions may also be expressed in terms of conditional expectations, and thus the powerful martingale theory tools apply, such as Gordin's [97] device that allowed Dedecker and Doukhan (2003) [43] to derive a sharp Donsker principle.

- The second class is that of noncausal processes such as two-sided linear processes for which specific techniques need to be developed. Moment inequalities are a main tool in this context.

In order to make this book useful to practitioners, we also develop some applications in the fields of Statistics, Stochastic Algorithms, Resampling, and Econometry. We also think that it is good to present here the notation for the concepts of weak dependence. Our aim in this book was to make it simple to read, and thus the mathematical level needed has been set as low as possible. The book may be used in different ways:

- First, this is a mathematical textbook aimed at fixing the notions in the area discussed. We do not intend to cover all the topics, but the book may be considered an introduction to weak dependence.

- Second, our main objective in this monograph is to propose models and tools for practitioners; hence the sections devoted to examples are really extensive.

- Finally, some of the applications already developed are also quoted for completeness.

A preliminary version of this joint book on weak dependence concepts was used in a course given by Paul Doukhan to the Latino Americana Escuela de Matemática in Merida (Venezuela). It was especially useful for the preparation of our manuscript that a graduate course in Merida (Venezuela) in September 2004 on this subject was based on these notes. The different contributors and authors of the present monograph participated in developing it jointly. We also want to thank the various coauthors of (published or not yet published) papers on the subject, namely Patrick Ango Nzé (Lille 3), Jean-Marc Bardet (Université Paris 1), Odile Brandière (Orsay), Alain Latour (Grenoble), Hélène Madre (Grenoble), Michael Neumann (Iena), Nicolas Ragache (INSEE), Mathieu Rosenbaum (Marne la Vallée), Gilles Teyssière (Göteborg), Lionel Truquet (Université Paris 1), Pablo Winant (ENS Lyon), Olivier Wintenberger (Université Paris 1), and Bernard Ycart (Grenoble). Even if all their work did not appear in those notes, they were really helpful for their conception. We also want to thank the various referees who provided us with helpful comments either for this monograph or for papers submitted for publication and related to weak dependence.

We now give some clarification concerning the origin of this notion of weak dependence. The seminal paper [67] was in fact submitted in 1996 and was part of the PhD dissertation of Sana Louhichi in 1998. The main tool developed in this work was combinatorial moment inequalities; analogous moment inequalities are also given in Bakhtin and Bulinski (1997) [8]. Another close definition of weak dependence was provided in a preprint by Bickel and Bühlmann (1995) [17] *anterior to* [67], also published in 1999 [18]. However, those authors aimed to work with the bootstrap; see Chapter 13 and Section 2.2 in [6]. The approach of Wu (2005) [188] detailed in Remark 3.1, based on $\mathbb{L}^2$-conditions for causal Bernoulli shifts, also yields interesting and sharp results.

This monograph is essentially built in four parts:

Definitions and models
In the first chapter, we make precise some issues and tools for investigating dependence: this is a *motivational* chapter. The second chapter introduces formally the notion of *weak dependence*. *Models* are then presented in a long third chapter. Indeed, in our mind, the richness of examples is at the core of the weak dependence properties.

Tools
Tools are given in two chapters (Chapters 4 and 5) concerned respectively with *noncausal* and *causal* properties. Tools are first used in the text for proving the forthcoming limit theorems, but they are essential for any type of further application. Two main tools may be found: *moment* bounds and *coupling* arguments. We also present specific *tightness criteria* adapted to work out empirical limit theorems.

Limit theorems
Laws of large numbers (and some applications), *central limit theorems*, *invariance principles*, *laws of the iterated logarithm*, and *empirical central limit theorems* are useful limit theorems in probability. They are precisely stated and worked out within Chapters 6–10 .

Applications
The end of the monograph is dedicated to applications. We first present in Chapter 11 the properties of the standard *nonparametric techniques*. After this, we consider some issues of *spectral estimation* in Chapter 12. Finally, Chapter 13 is devoted to some miscellaneous applications, namely *applications to econometrics*, the *bootstrap*, and *subsampling* techniques.

After the table of contents, a useful short *list of notation* allows rapid access to the main weak dependence coefficients and some useful notation.

Jérôme Dedecker, Paul Doukhan, Gabriel Lang, José R. León,
Sana Louhichi, and Clémentine Prieur

Contents

List of notations

We recall here the main specific or unusual notation used throughout this monograph.
As usual, $\#A$ denotes the cardinal of a finite set A, and $\mathbb{N}, \mathbb{Z}, \mathbb{R}, \mathbb{C}$ are the standard sets of calculus. For clarity and homogeneity, we note in this list:

- $a \preceq b$ or $a = \mathcal{O}(b)$ means that $a \le Cb$ for a constant $C > 0$ and $a, b \ge 0$,
- $a \ll b$ or $a = o(b)$ as $b \to 0$, for $a, b \ge 0$, means $\lim_{b \to 0} a/(a+b) = 1$,
- $a \wedge b$, $a \vee b$ are the minimum and the maximum of the numbers $a, b \ge 0$,
- $\mathcal{M}, \mathcal{U}, \mathcal{V}, \mathcal{A}, \mathcal{B}$ are σ-algebras, and $(\Omega, \mathcal{A}, \mathbb{P})$ is a probability space,
- $X, Y, Z, \ldots$ denote random variables (usually ξ, ζ are inputs),
- $\mathbb{L}^p(E, \mathcal{E}, m)$ are classes of measurable functions: $\|f\|_p = \left(\int_E |f(x)|^p dm(x)\right)^{\frac{1}{p}} < \infty$,
- F is a cumulative distribution function,
  Φ is the normal cumulative distribution function, and $\phi = \Phi'$ is its density,
- n is a time (space) delay, $r, s \in \mathbb{N}$ are "past-future" parameters, p is a moment order, and $\mathcal{F}, \mathcal{G}$ are function spaces.

$\alpha(\mathcal{U}, \mathcal{V})$	p. 4,	§ 1.2, eqn. (1.2.1),	*mixing coefficient*
$\widetilde{\alpha}(\mathcal{M}, X)$	p. 16,	§ 2.2.3, def. 2.5-1,	*dependence coefficient*
$\widetilde{\alpha}_r(n)$	p. 19,	§ 2.2.3, def. 2.2.15,	*dependence sequence*
BV	p. 11,	§ 2.1,	*bounded variation spaces*
BV_1			
$\beta(\mathcal{U}, \mathcal{V})$	p. 4,	§ 1.2, eqn. (1.2.4),	*mixing coefficient*
$\widetilde{\beta}(\mathcal{M}, X)$	p. 16,	§ 2.2.3, def. 2.5-2,	*dependence coefficient*
$\widetilde{\beta}_r(n)$	p. 19,	§ 2.2.3, def .2.2.15,	*dependence sequence*
$C_{n,r}$	p. 73,	§ 4.3, def. 4.1,	*moment coefficient sequence*
$c_{X,r}(n)$	p. 87,	§ 4.4.1, eqn. (4.4.7),	*vector moment sequence*
$c^\star_{X,r}(n)$	p. 87,	§ 4.4.1, eqn. (4.4.8),	*maximum moment sequence*
$\gamma_p(\mathcal{M}, X)$	p. 19,	§ 2.2.4, def. 2.2.16,	*dependence coefficient*
$\gamma_p(n)$	p. 19,	§ 2.2.4, def. 2.2.16,	*dependence sequence*
δ_n	p. 25,	§ 3.1.2, def. 3.1.10,	*Bernoulli shift coefficient*
$\epsilon(X, Y)$	p. 11,	§ 2.2, eqn. (2.2.1),	*general coefficient*

$\epsilon(n)$	pp. 4–11,	§§ 1.1–2.2, eqn. (1.1.3),	*coefficient sequence*
$\eta(n)$	p. 12,	§ 2.2.1, eqn. (2.2.3),	*dependence sequence*
f^{-1}	p. 18,	§ 2.2.3, eqn. (2.2.14),	*inverse of a monotonic f*
$(\mathcal{F}, \mathcal{G}, \Psi)$	p. 11,	§ 2.2, eqn. (2.2.1),	*weak dependence*
$(\mathcal{F}, \Psi)$			
$\phi(\mathcal{U}, \mathcal{V})$	p. 4,	§ 1.2, eqn. (1.2.3),	*mixing coefficient*
$\widetilde{\phi}(\mathcal{M}, X)$	p. 16,	§ 2.2.3, def. 2.5-3,	*dependence coefficient*
$\widetilde{\phi}_r(n)$	p. 19,	§ 2.2.3, def. 2.2.15,	*dependence sequence*
$\mathcal{I}$	p. 67,	§ 4.1,	*class of indicators*
$\kappa_p(n)$	p. 88,	§ 4.4.1, def. 4.4.9,	*cumulant sequence*
$\kappa(n)$	p. 12,	§ 2.2.1, eqn. (2.2.5),	*dependence sequence*
$\mathcal{L}$	p. 38,	§ 3.3,	*Perron Frobenius operator*
$\mathbb{L}^p$-NED	p. 6,	§ 1.3, def. 1.2,	*NED notion*
$\Lambda(\delta), \Lambda^{(1)}(\delta)$	p. 10,	§ 2.1,	*Lipschitz spaces*
$\lambda(n)$	p. 12,	§ 2.2.1, eqn. (2.2.4),	*dependence sequence*
$\mu_X(n)$	p. 5,	§ 1.2, eqn. (1.2.5),	*μ-mixing sequence*
$\mu_{X,r,s}(n)$	p. 5,	§ 1.2, eqn. (1.2.6),	*μ-mixing field*
$\psi(n)$	p. 5,	§ 1.3,	*$\mathbb{L}^p$-mixingale coefficient*
$Q_X(\cdot)$	p. 74,	§ 4.3, eqn. (4.3.3),	*quantile function*
$\rho(\mathcal{U}, \mathcal{V})$	p. 4,	§ 1.2, eqn. (1.2.2),	*mixing coefficient*
$\theta(n)$	p. 14,	§ 2.2.2, eqn. (2.2.7),	*dependence sequence*
$\theta_p(\mathcal{M}, X)$	p. 15,	§ 2.2.2, eqn. (2.2.8),	*dependence coefficient*
$\theta_{p,r}(n)$	p. 15,	§ 2.2.2, eqn. (2.2.9),	*dependence sequence*
$\tau_p(\mathcal{M}, X)$	p. 16,	§ 2.2.2, eqn. (2.2.12),	*coupling coefficient*
$\tau_{p,r}(n)$	p. 16,	§ 2.2.2, eqn. (2.4),	*coupling sequence*
$\zeta(n)$	p. 12,	§ 2.2.1, eqn. (2.2.6),	*dependence sequence*
$\omega_{p,n}$	p. 22,	§ 3.1, eqn. (3.1.3),	*shift increment coefficient*

Chapter 1

Introduction

This chapter is aimed to justify some of our choices and to provide a basic background of the other competitive notions like those linked to mixing conditions. In our mind mixing notions are not related to time series but really to σ-algebras. They are consequently more adapted to work in areas like Finance where history, that is the σ-algebra generated by the past is of a considerable importance.
Having in view the most elementary ideas, Doukhan and Louhichi (1999) [67] introduced the more adapted weak dependence condition developped in this monograph. This definition makes explicit the asymptotic independence between 'past' and 'future'; this means that the 'past' is progressively forgotten. In terms of the initial time series, 'past' and 'future' are elementary events given through finite dimensional marginals. Roughly speaking, for convenient functions f and g, we shall assume that

$$\mathrm{Cov}\,(f(\text{'past'}), g(\text{'future'}))$$

is small when the distance between the 'past' and the 'future' is sufficiently large. Such inequalities are significant only if the distance between indices of the initial time series in the 'past' and 'future' terms grows to infinity. The convergence is not assumed to hold uniformly on the dimension of the 'past' or 'future' involved. Another direction to describe the asymptotic behavior of certain time series is based on projective methods. It will be proved that this is coherent with the previous items.
Sections in this chapter first provide general considerations on independence, then we define classical mixing coefficients, mixingales and association to conclude with simple counterexamples.

1.1 From independence to dependence

We recall here some very basic facts concerning independence of random variables. Let $\mathbf{P}, \mathbf{F}$ be random variables defined on the same probability space

$(\Omega, \mathcal{A}, \mathbb{P})$ and taking values in measurable spaces $(E_{\mathbf{P}}, \mathcal{E}_{\mathbf{P}})$ and $(E_{\mathbf{F}}, \mathcal{E}_{\mathbf{F}})$. Independence of both random variables $\mathbf{P}, \mathbf{F}$ writes

$$\mathbb{P}(X \in A, Y \in B) = \mathbb{P}(X \in A)\mathbb{P}(Y \in B), \qquad \forall (A, B) \in \mathcal{E}_{\mathbf{P}} \times \mathcal{E}_{\mathbf{F}} \tag{1.1.1}$$

extending this identity by linearity to stepwise constant functions and using limits yields a formulation of this relation which looks more adapted for applications:

$$\text{Cov}(f(\mathbf{P}), g(\mathbf{F})) = 0, \qquad \forall (f, g) \in \mathbb{L}^\infty(E_{\mathbf{P}}, \mathcal{E}_{\mathbf{P}}) \times \mathbb{L}^\infty(E_{\mathbf{F}}, \mathcal{E}_{\mathbf{F}})$$

where, for instance, $\mathbb{L}^\infty(E_{\mathbf{P}}, \mathcal{E}_{\mathbf{P}})$ denotes the subspace of $\ell^\infty(E_{\mathbf{P}}, \mathbb{R})$ (the space of bounded and real valued functions), of measurable and bounded function $f : (E_{\mathbf{P}}, \mathcal{E}_{\mathbf{P}}) \to (\mathbb{R}, \mathcal{B}_{\mathbb{R}})$.
If the spaces $E_{\mathbf{P}}, E_{\mathbf{F}}$ are topological spaces endowed with their Borel σ-algebras (the σ-algebra generated by open sets) then it is sufficient to state

$$\text{Cov}(f(\mathbf{P}), g(\mathbf{F})) = 0, \qquad \forall (f, g) \in \mathcal{P} \times \mathcal{F} \tag{1.1.2}$$

where $\mathcal{P}, \mathcal{F}$ are dense subsets of the spaces of continuous functions $E_{\mathbf{P}} \to \mathbb{R}$ and $E_{\mathbf{F}} \to \mathbb{R}$. In order to qualify a simple topology on both spaces it will be convenient to assume that $E_{\mathbf{P}}$ and $E_{\mathbf{F}}$ are locally compact topological spaces and the density in the space of continuous functions will thus refer to uniform convergence of compact subsets of $E_{\mathbf{P}}, E_{\mathbf{F}}$.
From a general principle of economy, we always should wonder about the smallest classes $\mathcal{P}, \mathcal{F}$ possible. The more intuitive (necessary) condition for independence is governed by the idea of orthogonality.

In this idea we now develop some simple examples for which, however

$$\textbf{Orthogonality} \quad \Longrightarrow \quad \textbf{Independence.}$$

- **Bernoulli trials**
  If $E_{\mathbf{P}} = E_{\mathbf{F}} = \{0, 1\}$ are both two-points spaces, the random variables now follow Bernoulli distributions and independence follows form the simple orthogonality of $\text{Cov}(\mathbf{P}, \mathbf{F}) = 0$. Indeed from the standard bilinearity properties of the covariance, we also have $\text{Cov}(1-\mathbf{P}, \mathbf{F}) = 0$, $\text{Cov}(\mathbf{P}, 1-\mathbf{F}) = 0$ and $\text{Cov}(1-\mathbf{P}, 1-\mathbf{F}) = 0$ which respectively means that eqn. (1.1.1) holds if $(A, B) = (\{a\}, \{b\})$ with $(a, b) = (0, 0)$ or respectively $(1, 0)$, $(0, 1)$ and $(1, 1)$. The problem of this example is that in this case the σ-algebras generated by $\mathbf{P}, \mathbf{F}$ are very poor and this example will thus not fit the forthcoming case of 'important' past and future.
- **Gaussian vectors**
  If now $E_{\mathbf{P}} = \mathbb{R}^p$ and $E_{\mathbf{F}} = \mathbb{R}^q$, then if the vector $Z = (\mathbf{P}, \mathbf{F}) \in \mathbb{R}^{p+q}$ is Gaussian then its distribution only depends on its second order properties*,

*This only means that Z's distribution depends only on the expressions $\mathbb{E}Z_i$ and $\mathbb{E}Z_iZ_j$ for $1 \le i, j \le d$ if $Z = (Z_1, \ldots, Z_d)$.

hence coordinate functions are enough to determine independence. In more precise words two random vectors $\mathbf{P} = (P_1, \ldots, P_p)$ and $\mathbf{F} = (F_1, \ldots, F_q)$ are independent if and only if $\mathrm{Cov}(P_i, F_j) = 0$ for any $1 \le i \le p$ and any $1 \le j \le q$. This is a very simple characterization of independence which clearly does not extend to any category of random variables.

For example, let $X, Y \in \mathbb{R}$ be independent and symmetric random variables such that $\mathbb{P}(X = 0) = 0$ if we set $\mathbf{P} = X$ and $\mathbf{F} = \mathrm{sign}(X)Y$ then

$$\mathrm{Cov}(\mathbf{P}, \mathbf{F}) = \mathbb{E}(|X|Y) - \mathbb{E}X \cdot \mathbb{E}\mathbf{F} = \mathbb{E}|X| \cdot \mathbb{E}Y - \mathbb{E}X \cdot \mathbb{E}\mathbf{F} = 0$$

because X and Y are centered random variables even if those variables are not independent if the support of Y's distribution contains more than two points.

A simple situation of uncorrelated individually standard Gaussian random variables which are not independent is provided† with the couple $(\mathbf{P}, \mathbf{F}) = (N, RN)$ where $N \sim \mathcal{N}(0, 1)$ and R is a Rademacher random variable (that means $\mathbb{P}(R = 1) = \mathbb{P}(R = -1) = \frac{1}{2}$) independent of N.

- **Associated vectors (cf. § 1.4)**

  Again, we assume that $E_{\mathbf{P}} = \mathbb{R}^p$ and $E_{\mathbf{F}} = \mathbb{R}^q$, then the random vector $X = (\mathbf{P}, \mathbf{F}) \in \mathbb{R}^{p+q} = \mathbb{R}^d$ is called associated in case

  $$\mathrm{Cov}(h(X), k(X)) \ge 0$$

  for any measurable couple of functions $h, k : \mathbb{R}^d \to \mathbb{R}$ such that both $\mathbb{E}(h^2(X) + k^2(X)) < \infty$ and the partial functions $x_j \mapsto h(x_1, \ldots, x_d)$ and $x_j \mapsto k(x_1, \ldots, x_d)$ are non-decreasing for any choice of the remaining coordinates $x_1, \ldots, x_{j-1}, x_{j+1}, \ldots, x_d \in \mathbb{R}$ and any $1 \le j \le d$. A essential property of such vectors is that here too, orthogonality implies independence. This will be developed in a forthcoming chapter.

  An example of associated vectors is that of independent coordinates. Even if it looks uninteresting case in our dependent setting, this leads to much more involved examples of associated random vectors through monotonic functions.

  The class of such coordinatewise increasing functions is a cone of $\mathbb{L}^2(X)$, the class of functions such that $\mathbb{E}h^2(X) < \infty$, hence the set of associated distributions looks like a (very thin) cone of the space of distributions on $\mathbb{R}^d$.

  The same idea applies to Gaussian distributions which is even finite dimensional in the large set of laws on $\mathbb{R}^d$.

If now, we consider a time series $X = (X_n)_{n \in \mathbb{Z}}$ with values in a locally compact topological space E (typically $E = \mathbb{R}^d$) we may consider one variable $\mathbf{P}$ of the past and one variable $\mathbf{F}$ of the future:

$$\mathbf{P} = (X_{i_1}, \ldots, X_{i_u}), \qquad \mathbf{F} = (X_{j_1}, \ldots, X_{j_v}),$$

†In this case both variables are indeed centered with Normal distributions and $\mathbb{E}\{N(RN)\} = \mathbb{E}R \cdot \mathbb{E}N^2 = 0$ while $|N| = |RN|$ is not independent of itself since it is not *a.s.* a constant.

for $i_1 \leq i_2 \leq \cdots \leq i_u < j_1 \leq j_2 \leq \cdots \leq j_v, u, v \in \mathbb{N}^* = \{1, 2, \ldots\}$.
Independence of the time series X thus writes as the independence of $\mathbf{P}$ and $\mathbf{F}$ for any choices $i_1 \leq i_2 \leq \cdots \leq i_u < j_1 \leq j_2 \leq \cdots \leq j_v$. Independence of the times series up to time m, also called m-dependence, is now characterized as independence of $\mathbf{P}$ and $\mathbf{F}$ if $i_u + m \leq j_1$. Finally, asymptotic independence of past and future will thus be given by arbitrary asymptotics

$$\epsilon(r) = \sup_{d(\mathbf{P},\mathbf{F}) \geq r} \sup_{(f,g) \in \mathcal{F} \times \mathcal{G}} |\mathrm{Cov}(f(\mathbf{P}), g(\mathbf{F}))| \tag{1.1.3}$$

where $d(\mathbf{P}, \mathbf{F}) = j_1 - i_u$. The only problem of the previous definition is that the corresponding dependence coefficient should also be indexed by suitable multi-indices $(i_1, i_2, \ldots, i_u)$ and $(j_1, j_2, \ldots, j_v)$. This definition will be completed in chapter 2 by considering classes $\mathcal{P}_u \subset \mathcal{P}$ and $\mathcal{F}_v \subset \mathcal{F}$ and suprema as well with respect to ordered multi-indices $(i_1, i_2, \ldots, i_u)$ and $(j_1, j_2, \ldots, j_v)$ such that $j_1 - i_u \geq r$.

1.2 Mixing

Mixing conditions, as introduced by Rosenblatt (1956) [166] are weak dependence conditions in terms of the $\sigma-$algebras generated by a random sequence. In order to define such conditions we first introduce the conditions relative to sub-$\sigma-$algebras $\mathcal{U}, \mathcal{V} \subset \mathcal{A}$ on an abstract probability space $(\Omega, \mathcal{A}, \mathbb{P})$:

$$\alpha(\mathcal{U}, \mathcal{V}) = \sup_{U \in \mathcal{U}, V \in \mathcal{V}} |\mathbb{P}(U \cap V) - \mathbb{P}(U)\mathbb{P}(V)| \tag{1.2.1}$$

$$\rho(\mathcal{U}, \mathcal{V}) = \sup_{u \in \mathbb{L}^2(\mathcal{U}), v \in \mathbb{L}^2(\mathcal{V})} |\mathrm{Corr}(u, v)| \tag{1.2.2}$$

$$\phi(\mathcal{U}, \mathcal{V}) = \sup_{U \in \mathcal{U}, V \in \mathcal{V}} \left| \frac{\mathbb{P}(U \cap V)}{\mathbb{P}(U)} - \mathbb{P}(V) \right| \tag{1.2.3}$$

$$\beta(\mathcal{U}, \mathcal{V}) = \frac{1}{2} \sup_{\substack{I,\, J \geq 1 \\ (U_i)_{1 \leq i \leq I} \in \mathcal{U}^I, \\ (V_j)_{1 \leq j \leq J} \in \mathcal{V}^J}} \sum_{i=1}^{I} \sum_{j=1}^{J} |\mathbb{P}(U_i \cap V_j) - \mathbb{P}(U_i)\mathbb{P}(V_j)| \tag{1.2.4}$$

In the definition of β, the supremum is considered over all measurable partitions $(U_i)_{1 \leq i \leq I}$, $(V_j)_{1 \leq j \leq J}$ of Ω. The above coefficients are, respectively, Rosenblatt (1956) [166]'s strong mixing coefficient $\alpha(\mathcal{U}, \mathcal{V})$, Wolkonski and Rozanov (1959) [187]'s absolute regularity coefficient $\beta(\mathcal{U}, \mathcal{V})$, Kolmogorov and Rozanov (1960) [112]'s maximal correlation coefficient $\rho(\mathcal{U}, \mathcal{V})$, and Ibragimov (1962) [110]'s uniform mixing coefficient $\phi(\mathcal{U}, \mathcal{V})$. A more comprehensible formulation for β is written in terms of a norm in total variation

$$\beta(\mathcal{U}, \mathcal{V}) = \|\mathbb{P}_{\mathcal{U} \otimes \mathcal{V}} - \mathbb{P}_{\mathcal{U}} \otimes \mathbb{P}_{\mathcal{V}}\|_{TV}$$

here $\mathbb{P}_{\mathcal{U}}, \mathbb{P}_{\mathcal{V}}$ denote the restrictions of $\mathbb{P}$ to $\sigma-$fields $\mathcal{U}, \mathcal{V}$ and $\mathbb{P}_{\mathcal{U}\otimes\mathcal{V}}$ is a law on the product $\sigma-$fields defined of rectangles by $\mathbb{P}_{\mathcal{U}\otimes\mathcal{V}}(U, V) = \mathbb{P}(U, V)$. In case $\mathcal{U}, \mathcal{V}$ are generated by random variables U, V this may be written

$$\beta(\mathcal{U}, \mathcal{V}) = \|\mathbb{P}_{(U,V)} - \mathbb{P}_U \otimes \mathbb{P}_V\|_{TV}$$

as the total variation norm of distributions of (U, V) and $(U, V^\star)$ for some independent copy $V^\star$ of V. The Markov frame is however adapted to prove β-mixing since this condition holds under positive recurrence.
In fact any coefficient μ such that $\mu(\mathcal{U}, \mathcal{V}) \in [0, +\infty]$ is well defined and such that independence of $\mathcal{U}, \mathcal{V}$ implies $\mu(\mathcal{U}, \mathcal{V}) = 0$ may be considered as a mixing coefficient. Once a mixing coefficient has been chosen, the corresponding mixing condition is defined for random processes $(X_t)_{t\in\mathbb{Z}}$ and for random fields $(X_t)_{t\in\mathbb{Z}^d}$:

$$\mu_X(r) = \sup_{i\in\mathbb{Z}} c(\sigma(X_t, t \le i), \sigma(X_t, t \ge i + r)) \tag{1.2.5}$$

and the random process is called μ-mixing in case $\mu_X(r) \to_{r\to\infty} 0$. Here $\mu = \alpha, \beta, \phi$ or ρ thus yield the coefficient sequences $\alpha_X(r)$, $\beta_X(r)$, $\phi_X(r)$ or $\rho_X(r)$; many other coefficients may also be introduced.

For the more difficult case of random fields, one needs a more intricate definition. The one we propose depends on two additional integers, and the random field $(X_t)_{t\in\mathbb{Z}^d}$ is μ-mixing in case for any $u, v \in \mathbb{N}^\star$, $c_{X,u,v}(r) \to_{r\to\infty} 0$, where now

$$\mu_{X,a,b}(r) = \sup_{\#A=a,\#B=b,d(A,B)\ge r} c(\sigma(X_t, t \in A), \sigma(X_t, t \in B)) \tag{1.2.6}$$

the supremum is considered over finite subsets with cardinality u, v and at least r distant (where a metric has been fixed on $\mathbb{Z}^d$).
The following relations hold:

$$\phi - \text{mixing} \Rightarrow \left\{ \begin{array}{c} \rho - \text{mixing} \\ \beta - \text{mixing} \end{array} \right\} \Rightarrow \alpha - \text{mixing}$$

and no reverse implication holds in general.
Examples for such conditions to hold are investigated in Doukhan (1994) [61], and Rio (2000) [161] provides up-to-date results in this setting. We only quote here that those conditions are usually difficult to check.

1.3 Mixingales and Near Epoch Dependence

Definition 1.1 (Mc Leish (1975) [129], Andrews (1988) [3]). *Let $p \ge 1$ and let $(\mathcal{F}_n)_{n\in\mathbb{Z}}$ be an increasing sequence of σ-algebras. The sequence $(X_n, \mathcal{F}_n)_{n\in\mathbb{Z}}$ is* *called an* $\mathbb{L}^p$*-mixingale if there exist nonnegative sequences* $(c_n)_{n\in\mathbb{Z}}$ *and* $(\psi(n))_{n\in\mathbb{Z}}$ *such that* $\psi(n) \to 0$ *as* $n \to \infty$ *and for all integers* $n \in \mathbb{Z}$, $k \geq 0$,

$$\|X_n - \mathbb{E}(X_n \mid \mathcal{F}_{n+k})\|_p \leq c_n\psi(k+1), \tag{1.3.1}$$

$$\|\mathbb{E}(X_n \mid \mathcal{F}_{n-k})\|_p \leq c_n\psi(k). \tag{1.3.2}$$

This property of fading memory is easier to handle than the martingale condition. A more general concept is the near epoch dependence (NED) on a mixing process. Its definition can be found in Billingsley (1968) [20] who considered functions of $\phi-$mixing processes.

Definition 1.2 (Pötscher and Prucha (1991) [152]). *Let* $p \geq 1$. *We consider a c-mixing process (defined as in eqn. (1.2.5))* $(V_n)_{n\in\mathbb{Z}}$. *For any integers* $i \leq j$, *set* $\mathcal{F}_i^j = \sigma(V_i, \ldots, V_j)$. *The sequence* $(X_n, \mathcal{F}_n)_{n\in\mathbb{Z}}$ *is called an* $\mathbb{L}^p$*-NED process on the c-mixing process* $(V_n)_{n\in\mathbb{Z}}$ *if there exist nonnegative sequences* $(c_n)_{n\in\mathbb{Z}}$ *and* $(\psi(n))_{n\in\mathbb{Z}}$ *such that* $\psi(n) \to 0$ *as* $n \to \infty$ *and for all integers* $n \in \mathbb{Z}$, $k \geq 0$,

$$\left\|X_n - \mathbb{E}\left(X_n \mid \mathcal{F}_{n-k}^{n+k}\right)\right\|_p \leq c_n\psi(k).$$

This approach is developed in details in Pötscher and Prucha (1991) [152]. Functions of MA(∞) processes can be handled using NED concept. For instance, limit theorems can be deduced for sums of such Functions of MA(∞) processes. These previous definitions translate the fact that a k-period – ahead in the first case, both ahead and backwards in the second definition – projection is convergent to the unconditional mean. They are known to be satisfied by a wide class of models. For example, martingale differences can be described as $\mathbb{L}_1$-mixingale sequences, and linear processes with martingale difference innovations as well.

1.4 Association

The notion of association was introduced independently by Esary, Proschan and Walkup (1967) [85] and Fortuin, Kastelyn and Ginibre (1971) [87].
The motivations of those authors were radically different since the first ones were working in reliability theory and the others in mechanical statistics, and their condition is known as FKG inequality.

Definition 1.3. *The sequence* $(X_t)_{t\in\mathbb{Z}}$ *is associated, if for all coordinatewise increasing real-valued functions* h *and* k,

$$\mathrm{Cov}(h(X_t, t \in A), k(X_t, t \in B)) \geq 0$$

for all finite disjoint subsets A *and* B *of* $\mathbb{Z}$ *and if moreover*

$$\mathbb{E}\left(h(X_t, t \in A)^2 + k(X_t, t \in B)^2\right) < \infty.$$

This extends the positive correlation assumption to model the notion that two stochastic processes have a tendency to evolve in a similar way.
This definition is deeper than the simple positivity of the correlations. Besides the evident fact that it does not assume that the variances exist, one can easily construct orthogonal (hence positively correlated) sequences that do not have the association property. An important difference between the above conditions is that its uncorrelatedness implies independence of an associated sequence (Newman, 1984 [136]). Let for instance (ξ_k, η_k) be independent and i.i.d. $\mathcal{N}(0,1)$ sequences. Then the sequence $(X_n)_{n\in\mathbb{Z}}$ defined by $X_k = \xi_k(\eta_k - \eta_{k-1})$ is neither correlated nor independent, hence it is not an associated sequence. Heredity of association only holds under monotonic transformations. This unpleasant restriction will disappear under the assumption of weak dependence.
The following property of associated sequences was a guideline for the forthcoming definition of weak dependence. Association does not imply at all any mixing assumption[‡]. The forthcoming inequality (1.4.1) also contains the idea that weakly correlated associated sequences are also 'weakly dependent'. The following result provide a quantitative idea of the loss of association to independence:

Theorem 1.1 (Newman, 1984 [136]). *For a pair of measurable numeric functions (f, g) defined on $A \subset \mathbb{R}^k$, we write $f \ll g$ if both functions $g + f$ and $g - f$ are non-decreasing with respect to each argument. Let now X be any associated random vector with range in A. Then*

$$(f_i \ll g_i, \ \textit{for } i = 1, 2) \Rightarrow \Big(\big|\mathrm{Cov}(f_1(X), f_2(X))\big| \le \mathrm{Cov}(g_1(X), g_2(X))\Big).$$

This theorem follows simply from several applications of the definition to the coordinatewise non-decreasing functions $g_i - f_i$ and $g_i + f_i$. By an easy application of the above inequalities one can check that

$$|\mathrm{Cov}(f(X), g(Y))| \le \sum_{i=1}^{k}\sum_{j=1}^{l} \left\|\frac{\partial f}{\partial x_i}\right\|_\infty \left\|\frac{\partial g}{\partial y_j}\right\|_\infty \mathrm{Cov}(X_i, Y_j), \qquad (1.4.1)$$

for $\mathbb{R}^k$ or $\mathbb{R}^l$ valued associated random vectors X and Y and $\mathcal{C}^1$ functions f and g with bounded partial derivatives. For this, it suffices to note that $f \ll f_1$ if one makes use of Theorem 1.1 with $f_1(x_1, \ldots, x_p) = \sum_{i=1}^{p} \left\|\frac{\partial f}{\partial x_i}\right\|_\infty x_i$.
Denote by $\mathcal{R}(z)$ the real part of the complex number z. Theorem 1.1 can be extended to complex valued functions, up to a factor 2 in the left hand side of the above inequality (1.4.1). Indeed, we can set now $f \ll g$ if for any real number

‡E.g. Gaussian processes with nonnegative covariances are associated while this is well known that this condition does not implies mixing, see [61], page 62.

ω the mapping $t = (t_1, \ldots, t_k) \mapsto \mathcal{R}\left(g(t) + e^{i\omega(t_1+\cdots+t_k)} f(t)\right)$ is non-decreasing with respect to each argument. Also, for any real numbers $t_1, \ldots, t_k$,

$$\left|\mathbb{E}e^{i(t_1X_1+\cdots+t_kX_k)} - \mathbb{E}e^{it_1X_1} \ \cdots \ \mathbb{E}e^{it_kX_k}\right| \le 2\sum_{i=1}^{k}\sum_{j=1}^{k}|t_i||t_j|\mathrm{Cov}(X_i, X_j).$$

On the opposite side, negatively associated sequences of r.v.'s are defined by a similar relation than the aforementioned covariance inequality, except for the sense of this inequality. This property breaks the seemingly parallel definitions of positively and negatively associated sequences.

1.5 Nonmixing models

Finite moving averages $X_n = H(\xi_n, \xi_{n-1}, \ldots, \xi_{n-m})$ are trivially m-dependent. However this does not remain exact as $m \to \infty$. For example, the Bernoulli shift $X_n = H(\xi_n, \xi_{n-1}, \ldots)$ (with $H(x) = \sum_{k=0}^{\infty} 2^{-(k+1)}x_k$) is not mixing; this is an example of a Markovian, non-mixing sequence.

Indeed, its stationary representation writes $X_n = \sum_{k=0}^{\infty} 2^{-k-1}\xi_{n-k}$. Here ξ_{n-k} is the k-th digit in the binary expansion of the uniformly chosen number $X_n = 0.\xi_n\xi_{n-1}\cdots \in [0, 1]$. This proves that X_n is a deterministic function of X_0 which is the main argument to derive that such models are not mixing ([61], page 77, counterexample 2 or [2]); more precisely, as X_n is some deterministic function of X_0 the event $A = (X_0 \le \frac{1}{2})$ belongs both to the sigma algebras of the past $\sigma(X_t, t \le 0)$ an and the sigma algebras of the future $\sigma(X_t, t \ge n)$, hence with the notation in § 1.2,

$$\alpha(n) \ge |\mathbb{P}(A \cap A) - \mathbb{P}(A)\mathbb{P}(A)| = \frac{1}{2} - \frac{1}{4} = \frac{1}{4}.$$

The same arguments apply to the model described before of an autoregressive process with innovations taking p distinct values. The difference between two such independent processes of this type or $((-1)^n X_n)_n$ provide example of non-associated and non-mixing processes.

Assume now that more generally $\xi_j \sim b(s)$ follows a Bernoulli distribution with parameter $0 < s < 1$. Concentration properties then hold *e.g.* X_n is uniform if $s = \frac{1}{2}$, and it has a Cantor marginal distribution if $s = \frac{1}{3}$.

Much more stationary models may be in fact proved to be nonmixing; *e.g.* for integer valued models (3.6.2) this is simple to prove that $X_t = 0 \Rightarrow X_0 = 0$ and $\mathbb{P}(X_0 = 0) \in]0, 1[$. With stationarity this easily excludes this model to be strong mixing since, setting $\mathbb{P}(X_0 = 0) := p$,

$$\alpha(n) \ge \Big|\mathbb{P}\big((X_0 = 0) \cap (X_n = 0)\big) - \mathbb{P}(X_0 = 0)\mathbb{P}(X_n = 0)\Big| = p(1-p) > 0.$$

Chapter 2

Weak dependence

Many authors have used one of the two following type of dependence: on the one hand mixing properties, introduced by Rosenblatt (1956) [166], on the other hand martingales approximations or mixingales, following the works of Gordin (1969, 1973) [97], [98] and Mc Leisch (1974, 1975) [127], [129]. Concerning strongly mixing sequences, very deep and elegant results have been established: for recent works, we mention the books of Rio (2000) [161] and Bradley (2002) [30]. However many classes of time series do not satisfy any mixing condition as it is quoted *e.g.* in Eberlein and Taqqu (1986) [83] or Doukhan (1994) [61]. Conversely, most of such time series enter the scope of mixingales but limit theorems and moment inequalities are more difficult to obtain in this general setting.
Between those directions, Bickel and Bühlmann (1999) [18] and simultaneously Doukhan and Louhichi (1999) [67] introduced a new idea of weak dependence. Their notion of weak dependence makes explicit the asymptotic independence between 'past' and 'future'; this means that the 'past' is progressively forgotten. In terms of the initial time series, 'past' and 'future' are elementary events given through finite dimensional marginals. Roughly speaking, for convenient functions f and g, we shall assume that

$$\mathrm{Cov}\,(f(\text{`past'}), g(\text{`future'}))$$

is small when the distance between the 'past' and the 'future' is sufficiently large. Such inequalities are significant only if the distance between indices of the initial time series in the 'past' and 'future' terms grows to infinity. The convergence is not assumed to hold uniformly on the dimension of the 'past' or 'future' involved.
The main advantage is that such a kind of dependence contains lots of pertinent examples and can be used in various situations: empirical central limit theorems are proved in Doukhan and Louhichi (1999) [67] and Borovkova, Burton and

Dehling (2001) [25], while applications to Bootstrap are given by Bickel and Bühlmann (1999) [18] and Ango Nzé *et al.*(2002) [6] and to functional estimation (Coulon-Prieur & Doukhan, 2000 [40]).
In this chapter a first section introduces the function spaces necessary to define the various dependence coefficients of the second section. They are classified in separated subsections. We shall first consider noncausal coefficients and then their causal counterparts; in both cases the subjacent spaces are Lipschitz spaces. A further case associated to bounded variation spaces is provided in the following subsection. Projective measure of dependence are included in the last subsection.

2.1 Function spaces

In this section, we give the definitions of some function spaces used in this book.

- Let m be any measure on a measurable space $(\Omega, \mathcal{A})$. For any $p \geq 1$, we denote by $\mathbb{L}^p(m)$ the space of measurable functions f from Ω to $\mathbb{R}$ such that

$$\begin{aligned} \|f\|_{p,m} &= \left(\int |f(x)|^p m(dx)\right)^{1/p} < \infty, \\ \|f\|_{\infty,m} &= \inf\left\{M > 0 \,/\, m(|f| > M) = 0\right\} < \infty, \quad \text{for } p = \infty. \end{aligned}$$

  For simplicity, when no confusion can arise, we shall write $\mathbb{L}^p$ and $\|\cdot\|_p$ instead of $\mathbb{L}^p(m)$ and $\|\cdot\|_{p,m}$.

Let $\mathcal{X}$ be a Polish space and δ be some metric on $\mathcal{X}$ ($\mathcal{X}$ need not be Polish with respect to δ).

- Let $\Lambda(\delta)$ be the set of Lipschitz functions from $\mathcal{X}$ to $\mathbb{R}$ with respect to the distance δ. For $f \in \Lambda(\delta)$, denote by $\mathrm{Lip}\,(f)$, f's Lipschitz constant. Let

$$\Lambda^{(1)}(\delta) = \{f \in \Lambda(\delta) \;/\; \mathrm{Lip}\,(f) \leq 1\}.$$

- Let $(\Omega, \mathcal{A}, \mathbb{P})$ be a probability space. Let $\mathcal{X}$ be a Polish space and δ be a distance on $\mathcal{X}$. For any $p \in [1, \infty]$, we say that a random variable X with values in $\mathcal{X}$ is $\mathbb{L}^p$-integrable if, for some x_0 in $\mathcal{X}$, the real valued random variable $\delta(X, x_0)$ belongs to $\mathbb{L}^p(\mathbb{P})$.

Another type of function class will be used in this chapter: it is the class of functions with bounded variation on the real line. To be complete, we recall,

Definition 2.1. *A σ-finite signed measure is the difference of two positive σ-finite measures, one of them at least being finite. We say that a function h from $\mathbb{R}$ to $\mathbb{R}$ is σ-BV if there exists a σ-finite signed measure dh such that $h(x) = h(0) + dh([0, x[)$ if $x \geq 0$ and $h(x) = h(0) - dh([x, 0[)$ if $x \leq 0$ (h is left continuous). The function h is BV if the signed measure dh is finite.*

Recall also the Hahn-Jordan decomposition: for any σ-finite signed measure μ, there is a set D such that

$$\mu_+(A) = \mu(A \cap D) \geq 0, \qquad -\mu_-(A) = \mu(A \backslash D) \leq 0.$$

μ_+ and μ_- are mutually singular, one of them at least is finite and $\mu = \mu_+ - \mu_-$. The measure $|\mu| = \mu_+ + \mu_-$ is called the total variation measure for μ. The total variation of μ writes as $\|\mu\| = |\mu|(\mathbb{R})$.

Now we are in position to introduce

- BV_1 the space of BV functions $h : \mathbb{R} \to \mathbb{R}$ such that $\|dh\| \leq 1$.

2.2 Weak dependence

Let $(\Omega, \mathcal{A}, \mathbb{P})$ be a probability space and let $\mathcal{X}$ be a Polish space. Let

$$\mathcal{F} = \bigcup_{u \in \mathbb{N}^*} \mathcal{F}_u \quad \text{and} \quad \mathcal{G} = \bigcup_{u \in \mathbb{N}^*} \mathcal{G}_u\,,$$

where $\mathcal{F}_u$ and $\mathcal{G}_u$ are two classes of functions from $\mathcal{X}^u$ to $\mathbb{R}$.

Definition 2.2. *Let X and Y be two random variables with values in $\mathcal{X}^u$ and $\mathcal{X}^v$ respectively. If Ψ is some function from $\mathcal{F} \times \mathcal{G}$ to $\mathbb{R}_+$, define the $(\mathcal{F}, \mathcal{G}, \Psi)$-dependence coefficient $\epsilon(X, Y)$ by*

$$\epsilon(X, Y) = \sup_{f \in \mathcal{F}_u\, g \in \mathcal{G}_v} \frac{|\mathrm{Cov}(f(X), g(Y))|}{\Psi(f, g)}\,. \tag{2.2.1}$$

Let $(X_n)_{n \in \mathbb{Z}}$ be a sequence of $\mathcal{X}$-valued random variables. Let $\Gamma(u, v, k)$ be the set of (i, j) in $\mathbb{Z}^u \times \mathbb{Z}^v$ such that $i_1 < \cdots < i_u \leq i_u + k \leq j_1 < \cdots < j_v$. The dependence coefficient $\epsilon(k)$ is defined by

$$\epsilon(k) = \sup_{u,v} \sup_{(i,j) \in \Gamma(u,v,k)} \epsilon((X_{i_1}, \ldots, X_{i_u}), (X_{j_1}, \ldots, X_{j_v}))\,.$$

The sequence $(X_n)_{n \in \mathbb{Z}}$ is $(\mathcal{F}, \mathcal{G}, \Psi)$-dependent if the sequence $(\epsilon(k))_{k \in \mathbb{N}}$ tends to zero. If $\mathcal{F} = \mathcal{G}$ we simply denote this as $(\mathcal{F}, \Psi)$-dependence.

Remark 2.1. *Definition 2.2 above easily extends to general metric sets of indices T equipped with a distance δ (e.g. $T = \mathbb{Z}^d$ yields the case of random fields). The set $\Gamma(u, v, k)$ is then the set of (i, j) in $T^u \times T^v$ such that*

$$k = \min\left\{\delta(i_\ell, j_m) \,/\, 1 \leq \ell \leq u, 1 \leq m \leq v\right\}\,.$$

2.2.1 η, κ, λ and ζ-coefficients

In this section, we focus on the case where $\mathcal{F}_u = \mathcal{G}_u$. If f belongs to $\mathcal{F}_u$, we define $d_f = u$.
In a first time, $\mathcal{F}_u$ is the set of bounded functions from $\mathcal{X}^u$ to $\mathbb{R}$, which are Lipschitz with respect to the distance δ_1 on $\mathcal{X}^u$ defined by

$$\delta_1(x,y) = \sum_{i=1}^{u} \delta(x_i, y_i) . \tag{2.2.2}$$

In that case:

- the coefficient η corresponds to

$$\Psi(f,g) = d_f \|g\|_\infty \mathrm{Lip}\,(f) + d_g \|f\|_\infty \mathrm{Lip}\,(g) , \tag{2.2.3}$$

- the coefficient λ corresponds to

$$\Psi(f,g) = d_f \|g\|_\infty \mathrm{Lip}\,(f) + d_g \|f\|_\infty \mathrm{Lip}\,(g) + d_f d_g \mathrm{Lip}\,(f) \mathrm{Lip}\,(g) . \tag{2.2.4}$$

To define the coefficients κ and ζ, we consider for $\mathcal{F}_u$ the wider set of functions from $\mathcal{X}^u$ to $\mathbb{R}$, which are Lipschitz with respect to the distance δ_1 on $\mathcal{X}^u$, but which are not necessarily bounded. In that case we assume that the variables X_i are $\mathbb{L}^1$-integrable.

- the coefficient κ corresponds to

$$\Psi(f,g) = d_f d_g \mathrm{Lip}\,(f) \mathrm{Lip}\,(g) , \tag{2.2.5}$$

- the coefficient ζ corresponds to

$$\Psi(f,g) = \min(d_f, d_g) \mathrm{Lip}\,(f) \mathrm{Lip}\,(g) . \tag{2.2.6}$$

These coefficients have some hereditary properties. For example, let $h : \mathcal{X} \to \mathbb{R}$ be a Lipschitz function with respect to δ, then if the sequence $(X_n)_{n\in\mathbb{Z}}$ is η, κ, λ or ζ weakly dependent, then the same is true for the sequence $(h(X_n))_{n\in\mathbb{Z}}$. One can also obtain some hereditary properties for functions which are not Lipschitz on the whole space $\mathcal{X}$, as shown by Lemma 2.1 below, in the special case where $\mathcal{X} = \mathbb{R}^k$ equipped with the distance $\delta(x,y) = \max_{1\le i\le k} |x_i - y_i|$.

Proposition 2.1 (Bardet, Doukhan, León, 2006 [11]). *Let $(X_n)_{n\in\mathbb{Z}}$ be a sequence of $\mathbb{R}^k$-valued random variables. Let $p > 1$. We assume that there exists some constant $C > 0$ such that $\max_{1\le i\le k} \|X_i\|_p \le C$. Let h be a function from $\mathbb{R}^k$ to $\mathbb{R}$ such that $h(0) = 0$ and for $x, y \in \mathbb{R}^k$, there exist a in $[1, p[$ and $c > 0$ such that*

$$|h(x) - h(y)| \le c|x - y|(|x|^{a-1} + |y|^{a-1}) .$$

We define the sequence $(Y_n)_{n\in\mathbb{Z}}$ by $Y_n = h(X_n)$. Then,

- *if* $(X_n)_{n\in\mathbb{Z}}$ *is* η*-weak dependent, then* $(Y_n)_{n\in\mathbb{Z}}$ *also, and*

$$\eta_Y(n) = \mathcal{O}\left(\eta(n)^{\frac{p-a}{p-1}}\right);$$

- *if* $(X_n)_{n\in\mathbb{Z}}$ *is* λ*-weak dependent, then* $(Y_n)_{n\in\mathbb{Z}}$ *also, and*

$$\lambda_Y(n) = \mathcal{O}\left(\lambda(n)^{\frac{p-a}{p+a-2}}\right).$$

Remark 2.2. *The function* $h(x) = x^2$ *satisfies the previous assumptions with* $a = 2$. *This condition is satisfied by polynomials with degree* a.

Proof of Proposition 2.1. Let f and g be two real functions in $\mathcal{F}_u$ and $\mathcal{F}_v$ respectively. Denote $x^{(M)} = (x \wedge M) \vee (-M)$ for $x \in \mathbb{R}$. Now, for $x = (x_1, \ldots, x_k) \in \mathbb{R}^k$, we analogously denote $x^{(M)} = (x_1^{(M)}, \ldots, x_k^{(M)})$. Assume that (i, j) belong to the set $\Gamma(u, v, r)$ defined in Definition 2.2. Define $X_{\mathbf{i}} = (X_{i_1}, \ldots, X_{i_u})$ and $X_{\mathbf{j}} = (X_{j_1}, \ldots, X_{j_v})$. We then define functions $F : \mathbb{R}^{uk} \to \mathbb{R}$ and $G : \mathbb{R}^{vk} \to \mathbb{R}$ through the relations:

- $F(X_{\mathbf{i}}) = f(h(X_{i_1}), \ldots, h(X_{i_u}))$, $F^{(M)}(X_{\mathbf{i}}) = f(h(X_{i_1}^{(M)}), \ldots, h(X_{i_u}^{(M)}))$,
- $G(X_{\mathbf{j}}) = g(h(X_{j_1}), \ldots, h(X_{j_v}))$, $G^{(M)}(X_{\mathbf{j}}) = g(h(X_{j_1}^{(M)}), \ldots, h(X_{j_v}^{(M)}))$.

Then:

$$\begin{aligned}
|\mathrm{Cov}(F(X_{\mathbf{i}}), G(X_{\mathbf{j}}))| &\leq |\mathrm{Cov}(F(X_{\mathbf{i}}), G(X_{\mathbf{j}}) - G^{(M)}(X_{\mathbf{j}}))| \\
&\quad + |\mathrm{Cov}(F(X_{\mathbf{i}}), G^{(M)}(X_{\mathbf{j}}))| \\
&\leq 2\|f\|_\infty \mathbb{E}|G(X_{\mathbf{j}}) - G^{(M)}(X_{\mathbf{j}}))| \\
&\quad + 2\|g\|_\infty \mathbb{E}|F(X_{\mathbf{i}}) - F^{(M)}(X_{\mathbf{i}})| \\
&\quad + |\mathrm{Cov}(F^{(M)}(X_{\mathbf{i}}), G^{(M)}(X_{\mathbf{j}}))|
\end{aligned}$$

But we also have from the assumptions on h and Markov inequality,

$$\begin{aligned}
\mathbb{E}|G(X_{\mathbf{j}}) - G^{(M)}(X_{\mathbf{j}}))| &\leq \mathrm{Lip}\, g \sum_{l=1}^{v} \mathbb{E}|h(X_{j_l}) - h(X_{j_l}^{(M)})| \\
&\leq 2c\,\mathrm{Lip}\, g \sum_{l=1}^{v} \mathbb{E}\big(|X_{j_l}|^a \mathbf{1}_{|X_{j_l}|>M}\big), \\
&\leq 2c\, v\, \mathrm{Lip}\, g C^p M^{a-p}.
\end{aligned}$$

The same thing holds for F. Moreover, the functions $F^{(M)} : \mathbb{R}^{uk} \to \mathbb{R}$ and $G^{(M)} : \mathbb{R}^{vk} \to \mathbb{R}$ satisfy $\mathrm{Lip}\, F^{(M)} \leq 2cM^{a-1}\mathrm{Lip}\,(f)$ and $\mathrm{Lip}\, G^{(M)} \leq 2cM^{a-1}$

Lip (g), and $\|F^{(M)}\|_\infty \le \|f\|_\infty$, $\|G^{(M)}\|_\infty \le \|g\|_\infty$. Thus, from the definition of weak dependence of X and the choice of $\mathbf{i},\mathbf{j}$, we obtain respectively, if $M \ge 1$

$$\begin{aligned}\left|\mathrm{Cov}\left(F^{(M)}(X_{\mathbf{i}}), G^{(M)}(X_{\mathbf{j}})\right)\right| &\le 2c(u\mathrm{Lip}\,(f)\|g\|_\infty + v\mathrm{Lip}\,(g)\|f\|_\infty)M^{a-1}\eta(r),\\ &\le 2c(d_f\mathrm{Lip}\,(f)\|g\|_\infty + d_g\mathrm{Lip}\,(g)\|f\|_\infty)M^{a-1}\lambda(r)\\ &+ 4c^2 d_f d_g \mathrm{Lip}\,(f)\mathrm{Lip}\,(g)M^{2a-2}\lambda(r).\end{aligned}$$

Finally, we obtain respectively, if $M \ge 1$:

$$\begin{aligned}|\mathrm{Cov}(F(X_{\mathbf{i}}), G(X_{\mathbf{j}}))| &\le 2c(u\mathrm{Lip}\, f\|g\|_\infty + v\mathrm{Lip}\, g\|f\|_\infty)\\ &\qquad \times\left(M^{a-1}\eta(r) + 2C^pM^{a-p}\right),\\ &\le c(u\mathrm{Lip}\, f + v\mathrm{Lip}\, g + uv\mathrm{Lip}\, f\mathrm{Lip}\, g)\\ &\qquad \times(M^{2a-2}\lambda(r) + M^{a-p}).\end{aligned}$$

Choosing $M = \eta(r)^{1/(1-p)}$ and $M = \lambda(r)^{-1/(p+a-2)}$ respectively, we obtain the result. □

In the definition of the coefficients η, κ, λ and ζ, we assume some regularity conditions on $\mathcal{F}_u = \mathcal{G}_u$. In the case where the sequence $(X_n)_{n\in\mathbb{Z}}$ is an adapted process with respect to some increasing filtration $(\mathcal{M}_i)_{i\in\mathbb{Z}}$, it is often more suitable to work without assuming any regularity conditions on $\mathcal{F}_u$. In that case $\mathcal{G}_u$ is some space of regular functions and $\mathcal{F}_u \neq \mathcal{G}_u$. This last case is called the causal case. In the situations where both $\mathcal{F}_u$ and $\mathcal{G}_u$ are spaces of regular functions, we say that we are in the non causal case.

2.2.2 θ and τ-coefficients

Let $\mathcal{F}_u$ be the class of bounded functions from $\mathcal{X}_u$ to $\mathbb{R}$, and let $\mathcal{G}_u$ be the class of functions from $\mathcal{X}_u$ to $\mathbb{R}$ which are Lipschitz with respect to the distance δ_1 defined by (2.2.2). We assume that the variables X_i are $\mathbb{L}^1$-integrable.

- The coefficient θ corresponds to

$$\Psi(f, g) = d_g\|f\|_\infty \mathrm{Lip}\,(g)\,. \tag{2.2.7}$$

The coefficient θ has some hereditary properties. For example, Proposition 2.2 below gives hereditary properties similar to those given for the coefficients η and λ in Lemma 2.1.

Proposition 2.2. *Let $(X_n)_{n\in\mathbb{Z}}$ be a sequence of $\mathbb{R}^k$-valued random variables. We define the sequence $(Y_n)_{n\in\mathbb{Z}}$ by $Y_n = h(X_n)$. The assumptions on $(X_n)_{n\in\mathbb{Z}}$ and on h are the same as in Lemma 2.1. Then,*

- *if $(X_n)_{n\in\mathbb{Z}}$ is θ-weak dependent, $(Y_n)_{n\in\mathbb{Z}}$ also, and*

$$\theta_Y(n) = \mathcal{O}\left(\theta(n)^{\frac{p-a}{p-1}}\right).$$

The proof of Proposition 2.2 follows the same line as the proof of Proposition 2.1 and therefore is not detailed.
We shall see that the coefficient θ defined above belongs to a more general class of dependence coefficients defined through conditional expectations with respect to the filtration $\sigma(X_j, j \le i)$.

Definition 2.3. *Let $(\Omega, \mathcal{A}, \mathbb{P})$ be a probability space, and $\mathcal{M}$ be a σ-algebra of $\mathcal{A}$. Let $\mathcal{X}$ be a Polish space and δ a distance on $\mathcal{X}$. For any $\mathbb{L}^p$-integrable random variable X (see § 2.1) with values in $\mathcal{X}$, we define*

$$\theta_p(\mathcal{M}, X) = \sup\{\|\mathbb{E}(g(X)|\mathcal{M}) - \mathbb{E}(g(X))\|_p \,/\, g \in \Lambda^{(1)}(\delta)\}. \tag{2.2.8}$$

Let $(X_i)_{i\in\mathbb{Z}}$ be a sequence of $\mathbb{L}^p$-integrable $\mathcal{X}$-valued random variables, and let $(\mathcal{M}_i)_{i\in\mathbb{Z}}$ be a sequence of σ-algebras of $\mathcal{A}$. On $\mathcal{X}^l$, we consider the distance δ_1 defined by (2.2.2). The sequence of coefficients $\theta_{p,r}(k)$ is then defined by

$$\theta_{p,r}(k) = \max_{\ell \le r} \frac{1}{\ell} \sup_{(i,j)\in\Gamma(1,\ell,k)} \theta_p\left(\mathcal{M}_i, (X_{j_1}, \ldots, X_{j_\ell})\right). \tag{2.2.9}$$

When it is not clearly specified, we shall always take $\mathcal{M}_i = \sigma(X_k, k \le i)$.

The two preceding definitions are coherent as proved below.

Proposition 2.3. *Let $(X_i)_{i\in\mathbb{Z}}$ be a sequence of $\mathbb{L}^1$-integrable $\mathcal{X}$-valued random variables, and let $\mathcal{M}_i = \sigma(X_j, j \le i)$. According to the definition of $\theta(k)$ and to the definition 2.3, we have the equality*

$$\theta(k) = \theta_{1,\infty}(k). \tag{2.2.10}$$

Proof of Proposition 2.3. The fact that $\theta(k) \le \theta_{1,\infty}(k)$ is clear since, for any f in $\mathcal{F}_u$, g in $\mathcal{G}_v$, and any $(i,j) \in \Gamma(u,v,k)$,

$$\begin{aligned}
&\left|\mathrm{Cov}\left(\frac{f(X_{i_1},\ldots,X_{i_u})}{\|f\|_\infty}, \frac{g(X_{j_1},\ldots,X_{j_v})}{v\mathrm{Lip}\,(g)}\right)\right| \\
&\quad\le \frac{1}{v}\left\|\mathbb{E}\left(\frac{g(X_{j_1},\ldots,X_{j_v})}{\mathrm{Lip}\,(g)}\Big|\mathcal{M}_{i_u}\right) - \mathbb{E}\left(\frac{g(X_{j_1},\ldots,X_{j_v})}{\mathrm{Lip}\,(g)}\right)\right\|_1 \le \theta_{1,\infty}(k).
\end{aligned}$$

To prove the converse inequality, we first notice that

$$\theta(\mathcal{M}_i, (X_{j_1}, \ldots, X_{j_v}) = \lim_{k\to-\infty} \theta\left(\mathcal{M}_{k,i}, (X_{j_1}, \ldots, X_{j_v})\right), \tag{2.2.11}$$

where $\mathcal{M}_{k,i} = \sigma(X_j, k \le j \le i)$. Now, letting

$$f(X_k,\dots,X_i) = \operatorname{sign}\{\mathbb{E}(g(X_{j_1},\dots,X_{j_v})|\mathcal{M}_{k,i}) - \mathbb{E}(g(X_{j_1},\dots,X_{j_v}))\},$$

we have that, for (i,j) in $\Gamma(1,v,k)$ and g in $\Lambda^{(1)}(\delta_1)$,

$$\begin{aligned}\|\mathbb{E}(g(X_{j_1},\dots,X_{j_v})|\mathcal{M}_{k,i}) &- \mathbb{E}(g(X_{j_1},\dots,X_{j_v}))\|_1 \\ &= \operatorname{Cov}(f(X_k,\dots,X_i), g(X_{j_1},\dots,X_{j_v})) \le v\theta(k)\ .\end{aligned}$$

We infer that

$$\frac{1}{v}\theta(\mathcal{M}_{k,i},(X_{j_1},\dots,X_{j_v}) \le \theta(k)$$

and we conclude from (2.2.11) that $\theta_{1,\infty}(k) \le \theta(k)$. The proof is complete. □
Having in view the coupling arguments in § 5.3, we now define a variation of the coefficient (2.2.8) where we exchange the order of $\|.\|_p$ and the supremum. This is the same step as passing from $\alpha-$mixing to $\beta-$mixing, which is known to ensure nice coupling arguments (see Berbee, 1979 [16]).

Definition 2.4. *Let $(\Omega, \mathcal{A}, \mathbb{P})$ be a probability space, and $\mathcal{M}$ a σ-algebra of $\mathcal{A}$. Let $\mathcal{X}$ be a Polish space and δ a distance on $\mathcal{X}$. For any $\mathbb{L}^p-$integrable (see § 2.1)) $\mathcal{X}$-valued random variable X, we define the coefficient τ_p by:*

$$\tau_p(\mathcal{M}, X) = \left\| \sup_{g\in\Lambda^{(1)}(\delta)} \left\{ \int g(x)\mathbb{P}_{X|\mathcal{M}}(dx) - \int g(x)\mathbb{P}_X(dx) \right\} \right\|_p \tag{2.2.12}$$

where $\mathbb{P}_X$ is the distribution of X and $\mathbb{P}_{X|\mathcal{M}}$ is a conditional distribution of X given $\mathcal{M}$. We clearly have

$$\theta_p(\mathcal{M}, X) \le \tau_p(\mathcal{M}, X)\,. \tag{2.2.13}$$

Let $(X_i)_{i\in\mathbb{Z}}$ be a sequence of $\mathbb{L}^p$-integrable $\mathcal{X}$-valued random variables. The coefficients $\tau_{p,r}(k)$ are defined from τ_p as in (2.2.9).

2.2.3 $\tilde{\alpha}$, $\tilde{\beta}$ and $\tilde{\phi}$-coefficients.

In the case where $\mathcal{X} = (\mathbb{R}^d)^r$, we introduce some new coefficients based on indicator of quadrants. Recall that if x and y are two elements of $\mathbb{R}^d$, then $x \le y$ if and only if $x_i \le y_i$ for any $1 \le i \le d$.

Definition 2.5. *Let $X = (X_1,\dots,X_r)$ be a $(\mathbb{R}^d)^r$-valued random variable and $\mathcal{M}$ a σ-algebra of $\mathcal{A}$. For t_i in $\mathbb{R}^d$ and x in $\mathbb{R}^d$, let $g_{t_i,i}(x) = \mathbf{1}_{x\le t_i} - \mathbb{P}(X_i \le t_i)$. Keeping the same notations as in Definition 2.4, define for $t = (t_1,\dots,t_r)$ in $(\mathbb{R}^d)^r$,*

$$L_{X|\mathcal{M}}(t) = \int \prod_{i=1}^r g_{t_i,i}(x_i)\mathbb{P}_{X|\mathcal{M}}(dx) \quad \text{and} \quad L_X(t) = \mathbb{E}\prod_{i=1}^r g_{t_i,i}(X_i).$$

Define now the coefficients

1. $\tilde{\alpha}(\mathcal{M}, X) = \sup_{t \in (\mathbb{R}^d)^r} \|L_{X|\mathcal{M}}(t) - L_X(t)\|_1$.

2. $\tilde{\beta}(\mathcal{M}, X) = \left\| \sup_{t \in (\mathbb{R}^d)^r} |L_{X|\mathcal{M}}(t) - L_X(t)| \right\|_1$.

3. $\tilde{\phi}(\mathcal{M}, X) = \sup_{t \in (\mathbb{R}^d)^r} \|L_{X|\mathcal{M}}(t) - L_X(t)\|_\infty$.

Remark 2.3. *Note that if $r = 1, d = 1$ and $\delta(x, y) = |x - y|$, then, with the above notation,*

$$\tau_1(\mathcal{M}, X) = \int \|L_{X|\mathcal{M}}(t)\|_1 dt\,.$$

The proof of this equality follows the same lines than the proof of the coupling property of τ_1 (see Chapter 5, proof of Lemma 5.2).

In the definition of the coefficients θ and τ, we have used the class of functions $\Lambda^{(1)}(\delta)$. In the case where $d = 1$, we can define the coefficients $\tilde{\alpha}(\mathcal{M}, X)$, $\tilde{\beta}(\mathcal{M}, X)$ and $\tilde{\phi}(\mathcal{M}, X)$ with the help of bounded variation functions. This is the purpose of the following lemma:

Lemma 2.1. *Let $(\Omega, \mathcal{A}, \mathbb{P})$ be a probability space, $X = (X_1, \ldots, X_r)$ a $\mathbb{R}^r$-valued random variable and $\mathcal{M}$ a σ-algebra of $\mathcal{A}$. If f is a function in BV_1, let $f^{(i)}(x) = f(x) - \mathbb{E}(f(X_i))$. The following relations hold:*

1. $\tilde{\alpha}(\mathcal{M}, X) = \sup_{f_1, \ldots, f_r \in BV_1} \left\| \mathbb{E}\left(\prod_{i=1}^r f_i^{(i)}(X_i) \Big| \mathcal{M} \right) - \mathbb{E}\left(\prod_{i=1}^r f_i^{(i)}(X_i) \right) \right\|_1$.

2. $\tilde{\beta}(\mathcal{M}, X) = \left\| \sup_{f_1, \ldots, f_r \in BV_1} \left| \int \prod_{i=1}^r f_i^{(i)}(x_i) \left(\mathbb{P}_{X|\mathcal{M}} - \mathbb{P}_X \right)(dx) \right| \right\|_1$.

3. $\tilde{\phi}(\mathcal{M}, X) = \sup_{f_1, \ldots, f_r \in BV_1} \left\| \mathbb{E}\left(\prod_{i=1}^r f_i^{(i)}(X_i) | \mathcal{M} \right) - \mathbb{E}\left(\prod_{i=1}^r f_i^{(i)}(X_i) \right) \right\|_\infty$.

Remark 2.4. *For $r = 1$ and $d = 1$, the coefficient $\tilde{\alpha}(\mathcal{M}, X)$ was introduced by Rio (2000, equation 1.10c [161]) and used by Peligrad (2002) [140], while $\tau_1(\mathcal{M}, X)$ was introduced by Dedecker and Prieur (2004a) [45]. Let $\alpha(\mathcal{M}, \sigma(X))$, $\beta(\mathcal{M}, \sigma(X))$ and $\phi(\mathcal{M}, \sigma(X))$ be the usual mixing coefficients defined respectively by Rosenblatt (1956) [166], Rozanov and Volkonskii (1959) [187] and Ibragimov (1962) [110]. Starting from Definition 2.5 one can easily prove that*

$$\tilde{\alpha}(\mathcal{M}, X) \le 2\alpha(\mathcal{M}, \sigma(X)),\ \tilde{\beta}(\mathcal{M}, X) \le \beta(\mathcal{M}, \sigma(X)),\ \tilde{\phi}(\mathcal{M}, X) \le \phi(\mathcal{M}, \sigma(X)).$$

Proof of Lemma 2.1. Let f_i be a function in BV_1. Assume without loss of generality that $f_i(-\infty) = 0$. Then

$$f_i^{(i)}(x) = -\int (\mathbf{1}_{x \le t} - \mathbb{P}(X_i \le t))\, df_i(t).$$

Hence,

$$\int \prod_{i=1}^k f_i^{(i)}(x_i) \mathbb{P}_{X|\mathcal{M}}(dx) = (-1)^k \int \Big(\int \prod_{i=1}^k g_{t_i,i}(x_i) \mathbb{P}_{X|\mathcal{M}}(dx) \Big) \prod_{i=1}^k df_i(t_i),$$

and the same is true for $\mathbb{P}_X$ instead of $\mathbb{P}_{X|\mathcal{M}}$. From these inequalities and the fact that $|df_i|(\mathbb{R}) \le 1$, we infer that

$$\sup_{f_1,\dots,f_k \in BV_1} \Big| \int \prod_{i=1}^k f_i^{(i)}(x_i) \mathbb{P}_{X|\mathcal{M}}(dx) - \int \prod_{i=1}^k f_i^{(i)}(x_i) \mathbb{P}_X(dx) \Big| \le \sup_{t \in \mathbb{R}^r} |L_{X|\mathcal{M}}(t) - L_X(t)|.$$

The converse inequality follows by noting that $x \mapsto \mathbf{1}_{x \le t}$ belongs to BV_1. □
The following proposition gives the hereditary properties of these coefficients.

Proposition 2.4. *Let $(\Omega, \mathcal{A}, \mathbb{P})$ be a probability space, X an $\mathbb{R}^r$-valued, random variable and $\mathcal{M}$ a σ-algebra of $\mathcal{A}$. Let $g_1, \dots, g_r$ be any nondecreasing functions, and let $g(X) = (g_1(X_1), \dots, g_r(X_r))$. We have the inequalities $\tilde{\alpha}(\mathcal{M}, g(X)) \le \tilde{\alpha}(\mathcal{M}, X)$, $\tilde{\beta}(\mathcal{M}, g(X)) \le \tilde{\beta}(\mathcal{M}, X)$ and $\tilde{\phi}(\mathcal{M}, g(X)) \le \tilde{\phi}(\mathcal{M}, X)$. In particular, if F_i is the distribution function of X_i, we have $\tilde{\alpha}(\mathcal{M}, F(X)) = \tilde{\alpha}(\mathcal{M}, X)$, $\tilde{\beta}(\mathcal{M}, F(X)) = \tilde{\beta}(\mathcal{M}, X)$ and $\tilde{\phi}(\mathcal{M}, F(X)) = \tilde{\phi}(\mathcal{M}, X)$.*

Notations 2.1. For any distribution function F, we define the generalized inverse as

$$F^{-1}(x) = \inf \big\{ t \in \mathbb{R} \,/\, F(t) \ge x \big\}. \tag{2.2.14}$$

For any non-increasing càdlàg function $f : \mathbb{R} \to \mathbb{R}$ we analogously define the generalized inverse

$$f^{-1}(u) = \inf\{t / f(t) \le u\}.$$

Proof of Proposition 2.4. The fact that $\tilde{\alpha}(\mathcal{M}, g(X)) \le \tilde{\alpha}(\mathcal{M}, X)$ is immediate, from its definition. We infer that $\tilde{\alpha}(\mathcal{M}, F(X)) \le \tilde{\alpha}(\mathcal{M}, X)$. Applying the first result once more, we obtain that $\tilde{\alpha}(\mathcal{M}, F^{-1}(F(X))) \le \tilde{\alpha}(\mathcal{M}, F(X))$. To conclude, it suffices to note that $F^{-1} \circ F(X) = X$ almost surely, so that $\tilde{\alpha}(\mathcal{M}, X) \le \tilde{\alpha}(\mathcal{M}, F(X))$. Of course, the same arguments apply to $\tilde{\beta}(\mathcal{M}, X)$ and $\tilde{\phi}(\mathcal{M}, X)$. □
We now define the coefficients $\tilde{\alpha}_r(k)$, $\tilde{\beta}_r(k)$ and $\tilde{\phi}_r(k)$ for a sequence of σ-algebras and a sequence of $\mathbb{R}^d$-valued random variables.

Definition 2.6. *Let $(\Omega, \mathcal{A}, \mathbb{P})$ be a probability space. Let $(X_i)_{i\in\mathbb{Z}}$ be a sequence of $\mathbb{R}^d$-valued random variables, and let $(\mathcal{M}_i)_{i\in\mathbb{Z}}$ be a sequence of σ-algebras of $\mathcal{A}$. For $r \in \mathbb{N}^*$ and $k \geq 0$, define*

$$\tilde{\alpha}_r(k) = \max_{1\leq l\leq r} \sup_{(i,j)\in\Gamma(1,l,k)} \tilde{\alpha}(\mathcal{M}_i, (X_{j_1}, \ldots, X_{j_l})) . \tag{2.2.15}$$

The coefficients $\tilde{\beta}_r(k)$ and $\tilde{\phi}_r(k)$ are defined in the same way. When it is not clearly specified, we shall always take $\mathcal{M}_i = \sigma(X_k, k \leq i)$.

2.2.4 Projective measure of dependence

Sometimes, it is not necessary to introduce a supremum over a class of functions. We can work with the simple following projective measure of dependence

Definition 2.7. *Let $(\Omega, \mathcal{A}, \mathbb{P})$ be a probability space, and $\mathcal{M}$ a σ-algebra of $\mathcal{A}$. Let $p \in [1, \infty]$. For any $\mathbb{L}^p$−integrable real valued random variable define*

$$\gamma_p(\mathcal{M}, X) = \|\mathbb{E}(X|\mathcal{M}) - \mathbb{E}(X)\|_p. \tag{2.2.16}$$

Let $(X_i)_{i\in\mathbb{Z}}$ be a sequence of $\mathbb{L}^p$−integrable real valued random variables, and let $(\mathcal{M}_i)_{i\in\mathbb{Z}}$ be a sequence of σ-algebras of $\mathcal{A}$. The sequence of coefficients $\gamma_p(k)$ is then defined by

$$\gamma_p(k) = \sup_{i\in\mathbb{Z}} \gamma_p(\mathcal{M}_i, X_{i+k}) . \tag{2.2.17}$$

When it is not clearly specified, we shall always take $\mathcal{M}_i = \sigma(X_k, k \leq i)$.

Remark 2.5. *Those coefficients are defined in Gordin (1969) [97], if $p \geq 2$ and in Gordin (1973) [98] if $p = 1$. Mc Leish (1975a) [128] and (1975b) [129] uses these coefficients in order to derive various limit theorems. Let us notice that*

$$\gamma_p(\mathcal{M}, X) \leq \theta_p(\mathcal{M}, X) . \tag{2.2.18}$$

Chapter 3

Models

The chapter is organized as follows: we first introduce Bernoulli shifts, a very broad class of models that contains the major part of processes derived from a stationary sequence. As an example, we define the class of Volterra processes that are multipolynomial transformation of the stationary sequence. We will discuss the dependence properties of Bernoulli shifts, whether the initial is a dependent or independent sequence. When the innovation sequence is independent, we will distinguish between causal and non-causal processes. After these general properties, we focus on Markov models and some of their extensions, as well as dynamical systems which may be studied as Markov chains up to a time reversal. After this we shall consider LARCH(∞)-models which are built by a mix of definition of Volterra series and Markov processes and will provide an attractive class of non linear and non Markovian times series. To conclude, we consider associated processes and we review some other types of stationary processes or random fields which satisfy some weak dependence condition.

3.1 Bernoulli shifts

Definition 3.1. *Let $H : \mathbb{R}^{\mathbb{Z}} \to \mathbb{R}$ be a measurable function. Let $(\xi_n)_{n\in\mathbb{Z}}$ be a strictly stationary sequence of real-valued random variables. A Bernoulli shift with innovation process $(\xi_n)_{n\in\mathbb{Z}}$ is defined as*

$$X_n = H\left((\xi_{n-i})_{i\in\mathbb{Z}}\right), \qquad n \in \mathbb{Z}. \tag{3.1.1}$$

This sequence is strictly stationary.

Remark that the expression (3.1.1) is certainly not always clearly defined; as H is a function depending on an infinite number of arguments, it is generally given in form of a series, which is usually only defined in some $\mathbb{L}^p$ space. In order to define (3.1.1) in a general setting, we denote for any subset $J \subset \mathbb{Z}$:

$$H\left((\xi_{-j})_{j\in J}\right) = H\left((\xi_{-j}\,\mathbf{1}_{j\in J})_{j\in\mathbb{Z}}\right).$$

For finite subsets J this expression is generally well defined and simple to handle. In order to define such models in $\mathbb{L}^m$ we may assume

$$\sum_{n=1}^{\infty} \omega_{m,n} < \infty, \tag{3.1.2}$$

where, for some $m \geq 1$:

$$\omega_{m,n}^m = \mathbb{E}\left|H\left((\xi_{-j})_{|j|\leq n}\right) - H\left((\xi_{-j})_{|j|<n}\right)\right|^m. \tag{3.1.3}$$

This condition indeed proves that the sequence $H\left((\xi_{-j})_{|j|\leq n}\right)$ has the Cauchy property and thus converges in the Banach space $\mathbb{L}^m$ of the classes of random variables with a finite moment order m.

In fact the strict definition of the function H as an element of the space $\mathbb{L}^m(\mathbb{R}^Z, \mathcal{B}(\mathbb{R}^Z), \mu)$ is the following. Denote by μ the distribution of a process $\xi = (\xi_t)_{t\in\mathbb{Z}}$. The measure μ is a probability distribution on the measurable space $(\mathbb{R}^Z, \mathcal{B}(\mathbb{R}^Z))$. If as before we assume that ξ is stationary, that $S \subset \mathbb{R}$ is the support of the distribution of ξ_0, and that $\mathcal{S} \subset \mathbb{R}^{\mathbb{Z}}$ is the support of the distribution of the sequence ξ, then the random variable H defines a function over $\mathcal{S} \supset S^{(\mathbb{Z})}$ where $S^{(\mathbb{Z})}$ is the set of sequences with values 0 excepted for finitely many indices. Now given a function defined over $\mathbb{R}^{(\mathbb{Z})}$ the previous condition (3.1.2) ensures that such a function may be extended to a function $H \in \mathbb{L}^m(\mathbb{R}^Z, \mathcal{B}(\mathbb{R}^Z), \mu)$.

Dependence properties. No mixing properties have been derived for such models excepted for the simple case of m-dependent Bernoulli shifts, *i.e.* when H depends only on a finite number of variables.

3.1.1 Volterra processes

The most simple case of infinitely dependent Bernoulli shift is the infinite moving average process with independent innovations:

$$X_t = \sum_{-\infty}^{\infty} a_i \xi_{t-i} \tag{3.1.4}$$

This simple case is generalised by Volterra processes defined with use of polynomials of the innovation process. A Volterra process is a stationary process defined through a convergent Volterra expansion

$$X_t = v_0 + \sum_{k=1}^{\infty} V_{k;t}, \quad \text{where} \tag{3.1.5}$$

$$V_{k;t} = \sum_{i_1<\cdots<i_k} a_{k;i_1,\ldots,i_k} \xi_{t-i_1} \cdots \xi_{t-i_k}, \tag{3.1.6}$$

and v_0 denotes a constant and $(a_{k;i_1,\ldots,i_k})_{(i_1,\ldots,i_k)\in\mathbb{Z}^k}$ are real numbers for each $k \geq 1$. Let $p \geq 1$, then this expression converges in $\mathbb{L}^p$, provided that $\mathbb{E}|\xi_0|^p < \infty$ and the weights satisfy

$$\sum_{k=1}^{\infty} \sum_{i_1<\cdots<i_k} |a_{k;i_1,\ldots,i_k}|^p < \infty.$$

If the sequences $a_{k;i_1,\ldots,i_k} = 0$ as $i_1 < 0$ then the process is *causal* in the sense that X_t is measurable with respect to $\sigma\{\xi_i, i \leq t\}$. In this case t may be seen as the usual time $\sigma\{\xi_i, i \leq t\}$ denotes the history at epoch t.

Assume now that $p = 2$, $\mathbb{E}\,\xi_0 = 0$ and $\mathbb{E}\,\xi_i^2 = 1$, then the $k-$th order homogeneous chaotic processes $V_{k;t}$ are pairwise orthogonal, and it is thus enough to prove the existence of such homogeneous processes (3.1.6) in $\mathbb{L}^2$ in order to obtain the existence of the more general Volterra processes (3.1.5). Normal convergence of $V_{k;t}$ follows clearly from the convergence of the series defining its variance Γ_k^2

$$\Gamma_k^2 = \sum_{0\leq i_1<\cdots<i_{k-1}} a_{i_1,\ldots,i_{k-1},i_k}^2 < \infty.$$

For the general infinite order Volterra series (3.1.5), the corresponding variance is trivially related by orthogonality:

$$\Gamma^2 = \sum_{k=1}^{\infty} \Gamma_k^2 < \infty.$$

The formula defining Volterra processes can be generalized to expansions

$$X_t = v_0 + \sum_{k=1}^{\infty} V_{k;t}, \text{ where } V_{k;t} = \sum_{(i_1,\ldots,i_k)\in\mathbb{Z}^k} a_{k;i_1,\ldots,i_k} \xi_{t-i_1} \cdots \xi_{t-k}, \tag{3.1.7}$$

and v_0 denotes a constant and $(a_{k;i_1,\ldots,i_k})_{(i_1,\ldots,i_k)\in\mathbb{Z}^k}$ are real numbers for each $k \geq 1$. The major difference with the preceding definition is the fact that the indices in the product are not all different. Let $p \geq 1$, then the series converges in $\mathbb{L}^p$ provided that the weights satisfy

$$\sum_{k=1}^{\infty} \mathbb{E}|\xi_0|^{pk} \sum_{i_1\leq\cdots\leq i_k} |a_{k;i_1,\ldots,i_k}|^p < \infty. \tag{3.1.8}$$

The drawback of this generalization is the loss the properties of orthogonality and moments derived for the models (3.1.5). It is but possible to rewrite the process as a sum of orthogonal term by relaxing the condition of identical distribution and independence of the innovation process. Consider more general Volterra processes defined under the same condition (3.1.8) by

$$V_{k;t} = \sum_{j_1 < \cdots < j_k} a^{(k)}_{j_1,\dots,j_k} \xi^{(k,1)}_{t-j_1} \cdots \xi^{(k,k)}_{t-j_k} \tag{3.1.9}$$

For a fixed $k > 0$, the series $(\xi_t^{(k,l)})_{t\in\mathbb{Z}}$ are i.i.d. and mutually orthogonal for $l \le k$. Clearly, models (3.1.5) have this form but it is also interesting to see that models (3.1.7) may also be written as sums of such models. Consider an expansion (3.1.7), we may assume without loss of generality that $j_1 \le \cdots \le j_k$ and that $\mathbb{E}\xi_0 = 0$; we replace each power of an innovation variable by its decomposition on the Appell polynomial of the distribution of ξ_0. For example the squares will be replaced by

$$\xi_{t-i}^2 = (\xi_{t-i}^2 - \sigma^2) + \sigma^2 = A_2(\xi_{t-i}) + \sigma^2.$$

For higher order polynomials, recall that Appell polynomials (see *e.g.* Doukhan, 2002 [62]) are defined as $A_k(\xi_{t-i}) = \xi_{t-i}^k + \cdots$ in such a way that $\mathbb{E}A_k(\xi_t)P(\xi_t) = 0$ if the degree of the polynomial P is less than k. Replacing all the powers with the help of such Appell polynomial leads to a decomposition (3.1.9) in orthogonal terms.

Dependence properties. Note that such models may have no weak dependence properties, as in the case of simple moving averages, see Doukhan, Oppenheim and Taqqu (2003) [72] for a thorough survey of strongly dependent Volterra processes. No mixing property have been derived in the general case. The degenerated case of m dependence, when V_t depends only on the ξ_{t-i} for $i = 1$ to m, so that only a finite number of coefficients in each series are nonzero, satisfies any of the mixing properties. The mixing properties of causal linear processes corresponding to the term $V_{1;t}$ with $a_{i_1} = 0$ when $i_1 < 0$ were derived under the strong additional assumption that ξ_0's distribution admits a density which is itself an absolutely continuous function; see Doukhan, 1994 [61] for references, in this monograph the proof of mixing for non causal linear processes is not complete.

3.1.2 Noncausal shifts with independent inputs

Assume here that the shift is well defined and that the sequence of innovations (ξ_t) is i.i.d.

Dependence properties. In order to prove weak dependence properties, once the existence and measurability of the function H is ensured, it is sufficient to assume the sequence $\{\delta_r\}_{r\in\mathbb{N}}$ defined by:

$$\mathbb{E}\left|H\left(\xi_{t-j}, j\in\mathbb{Z}\right)-H\left(\xi_{t-j}\mathbf{1}_{|j|\le r}, j\in\mathbb{Z}\right)\right|=\delta_r \tag{3.1.10}$$

converges to 0 as r tends to infinity. Note that a simple bound for δ_r is $\delta_r \le \sum_{i>r}\omega_{1,i}$. The following elementary lemma is easily proved:

Lemma 3.1. *Bernoulli shifts are $\eta-$weakly dependent with*

$$\eta(r)\le 2\delta_{[r/2]}.$$

Proof. Let $X_n^{(s)}=H((\xi_{n-i}\mathbf{1}_{|i|\le s}))$. Clearly, the two sequences $(X_n^{(s)})_{n\le i}$ and $(X_n^{(s)})_{n\ge i+r}$ are independent if $r>2s$; now consider $\mathrm{Cov}(\mathbf{f},\mathbf{g})$ for the functions $\mathbf{f}=f(X_{i_1},\ldots,X_{i_u})$, $\mathbf{g}=g(X_{j_1},\ldots,X_{j_v})$, where f and g are bounded and $f,g\in\Lambda^{(1)}(|\cdot|_1)$ with $|\cdot|_1$ defined by (2.2.2). Let $i_1\le\cdots\le i_u$ and $j_1\le\cdots\le j_v$ such that $j_1-i_u>2s$. From the previous remark, $\mathbf{f}^{(s)}=f(X_{i_1}^{(s)},\ldots,X_{i_u}^{(s)})$ and $\mathbf{g}^{(s)}=g(X_{j_1}^{(s)},\ldots,X_{j_v}^{(s)})$ are independent, and consequently

$$\begin{aligned}
|\mathrm{Cov}(\mathbf{f},\mathbf{g})| &\le \left|\mathrm{Cov}(\mathbf{f}-\mathbf{f}^{(s)},\mathbf{g})\right|+\left|\mathrm{Cov}(\mathbf{f}^{(s)},\mathbf{g}-\mathbf{g}^{(s)})\right|\\
&\le 2\|g\|_\infty\mathbb{E}\left|\mathbf{f}-\mathbf{f}^{(s)}\right|+2\|f\|_\infty\mathbb{E}\left|\mathbf{g}-\mathbf{g}^{(s)}\right|\\
&\le 2\|g\|_\infty\mathrm{Lip}\,f\sum_{t=1}^{u}\mathbb{E}\left|X_{i_t}-X_{i_t}^{(s)}\right|+2\|f\|_\infty\mathrm{Lip}\,g\sum_{t=1}^{v}\mathbb{E}\left|X_{j_t}-X_{j_t}^{(s)}\right|\\
&\le 2(u\|g\|_\infty\mathrm{Lip}\,f+v\|f\|_\infty\mathrm{Lip}\,g)\delta_s\,. \qquad \square
\end{aligned}$$

The sequence $(\delta_k)_k$ is related to the modulus of uniform continuity of H. Under the following regularity conditions:

$$|H(u_i,\ i\in\mathbb{Z})-H(v_i,\ i\in\mathbb{Z})|\le\sum_{i\in\mathbb{Z}}a_i|u_i-v_i|^b,$$

for some non negative constants $(a_i)_{i\in\mathbb{Z}}, 0<b\le 1$ and if the sequence $(\xi_i)_{i\in\mathbb{Z}}$ has finite b-th order moment, then $\delta_k\le\sum_{|i|>k}a_iE|\xi_i|^b$.

Recall here that processes can be η-weakly dependent and nonmixing, see § 1.5.

3.1.3 Noncausal shifts with dependent inputs

The condition of independent inputs ξ may be relaxed. *E.g.* in eqn. (3.1.4), instead of independence, assume that the sequence $(\xi_n)_{n\in\mathbb{Z}}$ is η_ξ-weak dependent

then the process $(X_n)_{n\in\mathbb{Z}}$ is η-weak dependent with $\eta(r) \le \eta_\xi(r/2) + \delta_{r/2}$. Such an heredity property of weak dependence is unknown under mixing. A general statement of this property is provided below in lemma 3.3. Let us now note by $(\xi_i)_{i\in\mathbb{Z}}$ a weakly dependent innovation process. The coefficient λ is proved to very useful to study Bernoulli shifts $X_n = H(\xi_{n-j}, j \in \mathbb{Z})$ with weakly dependent innovation process $(\xi_i)_i$ from the forthcoming lemma (see Doukhan and Wintenberger, 2005 [77]).
Let $H : \mathbb{R}^{\mathbb{Z}} \to \mathbb{R}$ be a measurable function and $X_n = H(\xi_{n-i}, i \in \mathbb{Z})$. In order to define X_n, we assume that H satisfies: for each $s \in \mathbb{Z}$, if $x, y \in \mathbb{R}^{\mathbb{Z}}$ satisfy $x_i = y_i$ for each index $i \neq s$

$$|H(x) - H(y)| \le b_s(\sup_{i\neq s} |x_i|^l \vee 1)|x_s - y_s| \tag{3.1.11}$$

where z is defined by $z_s = 0$ and $z_i = x_i = y_i$ for each $i \neq s$. This assumption is stronger than in the case of independent innovations (see equation (3.1.10)). The following lemma proves the existence of such models:

Lemma 3.2. *Let $X_n = H(\xi_{n-i}, i \in \mathbb{Z})$ be a Bernoulli shift such that $H : \mathbb{R}^{\mathbb{Z}} \to \mathbb{R}$ satisfies the condition (3.1.11) with $l \ge 0$ and some sequence $b_s \ge 0$ such that $\sum_s |s| b_s < \infty$. Assume that $\mathbb{E}|\xi_0|^{m'} < \infty$ with $lm + 1 < m'$ for some $m > 2$. Then $X_n = H(\xi_{n-i}, i \in \mathbb{Z})$ is a strongly stationary process, well defined in $\mathbb{L}^m$.*

The existence of example (3.1.4) was stated without proof, we now precise more involved examples of Bernoulli shifts with dependent innovations:

Example 3.1 (Volterra models with dependent inputs.). *Consider*

$$H(x) = \sum_{k=0}^{K} \sum_{j_1,\ldots,j_k} a^{(k)}_{j_1,\ldots,j_k} x_{j_1} \cdots x_{j_k},$$

then if x, y are as in eqn. (3.1.11):

$$H(x) - H(y) = \sum_{\substack{1 \le u \le k \le K \\ j_1,\ldots,j_{u-1} \\ j_{u+1},\ldots,j_k}} a^{(k)}_{j_1,\ldots,j_{u-1},s,j_{u+1},\ldots,j_k} x_{j_1} \cdots x_{j_{u-1}} (x_s - y_s) x_{j_{u+1}} \cdots x_{j_k}.$$

From the triangular inequality we derive that suitable constants in condition (3.1.11) may be chosen as $l = K - 1$ and

$$b_s = \sum_{k=1}^{K} \sum\nolimits^{(k,s)} |a^{(k)}_{j_1,\ldots,j_k}|,$$

where $\sum^{(k,s)}$ stands for the sums over all indices in $\mathbb{Z}^k$ and one of the indices $j_1, \ldots, j_k$ takes the value s.

Example 3.2 (Uniform Lipschitz Bernoulli shifts). *Assume that the condition (3.1.11) holds with $l = 0$, then the previous result still holds. An example of such a situation is the case of LARCH(∞) non-causal processes with bounded ($m' = +\infty$) and dependent stationary innovations.*

Proof of lemma 3.2. We first prove the existence of Bernoulli shift with dependent innovations in $\mathbb{L}^1$. The same proof leads to the existence in $\mathbb{L}^m$ for all $m \geq 1$ such that $lm + 1 \leq m'$. Here we set $\xi^{(s)} = (\xi_{-i}\mathbf{1}_{|i|<s})_{i\in\mathbb{Z}}$ and $\xi_+^{(s)} = (\xi_{-i}\mathbf{1}_{-s<i\leq s})_{i\in\mathbb{Z}}$ for $i \in \mathbb{Z} \cup \{\infty\}$. In order to prove the existence of Bernoulli shift with dependent innovations, we show that $H(\xi^{(\infty)})$ is the sum of a normally convergent series in $\mathbb{L}^1$. Then formally

$$X_0 = H(\xi^{(\infty)}) = H(0) + (H(\xi^{(1)}) - H(0)) + \sum_{s=1}^{\infty}\left((H(\xi^{(s+1)}) - H(\xi_+^{(s)})) + (H(\xi_+^{(s)}) - H(\xi^{(s)}))\right).$$

With (3.1.11) we obtain

$$\begin{aligned} |H(\xi^{(1)}) - H(0)| &\leq b_0|\xi_0|, \\ |H(\xi^{(s+1)}) - H(\xi_+^{(s)})| &\leq b_{-s}(\sup_{-s<i\leq s} |\xi_{-i}|^l \vee 1)|\xi_{-s}|, \\ |H(\xi_+^{(s)}) - H(\xi^{(s)})| &\leq b_s(\sup_{|i|<s} |\xi_{-i}|^l \vee 1)|\xi_s|. \end{aligned}$$

Hölder inequality yields

$$\mathbb{E}\left|H(\xi^{(1)}) - H(0)\right| + \sum_{s=1}^{\infty}\mathbb{E}\left|H(\xi^{(s+1)}) - H(\xi_+^{(s)})\right| + \mathbb{E}\left|H(\xi_+^{(s)}) - H(\xi^{(s)})\right| \leq \sum_{i\in\mathbb{Z}} 2|i|b_i(\|\xi_0\|_1 + \|\xi_0\|_{l+1}^{l+1}). \quad (3.1.12)$$

Hence assumption $l + 1 \leq m'$ with $\sum_{i\in\mathbb{Z}} |i|b_i < \infty$ together imply that the variable $H(\xi)$ is well defined. The same way proves that the process $X_n = H(\xi_{n-i}, i \in \mathbb{Z})$ is a well defined process in $\mathbb{L}^1$ and that it is strongly stationary. We can extend this result in $\mathbb{L}^m$ for all $m \geq 1$ such that $lm + 1 \leq m'$.

Dependence properties. Such models are proved to exhibit either λ- or η-weak dependence properties, as described below.

Lemma 3.3. *Assume that the conditions of lemma 3.2 are satisfied*

- *if the innovation process $(\xi_i)_{i\in\mathbb{Z}}$ is λ-weakly dependent (with coefficients $\lambda_\xi(r)$), then X_n is λ-weakly dependent with*

$$\lambda(k) = c \inf_{r\le[k/2]} \Big(\sum_{i\ge r} |i| b_i\Big) \vee \left((2r+1)^2 \lambda_\xi(k-2r)^{\frac{m'-1-l}{m'-1+l}}\right).$$

- *if the innovation process $(\xi_i)_{i\in\mathbb{Z}}$ is η-weakly dependent (with coefficients $\eta_\xi(r)$) then X_n is η-weakly dependent and there exists a constant $c>0$ such that*

$$\eta(k) = c \inf_{r\le[k/2]} \left(\sum_{i\ge r} |i| b_i\right) \vee \left((2r+1)^{1+\frac{l}{m'-1}} \eta_\xi(k-2r)^{\frac{m'-2}{m'-1}}\right).$$

Because Bernoulli shifts of κ-weak dependent innovations are neither κ- nor η-weakly dependent, the case of κ dependent innovation is here included in that of λ dependent inputs.
The proof of lemma 3.2 will be given below. If the weak dependence coefficients of ξ are regularly decreasing, it is easy to explicit the decay of the weak dependence coefficients of X:

Proposition 3.1. *Here $\lambda>0$ and $\eta>0$ are constants which can differ in each case.*

- *If $b_i = \mathcal{O}\left(i^{-b}\right)$ for some $b>2$ and $\lambda_\xi(i) = \mathcal{O}\left(i^{-\lambda}\right)$, resp. $\eta_\xi(i) = \mathcal{O}\left(i^{-\eta}\right)$ (as $i\uparrow\infty$) then from a simple calculation, we optimize both terms in order to prove that $\lambda(k) = \mathcal{O}\left(k^{-\lambda(1-\frac{2}{b})\frac{m'-1-l}{m'-1+l}}\right)$, resp. $\eta(k) = \mathcal{O}\left(k^{-\eta\frac{(b-2)(m'-2)}{(b-1)(m'-1)-l}}\right)$. Note that in the case $m'=\infty$ this exponent is arbitrarily close to λ for large values of $b>0$ and takes all possible values between 0 and λ.*

- *If $b_i = \mathcal{O}\left(e^{-ib}\right)$ for some $b>0$ and $\lambda_\xi(i) = \mathcal{O}\left(e^{-i\lambda}\right)$, resp. $\eta_\xi(i) = \mathcal{O}\left(e^{-i\eta}\right)$ (as $i\uparrow\infty$) we have $\lambda(k) = \mathcal{O}\left(k^2 e^{-\lambda k\frac{b(m'-1-l)}{b(m'-1+l)+2\eta(m'-1-l)}}\right)$, resp. $\eta(k) = \mathcal{O}\left(k^{\frac{m'-1-l}{m'-1}} e^{-\eta k\frac{b(m'-2)}{b(m'-1)+2\eta(m'-2)}}\right)$. The geometrical decays for both $(b_i)_i$ and coefficients of the innovations ensure the geometric decay of the weakly dependence coefficient of the Bernoulli shift.*

- *If the Bernoulli shift coefficients have a geometric decay, say $b_i = \mathcal{O}\left(e^{-ib}\right)$ and $\lambda_\xi(i) = \mathcal{O}\left(i^{-\lambda}\right)$, resp. $\eta_\xi(i) = \mathcal{O}\left(i^{-\eta}\right)$ (as $i\uparrow\infty$) we find*

$$\lambda(k) = \mathcal{O}\left((\log k)^2 k^{-\lambda\frac{m'-1-l}{m'-1+l}}\right), \textit{ resp. } \eta(k) = \mathcal{O}\left((\log k)^{1+\frac{l}{m'-1}} k^{-\eta\frac{m'-2}{m'-1}}\right).$$

If $m' = \infty$ this means that we only lose at most a factor $\log^2 k$ with respect to the dependence coefficients of the input dependent series $(\xi_i)_i$.

Proof of lemma 3.3. We exhibit some Lipschitz function by using a convenient truncation. Write $\overline{\xi} = \xi \vee (-T) \wedge T$ for a truncation T which will be precisely stated later. As in the proof of Lemma 3.1, we denote by $X_n^{(r)} = H((\xi_{n-i}\mathbf{1}_{|i|\le r}))$ and $\overline{X}_n^{(r)} = H((\overline{\xi}_{n-i}\mathbf{1}_{|i|\le s}))$. Furthermore, for any $k \ge 0$ and any $(u+v)$-tuples such that $s_1 < \cdots < s_u \le s_u + k \le t_1 < \cdots < t_v$, set $X_{\mathbf{s}} = (X_{s_1}, \ldots, X_{s_u})$, $X_{\mathbf{t}} = (X_{t_1}, \ldots, X_{t_v})$ and $\overline{X}_{\mathbf{s}}^{(r)} = (\overline{X}_{s_1}^{(r)}, \ldots, \overline{X}_{s_u}^{(r)})$, $\overline{X}_{\mathbf{t}}^{(r)} = (\overline{X}_{t_1}^{(r)}, \ldots, \overline{X}_{t_v}^{(r)})$. Then we have for all f, g satisfying $\|f\|_\infty, \|g\|_\infty \le 1$ and $\mathrm{Lip}\, f + \mathrm{Lip}\, g < \infty$:

$$\begin{aligned} |\mathrm{Cov}(f(X_{\mathbf{s}}), g(X_{\mathbf{t}}))| &\le |\mathrm{Cov}(f(X_{\mathbf{s}}) - f(\overline{X}_{\mathbf{s}}^{(r)}), g(X_{\mathbf{t}}))| & (3.1.13)\\ &+ |\mathrm{Cov}(f(\overline{X}_{\mathbf{s}}^{(r)}), g(X_{\mathbf{t}}) - g(\overline{X}_{\mathbf{t}}^{(r)}))| & (3.1.14)\\ &+ |\mathrm{Cov}(f(\overline{X}_{\mathbf{s}}^{(r)}), g(\overline{X}_{\mathbf{t}}^{(r)}))|. & (3.1.15)\end{aligned}$$

Using that $\|g\|_\infty \le 1$, the term (3.1.13) in the sum is bounded by

$$\begin{aligned} &2\mathrm{Lip}\, f \cdot \mathbb{E}\Big|\sum_{i=1}^{u} (X_{s_i} - \overline{X}_{s_i}^{(r)})\Big| \\ &\qquad \le 2\, u\, \mathrm{Lip}\, f\Big(\max_{1\le i\le u} \mathbb{E}|X_{s_i} - X_{s_i}^{(r)}| + \max_{1\le i\le u} \mathbb{E}|X_{s_i}^{(r)} - \overline{X}_{s_i}^{(r)}|\Big).\end{aligned}$$

With the same arguments as in the proof of the existence of $H(\xi^{(\infty)})$ (see equation (3.1.12)), the first term in the right hand side is bounded by $(\|\xi_0\|_1 + \|\xi_0\|_{l+1}^{l+1}) \sum_{i\ge s} 2|i|b_i$. Notice now that if x, y are sequences with $x_i = y_i = 0$ if $|i| \ge r$ then a repeated application of the previous inequality (3.1.11) yields

$$|H(x) - H(y)| \le L(\|x\|_\infty^l \vee \|y\|_\infty^l \vee 1)\|x - y\| \qquad (3.1.16)$$

where $L = \sum_{i\in\mathbb{Z}} |i|b_i < \infty$. The second term is bounded by using (3.1.16):

$$\begin{aligned} \mathbb{E}\left|X_{s_i}^{(r)} - \overline{X}_{s_i}^{(r)}\right| &= \mathbb{E}\left|H\left(\xi^{(r)}\right) - H\left(\overline{\xi}^{(r)}\right)\right| \\ &\le L\mathbb{E}\left(\left(\max_{-r\le i\le r} |\xi_i|\right)^l \sum_{-r\le j\le r} \{|\xi_j|\mathbf{1}_{\xi_j\ge T}\}\right) \\ &\le L(2r+1)^2\mathbb{E}\left(\max_{-r\le i,j\le r} |\xi_i|^l\{|\xi_j|\mathbf{1}_{|\xi_j|\ge T}\}\right) \\ &\le L(2r+1)^2\|\xi_0\|_{m'}^{m'} T^{l+1-m'}\end{aligned}$$

The term (3.1.14) is analogously bounded. We write (3.1.15) as

$$\left|\mathrm{Cov}(\overline{F}^{(r)}(\xi_{s_i+j}, 1 \le i \le u, |j| \le r), \overline{G}^{(r)}(\xi_{t_i+j}, 1 \le i \le v, |j| \le r)\right|,$$

where $\overline{F}^{(r)} : \mathbb{R}^{u(2r+1)} \to \mathbb{R}$ and $\overline{G}^{(r)} : \mathbb{R}^{u(2r+1)} \to \mathbb{R}$. If $r \le [k/2]$, assume that ξ is η weakly dependent (resp. λ-weak dependent) to bound this covariance by $\psi(\mathrm{Lip}\,\overline{F}^{(r)}, \mathrm{Lip}\,\overline{G}^{(r)}, u(2r+1), v(2r+1))\epsilon_{k-2r}$, where $\psi(u,v,a,b) = uvab$ and $\epsilon_i = \eta_i$ (resp. $\psi(u,v,a,b) = uvab + ua + vb$ and $\epsilon_i = \lambda_i$). Let $x = (x_1, \ldots, x_u)$ and $y = (y_1, \ldots, y_u)$ where $x_i, y_i \in \mathbb{R}^{2r+1}$; a bound for $\mathrm{Lip}\,\overline{F}^{(r)}$ writes as the supremum over sequences x, y of:

$$\frac{|f(H(\overline{x}_{s_i+l}, 1 \le i \le u, |l| \le r) - f(H(\overline{y}_{s_i+l}, 1 \le i \le u, |l| \le r)|}{\sum_{j=1}^{u} \|x_j - y_j\|}.$$

Using (3.1.16), we have:

$$\begin{aligned}
|\overline{F}^{(r)}(x) - \overline{F}^{(r)}(y)| &\le \mathrm{Lip}\, f L \sum_{i=1}^{u} \left(\|\overline{x}_{s_i}\|_\infty \vee \|\overline{y}_{s_i}\|_\infty \vee 1\right)^l \|\overline{x}_{s_i} - \overline{y}_{s_i}\| \\
&\le \mathrm{Lip}\, f L T^l \sum_{i=1}^{u} \sum_{-r \le l \le r} |x_{s_i+l} - y_{s_i+l}|.
\end{aligned}$$

Hence $\mathrm{Lip}\, F^{(r)} \le \mathrm{Lip}\, f \cdot L \cdot T^l$ and, similarly, $\mathrm{Lip}\, G^{(r)} \le \mathrm{Lip}\, g \cdot L \cdot T^l$.
• Under η-weak dependence, we bound the covariance as:

$$\begin{aligned}
|\mathrm{Cov}(f(X_{\mathbf{s}}), g(X_{\mathbf{t}}))| &\le (u\mathrm{Lip}\, f + v\mathrm{Lip}\, g) \\
&\quad \times \left[4 \sum_{i \ge r} |i| b_i (\|\xi_0\|_1 + \|\xi_0\|_{l+1}^{l+1}) \right. \\
&\quad \left. + (2r+1)L\left((2r+1)2\|\xi_0\|_{m'}^{m'} T^{l+1-m'} + T^l \eta_\xi(k-2r)\right)\right]
\end{aligned}$$

We then fix the truncation $T^{m'-1} = 2(2r+1)\|\xi_0\|_{m'}^{m'}/\eta_\xi(k-2r)$ to obtain the result of the lemma 3.3 in the η-weak dependent case.
• Under λ-weak dependence, we obtain:

$$\begin{aligned}
|\mathrm{Cov}(f(X_{\mathbf{s}}), g(X_{\mathbf{t}}))| &\le (u\mathrm{Lip}\, f + v\mathrm{Lip}\, g + uv\mathrm{Lip}\, f\mathrm{Lip}\, g) \\
&\quad \times \left(\left\{4 \sum_{i \ge r} |i| b_i (\|\xi_0\|_1 + \|\xi_0\|_{l+1}^{l+1}) \right.\right. \\
&\quad \left. + (2r+1)L\left(2(2r+1)T^{l+1-m'}\|\xi_0\|_{m'}^{m'} + T^l \lambda_\xi(k-2r)\right)\right\} \\
&\quad \left. \vee \left\{(2r+1)^2 L^2 T^{2l} \lambda_\xi(k-2r)\right\}\right)
\end{aligned}$$

With the truncation such that $T^{l+m'-1} = 2\|\xi_0\|_{m'}^{m'}/(L\lambda_\xi(k-2r))$, we obtain the result of the lemma 3.3 in the present η-weak dependent case. □

3.1.4 Causal shifts with independent inputs

Let $(\xi_i)_{i\in\mathbb{Z}}$ be a stationary sequence of random variables with values in a measurable space $\mathcal{X}$. Assume that there exists a function H defined on a subset of $\mathcal{X}^{\mathbb{N}}$, with values in $\mathbb{R}$ and such that $H(\xi_0, \xi_{-1}, \xi_{-2}, \ldots)$ is defined almost surely. The stationary sequence $(X_n)_{n\in\mathbb{Z}}$ defined by

$$X_n = H(\xi_n, \xi_{n-1}, \xi_{n-2}, \ldots) \tag{3.1.17}$$

is called a causal function of $(\xi_i)_{i\in\mathbb{Z}}$.
In this section, we assume that $(\xi_i)_{i\in\mathbb{Z}}$ is i.i.d. In this causal case, another way to define a coupling coefficient is to consider a non increasing sequence $(\tilde{\delta}_{p,n})_{n\geq 0}$ (p may be infinite) such that

$$\tilde{\delta}_{p,n} \geq \|X_n - \tilde{X}_n\|_p, \tag{3.1.18}$$

where $\tilde{X}_t = H(\tilde{\xi}_t, \tilde{\xi}_{t-1}, \tilde{\xi}_{t-2}, \ldots)$, $\tilde{\xi}_n = \xi_n$ if $n > 0$ and $\tilde{\xi}_n = \xi'_n$ for $n \leq 0$ for an independent copy $(\xi'_t)_{t\in\mathbb{Z}}$ of $(\xi_t)_{t\in\mathbb{Z}}$. Here $\tilde{X}_t$ has the same distribution as X_t and is independent of $\mathcal{M}_0 = \sigma(X_i, i \leq 0)$.

Dependence properties. In this section, we shall use the results of chapter 5 to give upper bounds for the coefficients $\theta_{p,\infty}(n)$, $\tau_{p,\infty}(n)$, $\tilde{\alpha}_k(n)$, $\tilde{\beta}_k(n)$ and $\tilde{\phi}_k(n)$. More precisely, we have that

1. $\theta_{p,\infty}(n) \leq \tau_{p,\infty}(n) \leq \tilde{\delta}_{p,n}$.

2. Assume that X_0 has a continuous distribution function with modulus of continuity w. Let $g_p(y) = y(w(y))^{1/p}$. For any $1 \leq p < \infty$, we have

$$\tilde{\alpha}_k(n) \leq \tilde{\beta}_k(n) \leq 2k\Big(\frac{\tilde{\delta}_{p,n}}{g_p^{-1}(\tilde{\delta}_{p,n})}\Big)^{p}.$$

   In particular, if X_0 has a density bounded by K, we obtain the inequality $\tilde{\beta}_k(n) \leq 2k(K\tilde{\delta}_{p,n})^{p/(p+1)}$.

3. Assume that X_0 has a continuous distribution function, with modulus of uniform continuity w. Then

$$\tilde{\alpha}_k(n) \leq \tilde{\beta}_k(n) \leq \tilde{\phi}_k(n) \leq k\, w(\tilde{\delta}_{\infty,n}).$$

4. For $\tilde{\phi}_k(n)$ it is sometimes interesting to use the coefficient

$$\delta'_{p,n} = \|\mathbb{E}(|X_n - X_n^*|^p|\mathcal{M}_0)\|_\infty^{1/p}.$$

   With the same notations as in point 2, $\tilde{\phi}_k(n) \leq 2k\Big(\frac{\delta'_{p,n}}{g_p^{-1}(\delta'_{p,n})}\Big)^{p}$.

The point 1. can be proved by using Lemma 5.3. The points 2. 3. and 4. can be proved as the point 3. of Lemma 5.1, by using Proposition 5.1 and Markov inequality.

Application (causal linear processes). In that case $X_n = \sum_{j\geq 0} a_j \xi_{n-j}$. One can take $\delta'_{p,n} = 2\|\xi_0\|_p \sum_{j\geq n} |a_j|$. For $p=2$ and $\mathbb{E}\xi_0 = 0$ one may set

$$\delta'_{2,n} = \Big(2\mathbb{E}\xi_0^2 \sum_{j\geq n} a_j^2\Big)^{\frac{1}{2}}.$$

For instance, if $a_i = 2^{-i-1}$ and $\xi_0 \sim \mathcal{B}(1/2)$, then $\delta_{i,\infty} \leq 2^{-i}$. Since X_0 is uniformly distributed over $[0,1]$, we have $\tilde{\phi}_1(i) \leq 2^{-i}$. Recall that this sequence is not strongly mixing (see section 1.5).

Remark 3.1. *By interpreting causal Bernoulli shifts as physical systems, denoted $X_t = g(\ldots, \epsilon_{t-1}, \epsilon_t)$ Wu (2005) [188] introduces physical dependence coefficients quantifying the dependence of outputs (X_t) on inputs (ϵ_t). He considers the nonlinear system theory's coefficient*

$$\bar{\delta}_t = \|g(\ldots, \epsilon_0, \ldots, \epsilon_{t-1}, \epsilon_t) - g(\ldots, \epsilon_{-1}, \epsilon'_0, \ldots, \epsilon_{t-1}, \epsilon_t)\|_2$$

with ϵ' an independent copy of ϵ. This provides a sharp framework for the study of the question of CLT random processes and shed new light on a variety of problems including estimation of linear models with dependent errors in Wu (2006) [191], nonparametric inference of time series in Wu (2005) [192], representations of sample quantiles (Wu 2005 [189]) and spectral estimation (Wu 2005 [190]) among others. This specific $\mathbb{L}^2$-formulation is rather adapted to CLT and it is not directly possible to compare it with τ-dependence because coupling is given here with only one element in the past. Justification of the Bernoulli shift representation follows from Ornstein (1973) [138].

3.1.5 Causal shifts with dependent inputs

In this section, the innovations are not required to be i.i.d., but the method introduced in the preceding section still works. More precisely, assume that there exists a stationary sequence $(\xi'_i)_{i\in\mathbb{Z}}$ distributed as $(\xi_i)_{i\in\mathbb{Z}}$ and independent of $(\xi_i)_{i\leq 0}$. Define $\tilde{X}_n = H(\xi'_n, \xi'_{n-1}, \xi'_{n-2}, \ldots)$. Clearly $\tilde{X}_n$ is independent of $\mathcal{M}_0 = \sigma(X_i, i \leq 0)$ and distributed as X_n. Hence one can apply the result of Lemma 5.3: if $(\tilde{\delta}_{p,n})_{n\geq 0}$ is a non increasing sequence satisfying (3.2.2), then the upper bounds 1. 2. 3. and 4. of the preceding section hold.
In particular, these results apply to the case where the sequence $(\xi_i)_{i\in\mathbb{Z}}$ is β-mixing. According to Theorem 4.4.7 in Berbee (1979) [16], if Ω is rich enough,

there exists $(\xi'_i)_{i\in\mathbb{Z}}$ distributed as $(\xi_i)_{i\in\mathbb{Z}}$ and independent of $(\xi_i)_{i\le 0}$ such that $\mathbb{P}(\xi_i \neq \xi'_i \text{ for some } i \ge k) = \beta(\sigma(\xi_i, i \le 0), \sigma(\xi_i, i \ge k))$.

Application (causal linear processes). In that case $X_n = \sum_{j\ge 0} a_j \xi_{n-j}$. For any $p \ge 1$, we can take the non increasing sequence $(\tilde{\delta}_{p,n})_{n\ge 0}$ such that

$$\tilde{\delta}_{p,n} \ge \|\xi_0 - \xi'_0\|_p \sum_{j\ge n} |a_j| + \sum_{j=0}^{n-1} |a_j| \|\xi_{i-j} - \xi'_{i-j}\|_p \ge \sum_{j\ge 0} |a_j| \|\xi_{i-j} - \xi'_{i-j}\|_p \,.$$

From Proposition 2.3 in Merlevède and Peligrad (2002) [130], one can take

$$\tilde{\delta}_{p,n} \ge \|\xi_0 - \xi'_0\|_p \sum_{j\ge n} |a_j| + \sum_{j=0}^{n-1} |a_j| \Big(2^p \int_0^{\beta(\sigma(\xi_k, k\le 0), \sigma(\xi_k, k\ge i-j))} Q^p_{\xi_0}(u) \Big)^{1/p} du \,,$$

where Q_{ξ_0} is the generalized inverse of the tail function $x \mapsto \mathbb{P}(|\xi_0| > x)$ (see Lemma 5.1 for the precise definition).

3.2 Markov sequences

Let $(X_n)_{n\ge 1-d}$ be sequence of random variables with values in a Banach space $(\mathbb{B}, \|\cdot\|)$. Assume that X_n satisfies the recurrence equation

$$X_n = F(X_{n-1}, \ldots, X_{n-d}; \xi_n). \tag{3.2.1}$$

where F is a measurable function with values in $\mathbb{B}$, the sequence $(\xi_n)_{n>0}$ is i.i.d. and $(\xi_n)_{n>0}$ is independent of $(X_0, \ldots, X_{d-1})$. Note that if X_n satisfies (3.2.1) then the random variable $Y_n = (X_n, \ldots, X_{n-d+1})$ defines a Markov chain such that $Y_n = M(Y_{n-1}; \xi_n)$ with

$$M(x_1, \ldots, x_d; \xi) = (F(x_1, \ldots, x_d; \xi), x_1, \ldots, x_{d-1}).$$

Dependence properties. Assume that $(X_n)_{n\le d-1}$ is a stationary solution to (3.2.1). As previously, let $Y_0 = (X_0, \ldots, X_{1-d})$, and let $\tilde{Y}_0 = (\tilde{X}_0, \ldots, \tilde{X}_{1-d})$ be and independent vectors with the same law as Y_0 (that is a distribution invariant by M). Let then $\tilde{X}_n = F(\tilde{X}_{n-1}, \ldots, \tilde{X}_{n-d}; \xi_n)$. Clearly, for $n > 0$, $\tilde{X}_n$ is distributed as X_n and independent of $\mathcal{M}_0 = \sigma(X_i, 1-d \le i \le 0)$. As in the preceding sections, let $(\tilde{\delta}_{p,n})_{n\ge 0}$ (p may be infinite) be a non increasing sequence such that

$$\tilde{\delta}_{p,n} \ge (\mathbb{E}\|X_n - \tilde{X}_n\|^p)^{1/p} \,. \tag{3.2.2}$$

Applying Lemma 5.3, we infer that

$$\theta_{p,\infty}(n) \le \tau_{p,\infty}(n) \le \tilde{\delta}_{p,n} \,.$$

For the coefficients $\tilde{\alpha}_k(n)$, $\tilde{\beta}_k(n)$ and $\tilde{\phi}_k(n)$ in the case $\mathbb{B} = \mathbb{R}$, the upper bounds 2., 3., and 4. of section 3.1.4 also hold under the same conditions on the distribution function of X_0.
Assume that, for some $p \geq 1$, the function F satisfies

$$(\mathbb{E}\|F(x;\xi_1) - F(y;\xi_1)\|^p)^{\frac{1}{p}} \leq \sum_{i=1}^{d} a_i \|x_i - y_i\|, \qquad \sum_{i=1}^{d} a_i < 1. \tag{3.2.3}$$

Then one can prove that the coefficient $\tau_{p,\infty}(n)$ decreases at an exponential rate. Indeed, we have that

$$\tilde{\delta}_{p,n} \leq \sum_{i=1}^{d} a_i \tilde{\delta}_{p,n-i}.$$

For two vectors x, y in $\mathbb{R}^d$, we write $x \leq y$ if $x_i \leq y_i$ for any $1 \leq i \leq d$. Using this notation, we have that

$$(\tilde{\delta}_{p,n}, \dots, \tilde{\delta}_{p,n-d+1})^t \leq A(\tilde{\delta}_{p,n-1}, \dots, \tilde{\delta}_{p,n-d})^t,$$

with the matrix A equal to

$$\begin{pmatrix} a_1 & a_2 & \cdot & \cdot & \cdot & a_{d-1} & a_d \\ 1 & 0 & \cdot & \cdot & \cdot & 0 & 0 \\ 0 & 1 & \cdot & \cdot & \cdot & 0 & 0 \\ \cdot & \cdot & \cdot & \cdot & \cdot & \cdot & \cdot \\ \cdot & \cdot & \cdot & \cdot & \cdot & \cdot & \cdot \\ \cdot & \cdot & \cdot & \cdot & \cdot & \cdot & \cdot \\ 0 & 0 & \cdot & \cdot & \cdot & 1 & 0 \end{pmatrix}$$

Iterating this inequality, we obtain that

$$(\tilde{\delta}_{p,n}, \dots, \tilde{\delta}_{p,n-d+1})^t \leq A^n(\tilde{\delta}_{p,0}, \dots, \tilde{\delta}_{p,1-d})^t.$$

Since $\sum_{i=1}^{d} a_i < 1$, the matrix A has a spectral radius strictly smaller than 1. Hence, we obtain that there exists $C > 0$ and ρ in $[0, 1[$ such that $\tilde{\delta}_{p,n} \leq C\rho^n$. Consequently

$$\theta_{p,\infty}(n) \leq \tau_{p,\infty}(n) \leq C\rho^n.$$

If $\mathbb{B} = \mathbb{R}$ and the condition (3.2.3) holds for $p = 1$, and if the distribution function F_X of X_0 is such that $|F_X(x) - F_X(y)| \leq K|x - y|^\gamma$ for some γ in $]0, 1]$, then we have the upper bound

$$\tilde{\alpha}_k(n) \leq \tilde{\beta}_k(n) \leq 2kK^{1/(\gamma+1)}C^{\gamma/(\gamma+1)}\rho^{n\gamma/(\gamma+1)}.$$

If the condition (3.2.3) holds for $p = \infty$, then

$$\tilde{\alpha}_k(n) \leq \tilde{\beta}_k(n) \leq \tilde{\phi}_k(n) \leq kKC^\alpha\rho^{\gamma n}.$$

We give below some examples of the general situation described in this section.

3.2.1 Contracting Markov chain.

For simplicity, let $\mathbb{B} = \mathbb{R}^d$, and let $\|\cdot\|$ be a norm on $\mathbb{R}^d$. Let F be a $\mathbb{R}^d$ valued function and consider the recurrence equation

$$X_n = F(X_{n-1}; \xi_n). \tag{3.2.4}$$

Assume that

$$A^p = \mathbb{E}\|F(0;\xi_1)\|^p < \infty \quad \text{and} \quad \mathbb{E}\|F(x,\xi_1) - F(y,\xi_1)\|^p \le a^p\|x-y\|^p, \tag{3.2.5}$$

for some $a < 1$ and $p \ge 1$. Duflo (1996) [81] proves that condition (3.2.5) implies that the Markov chain $(X_i)_{i\in\mathbb{N}}$ has a stationary law μ with finite moment of order p. In the sequel, we suppose that μ is the distribution of X_0 (*i.e.* the Markov chain is stationary).
Bougerol (1993) [27] and Diaconis and Friedmann (1999) [60] provide a wide variety of examples of stable Markov chains, see also Ango-Nzé and Doukhan (2002) [7].

Dependence properties. Mixing properties may be derived for Markov chains (see the previous references and Mokkadem (1990) [132]), but this property always need an additional regularity assumption on the innovations, namely the innovations must have some absolutely continuous component. By contrast, no assumption on the distribution of ξ_0 is necessary to obtain a geometrical decay of the coefficient $\tau_{p,\infty}(n)$. More precisely, arguing as in the previous section, one has the upper bounds: if $\tilde{X}_0$ is independent of X_0 and distributed as X_0,

$$\theta_{p,\infty}(n) \le \tau_{p,\infty}(n) \le \|\tilde{X}_0 - X_0\|_p \, a^n.$$

In the same way, if each component of X_0 has a distribution function which is Hölder, then the coefficients $\alpha_k(n)$ and $\beta_k(n)$ decrease geometrically (see lemma 5.1).

Let us show now that contractive Markov chains can be represented as Bernoulli shifts in a general situation when X_t and ζ_t take values in Euclidean spaces $\mathbb{R}^d$ and $\mathbb{R}^D$, respectively, $d, D \ge 1$ with $\|\cdot\|$ denoting indifferently a norm on $\mathbb{R}^D$ or on $\mathbb{R}^d$. Any homogeneous Markov chain X_t may also be represented as solution of a recurrence equation

$$X_n = F(X_{n-1}, \xi_n) \tag{3.2.6}$$

where $F(u,z)$ is a measurable function and $(\xi_n)_{n>0}$ is an i.i.d. sequence independent of X_0, see *e.g.* Kallenberg (1997, Proposition 7.6) [111].

Proposition 3.2 (Stable Markov chains as Bernoulli shifts). *The stationary iterative models (3.2.6) are Bernoulli shifts (3.1.17) if condition (3.2.5) holds.*

Proof. We denote by μ the distribution of $\zeta = (\zeta_n)_{n\in\mathbb{N}}$ on $\left(\mathbb{R}^D\right)^{\mathbb{N}}$. We define the space $\mathbb{L}^p(\mu, \mathbb{R}^d)$ of $\mathbb{R}^d$-valued Bernoulli shifts functions H such that $H(\zeta_0, \zeta_1, \ldots) \in \mathbb{L}^p$. We shall prove as in Doukhan and Truquet (2006) [76] that the operator of $\mathbb{L}^p(\mu, \mathbb{R}^d)$ $\Phi : H \mapsto K$ with $K(\zeta_0, \zeta_1, \ldots) = M(H(\zeta_1, \zeta_2, \ldots), \zeta_0)$ satisfies the contraction principle. Then, Picard fixed point theorem will allows to conclude.
We first mention that the condition (3.2.5) implies $\|M(x, \zeta_0)\|_p \le A + a\|x\|$ hence with independence of the sequence ζ this yields $\|K\|_p \le A + a\|H\|_p$; thus $\Phi(\mathbb{L}^p(\mu, \mathbb{R}^d)) \subset \mathbb{L}^p(\mu, \mathbb{R}^d)$. Now for $H, H' \in \mathbb{L}^p(\mu, \mathbb{R}^d)$ we also derive with analogous arguments that $\|\Phi(H) - \Phi(H')\|_p \le a\|H - H'\|_p$. □

Remark 3.2. *It is also possible to derive the Bernoulli shift representation through a recursive iteration in the autoregressive formula.*

3.2.2 Nonlinear AR(d) models

For simplicity, let $\mathbb{B} = \mathbb{R}$. Autoregressive models of order d are models such that:

$$X_n = r(X_{n-1}, \ldots, X_{n-d}) + \xi_n \,. \tag{3.2.7}$$

In such a case, the function F is given by

$$F(u_1, \ldots, u_d, \xi) = r(u_1, \ldots, u_d) + \xi,$$

Assume that $\mathbb{E}|\xi_1|^p < \infty$ and that

$$|r(u_1, \ldots, u_d) - r(v_1, \ldots, v_d)| \le \sum_{i=1}^d a_i |u_i - v_i|$$

for some $a_1, \ldots, a_d \ge 0$ such that $a = \sum_{i=1}^d a_i < 1$. Then the condition (3.2.3) holds, and we infer from section 3.2 that the coefficients $\tau_{p,\infty}(n)$ decrease exponentially fast.

3.2.3 ARCH-type processes

For simplicity, let $\mathbb{B} = \mathbb{R}$. Let

$$F(u, z) = A(u) + B(u)z \tag{3.2.8}$$

for suitable Lipschitz functions $A(u), B(u), u \in \mathbb{R}$. The corresponding iterative model (3.2.6) satisfies (3.2.5) if

$$a = \mathrm{Lip}\,(A) + \|\xi_1\|_p \,\mathrm{Lip}\,(B) < 1.$$

If $p = 2$ and $\mathbb{E}(\xi_n) = 0$, it is sufficient to assume that

$$a = \sqrt{(\mathrm{Lip}\,(A))^2 + \mathbb{E}(\xi_1^2)\,(\mathrm{Lip}\,(B))^2} < 1.$$

Some examples of iterative Markov processes given by functions of type (3.2.8) are:
• nonlinear AR(1) processes (case $B \equiv 1$);
• stochastic volatility models (case $A \equiv 0$);
• classical ARCH(1) models (case $A(u) = \alpha u$, $B(u) = \sqrt{\beta + \gamma u^2}$, $\alpha, \beta, \gamma \geq 0$).
In the last example, the inequality (3.2.5) holds for $p = 2$ with $a^2 = \alpha^2 + \mathbb{E}\xi_0^2\gamma$. A general description of these models can be found in Section 3.4.2.

3.2.4 Branching type models

Here $\mathbb{B} = \mathbb{R}$ and ξ_n is $\mathbb{R}^D$-valued. Let $\xi_n = \left(\xi_n^{(1)}, \dots, \xi_n^{(D)}\right)$. Let now $A_1, \dots, A_D$ be Lipschitz functions from $\mathbb{R}$ to $\mathbb{R}$, and let

$$F\left(u, (z^{(1)}, \dots, z^{(D)})\right) = \sum_{j=1}^{D} A_j(u) z^{(j)},$$

for $(z^{(1)}, \dots, z^{(D)}) \in \mathbb{R}^D$. For such functions F, if $\mathbb{E}(\xi_1^{(i)}\xi_1^{(j)}) = 0$ for $i \neq j$, the relation (3.2.5) holds with $p = 2$ if

$$a^2 = \sum_{j=1}^{D} (\mathrm{Lip}\,(A_j))^2\, \mathbb{E}\big((\xi_0^{(j)})^2\big) < 1.$$

Some examples of this situation are

- If $D = 2$, and $\xi_1^{(1)} \sim b(\bar{p})$ is a Bernoulli variable independent of a centered variable $\xi_1^{(2)} \in \mathbb{L}^2$ and $A_1(u) = u, A_2(u) = 1$ then the previous relations hold if $\overline{p} < 1$.

- If $D = 3$, $\xi_1^{(1)} = 1 - \xi_1^{(2)} \sim b(\overline{p})$ is independent of a centered variable $\xi_1^{(3)} \in \mathbb{L}^2$, then one obtains usual threshold models if $A_3 \equiv 1$.
  This only means that $X_n = F_n(X_{n-1}) + \xi_n^{(3)}$ where F_n is an i.i.d. sequence, independent of the sequence $(\xi_n^{(3)})_{n\geq 1}$, and such that $F_n = A_1$ with probability $\overline{p}$ and $F_n = A_2$, else.
  The condition (3.2.5) with $p = 2$ writes here

$$a = \overline{p}\,(\mathrm{Lip}\,(A_1))^2 + (1 - \overline{p})\,(\mathrm{Lip}\,(A_2))^2 < 1.$$

3.3 Dynamical systems

Let $I = [0,1]$, T be a map from I to I and define $X_i = T^i$. If μ is invariant by T, the sequence $(X_i)_{i\geq 0}$ of random variables from (I,μ) to I is strictly stationary. Denote by $\|g\|_{1,\lambda}$ the $\mathbb{L}^1$-norm with respect to the Lebesgue measure λ on I and by $\|\nu\| = |\nu|(I)$ the total variation of ν.

Covariance inequalities. In many interesting cases, one can prove that, for any BV function h and any k in $\mathbb{L}^1(I,\mu)$,

$$|\mathrm{Cov}(h(X_0), k(X_n))| \leq a_n \|k(X_n)\|_1 (\|h\|_{1,\lambda} + \|dh\|)\,, \tag{3.3.1}$$

for some non increasing sequence a_n tending to zero as n tends to infinity. Note that if (3.3.1) holds, then

$$\begin{aligned} |\mathrm{Cov}(h(X_0), k(X_n))| &= |\mathrm{Cov}(h(X_0) - h(0), k(X_n))| \\ &\leq a_n \|k(X_n)\|_1 (\|h - h(0)\|_{1,\lambda} + \|dh\|)\,. \end{aligned}$$

Since $\|h - h(0)\|_{1,\lambda} \leq \|dh\|$, we obtain that

$$|\mathrm{Cov}(h(X_0), k(X_n))| \leq 2a_n \|k(X_n)\|_1 \|dh\|\,. \tag{3.3.2}$$

The associated Markov chain. Define the operator $\mathcal{L}$ from $\mathbb{L}^1(I,\lambda)$ to $\mathbb{L}^1(I,\lambda)$ *via* the equality

$$\int_0^1 \mathcal{L}(h)(x) k(x) \lambda(dx) = \int_0^1 h(x)(k \circ T)(x)\lambda(dx)$$

where $h \in \mathbb{L}^1(I,\lambda)$ and $k \in \mathbb{L}^\infty(I,\lambda)$. The operator $\mathcal{L}$ is called the Perron-Frobenius operator of T. Assume that μ is absolutely continuous with respect to the Lebesgue measure, with density f_μ. Let I^* be the support of μ (that is $(I^*)^c$ is the largest open set in I such that $\mu((I^*)^c) = 0$) and choose a version of f_μ such that $f_\mu > 0$ on I^* and $f_\mu = 0$ on $(I^*)^c$. Note that one can always choose $\mathcal{L}$ such that $\mathcal{L}(f_\mu h)(x) = \mathcal{L}(f_\mu h)(x)\mathbf{1}_{f_\mu(x)>0}$. Define a Markov kernel associated to T by

$$K(h)(x) = \frac{\mathcal{L}(f_\mu h)(x)}{f_\mu(x)}\mathbf{1}_{f_\mu(x)>0} + \mu(h)\mathbf{1}_{f_\mu(x)=0}. \tag{3.3.3}$$

It is easy to check (see for instance Barbour *et al.*(2000) [9]) that $(X_0, X_1, \ldots, X_n)$ has the same distribution as $(Y_n, Y_{n-1}, \ldots, Y_0)$ where $(Y_i)_{i\geq 0}$ is a stationary Markov chain with invariant distribution μ and transition kernel K.

Spectral gap. In many interesting cases, the spectral analysis of $\mathcal{L}$ in the Banach space of BV-functions equipped with the norm $\|h\|_v = \|dh\| + \|h\|_{1,\lambda}$

can be done by using the Theorem of Ionescu-Tulcea and Marinescu (see Lasota and Yorke (1974) [115]): assume that 1 is a simple eigenvalue of $\mathcal{L}$ and that the rest of the spectrum is contained in a closed disk of radius strictly smaller than one. Then there exists an unique T-invariant absolutely continuous probability μ whose density f_μ is BV, and

$$\mathcal{L}^n(h) = \lambda(h) f_\mu + \Psi^n(h) \tag{3.3.4}$$

with $\Psi(f_\mu) = 0$ and $\|\Psi^n(h)\|_v \le D\rho^n \|h\|_v$, for some $0 \le \rho < 1$ and $D > 0$. Assume moreover that

$$\left\| \frac{1}{f_\mu} \mathbf{1}_{f_\mu > 0} \right\|_v = \gamma < \infty. \tag{3.3.5}$$

Starting from (3.3.3), we have that

$$K^n(h) = \mu(h) + \frac{\Psi^n(h f_\mu)}{f_\mu} \mathbf{1}_{f_\mu > 0}.$$

Let $\|\cdot\|_{\infty,\lambda}$ be the essential sup with respect to λ. Taking $C_1 = 2D\gamma(\|df_\mu\| + 1)$, we obtain $\|K^n(h) - \mu(h)\|_{\infty,\lambda} \le C_1 \rho^n \|h\|_v$.
This estimate implies (3.3.1) with $a_n = C_1\rho^n$. Indeed,

$$\begin{aligned}
|\mathrm{Cov}(h(X_0), k(X_n))| &= |\mathrm{Cov}(h(Y_n), k(Y_0))| \\
&\le \|k(Y_0)(\mathbb{E}(h(Y_n)|\sigma(Y_0)) - \mathbb{E}(h(Y_n)))\|_1 \\
&\le \|k(Y_0)\|_1 \|K^n(h) - \mu(h)\|_{\infty,\lambda} \\
&\le C_1 \rho^n \|k(Y_0)\|_1 (\|dh\| + \|h\|_{1,\lambda}) .
\end{aligned}$$

Moreover, we also have that

$$\begin{aligned}
\|dK^n(h)\| = \|dK^n(h - h(0))\| &\le 2\gamma \|\Psi^n(f_\mu(h - h(0)))\|_v \\
&\le 8D\rho^n \gamma (1 + \|df_\mu\|) \|dh\| . \qquad (3.3.6)
\end{aligned}$$

Dependence properties. If (3.3.2) holds, the upper bound

$$\tilde{\phi}(\sigma(X_n), X_0) \le 2a_n$$

follows from the following lemma.

Lemma 3.4. *Let $(\Omega, \mathcal{A}, \mathbb{P})$ be a probability space, X a real-valued random variable and $\mathcal{M}$ a σ-algebra of $\mathcal{A}$. We have the equality*

$$\tilde{\phi}(\mathcal{M}, X) = \sup\left\{ |\mathrm{Cov}(Y, h(X))| \,/\, Y \text{ is } \mathcal{M}\text{-measurable}, \ \|Y\|_1 \le 1 \text{ and } h \in BV_1 \right\}.$$

Proof of Lemma 3.4. Write first $|\mathrm{Cov}(Y, h(X))| = |\mathbb{E}(Y(\mathbb{E}(h(X)|\mathcal{M}) - \mathbb{E}(h(X))))|$. For any positive ε, there exists A_ε in $\mathcal{M}$ such that $\mathbb{P}(A_\varepsilon) > 0$ and for any ω in A_ε,

$$|\mathbb{E}(h(X)|\mathcal{M})(\omega) - \mathbb{E}(h(X))| > \|\mathbb{E}(h(X)|\mathcal{M}) - \mathbb{E}(h(X))\|_\infty - \varepsilon.$$

Define the random variable

$$Y_\varepsilon := \frac{\mathbf{1}_{A_\varepsilon}}{\mathbb{P}(A_\varepsilon)} \operatorname{sign}\left(\mathbb{E}(h(X)|\mathcal{M}) - \mathbb{E}(h(X))\right).$$

Y_ε is $\mathcal{M}$-measurable, $\mathbb{E}|Y_\varepsilon| = 1$ and

$$|\mathrm{Cov}(Y_\varepsilon, h(X))| \geq \|\mathbb{E}(h(X)|\mathcal{M}) - \mathbb{E}(h(X))\|_\infty - \varepsilon.$$

This is true for any positive ε, we infer from Definition 2.5 that

$$\tilde{\phi}(\mathcal{M}, X) \leq \sup\{|\mathrm{Cov}(Y, h(X))| \,/\, Y \text{ is } \mathcal{M}\text{-measurable, } \|Y\|_1 \leq 1 \text{ and } h \in BV_1\}.$$

The converse inequality follows straightforwardly from Definition 2.5. □

Now, if (3.3.4) and (3.3.5) hold, we have that: for any $n \geq i_l > \cdots > i_1 \geq 0$,

$$\tilde{\phi}(\sigma(X_k, k \geq n), X_{n-i_1}, \ldots, X_{n-i_l}) \leq C(l)\rho^{i_1},$$

for some positive constant $C(l)$. This is a consequence of the following lemma by using the upper bound (3.3.6).

Lemma 3.5. *Let $(Y_i)_{i\geq 0}$ be a real-valued Markov chain with transition kernel K. Assume that there exists a constant C such that*

$$\text{for any } BV \text{ function } f \text{ and any } n > 0, \quad \|dK^n(f)\| \leq C\|df\|. \tag{3.3.7}$$

Then, for any $i_l > \cdots > i_1 \geq 0$,

$$\tilde{\phi}(\sigma(Y_k), Y_{k+i_1}, \ldots, Y_{k+i_l}) \leq (1 + C + \cdots + C^{l-1})\tilde{\phi}(\sigma(Y_k), Y_{k+i_1}).$$

Consequently, if (3.3.4) and (3.3.5) hold, the coefficients $\tilde{\phi}_k(i)$ of the associated Markov chain $(Y_i)_{i\geq 0}$ satisfy: for any $k > 0$,

$$\tilde{\phi}_k(i) \leq C(k)\rho^i.$$

Proof of Lemma 3.5. We only give the proof for two points $i_1 = i$ and $i_2 = j$, the general case being similar. Let $f_k(x) = f(x) - \mathbb{E}(f(Y_k))$. We have, almost surely,

$$\mathbb{E}(f_{k+i}(Y_{k+i})g_{k+j}(Y_{k+j})|Y_k) - \mathbb{E}(f_{k+i}(Y_{k+i})g_{k+j}(Y_{k+j})) =$$
$$\mathbb{E}(f_{k+i}(Y_{k+i})(K^{j-i}(g))_{k+i}(Y_{k+i})|Y_k) - \mathbb{E}(f_{k+i}(Y_{k+i})(K^{j-i}(g))_{k+i}(Y_{k+i})).$$

Let f and g be two functions in BV_1. It is easy to see that

$$\begin{aligned}\|d((K^{j-i}(g))_{k+i}f_{k+i})\| &\leq \|df_{k+i}\| \, \|(K^{j-i}(g))_{k+i}\|_\infty \\ &\quad + \|d(K^{j-i}(g))_{k+i}\| \, \|f_{k+i}\|_\infty \\ &\leq (1 + \|d(K^{j-i}(g))_{k+i}\|).\end{aligned}$$

Hence, applying (3.3.7), the function $(K^{j-i}(g))_{k+i}f_{k+i}/(1+C)$ belongs to BV_1. The result follows from Definition 2.5. □

Application: uniformly expanding maps. A large class of uniformly expanding maps T is given in Broise (1996) [31], Section 2.1, page 11. If Broise's conditions are satisfied and if T is mixing in the ergodic-theoretic sense, then the Perron-Frobenius operator $\mathcal{L}$ satisfies the assumption (3.3.4). Let us recall some well know examples (see Section 2.2 in Broise):

1. $T(x) = \beta x - [\beta x]$ for $\beta > 1$. These maps are called β-transformations.
2. I is the finite union of disjoints intervals $(I_k)_{1 \leq k \leq n}$, and $T(x) = a_k x + b_k$ on I_k, with $|a_k| > 1$.
3. $T(x) = a(x^{-1} - 1) - [a(x^{-1} - 1)]$ for some $a > 0$. For $a = 1$, this transformation is known as the Gauss map.

Remark 3.3. Expanding maps with a neutral fixed point. *For some maps which are non uniformly expanding, in the sense that there exists a point for which the right (or left) derivative of T is equal to 1, Young (1999) [194] gives some sharp upper bounds for the covariances of Hölder functions of T^n. For instance, let us consider the maps introduced by Liverani et al. (1999) [123]:*

$$\text{for } 0 < \gamma < 1, \quad T(x) = \begin{cases} x(1 + 2^\gamma x^\gamma) & \text{if } x \in [0, 1/2[\\ 2x - 1 & \text{if } x \in [1/2, 1], \end{cases}$$

for which there exists a unique invariant probability μ. Contrary to uniformly expanding maps, these maps are not $\tilde{\phi}$-dependent. For $\lambda \in]0, 1]$, let $\delta_\lambda(x, y) = |x - y|^\lambda$ and let $\theta_1^{(\lambda)}$ be the coefficient associated to the distance δ_λ (see definition 2.3). Starting from the upper bounds given by Young, one can prove that there exist some positive constants $C_1(\lambda, \gamma)$ and $C_2(\lambda, \gamma)$ such that

$$\frac{C_1(\lambda, \gamma)}{n^{\frac{\gamma-1}{\gamma}}} \leq \theta_1^{(\lambda)}(\sigma(T^n), T) \leq \frac{C_2(\lambda, \gamma)}{n^{\frac{\gamma-1}{\gamma}}}.$$

Approximating the indicator function $f_x(t) = \mathbf{1}_{x \leq t}$ by λ-Hölder functions for small enough λ, one can prove that for any $\epsilon > 0$, there exist some positive constant $C_3(\epsilon, \gamma)$ such that

$$\frac{C_1(1, \gamma)}{n^{\frac{\gamma-1}{\gamma}}} \leq \tilde{\alpha}(\sigma(T^n), T) \leq \frac{C_3(\epsilon, \gamma)}{n^{\frac{\gamma-1}{\gamma}-\epsilon}}.$$

3.4 Vector valued LARCH(∞) processes

A vast literature is devoted to the study of conditionally heteroskedastic models. One of the best-known model is the GARCH model (Generalized Autoregressive Conditionally Heteroskedastic) introduced by Engle (1982) [84] and Bollerslev (1986) [23]. A usual GARCH(p,q) model can be written:

$$r_t = \sigma_t \xi_t, \quad \sigma_t^2 = \alpha_0 + \sum_{i=1}^{p} \beta_i \sigma_{t-i}^2 + \sum_{j=1}^{q} \alpha_j r_{t-j}^2$$

where $\alpha_0 \geq 0,\ \beta_i \geq 0,\ \alpha_j \geq 0,\ p \geq 0,\ q \geq 0$ are the model's parameters and the ξ_t are i.i.d.
If the β_i are null, we have an ARCH(q) model which can be extended in LARCH(∞) model (see Robinson, 1991 [165], Giraitis, Kokozska and Leipus, 2000 [92]). These models are often used in finance because their properties are close to the properties observed on empirical financial data such as volatility clustering, white noise behaviour or autocorrelation of the squares of those series. To reproduce other properties of the empirical data, such as leverage effect, a lot of extensions of the GARCH model have been introduced: EGARCH, TGARCH...
A simple equation in terms of a vector valued process allows simplifications in the definition of various ARCH type models Let $(\xi_t)_{t\in\mathbb{Z}}$ be an i.i.d. sequence of random $d \times m$-matrices, $(a_j)_{j\in\mathbb{N}^*}$ be a sequence of $m \times d$ matrices, and a be a vector in $\mathbb{R}^m$. A vector valued LARCH(∞) model is a solution of the recurrence equation

$$X_t = \xi_t \left(a + \sum_{j=1}^{\infty} a_j X_{t-j} \right) \tag{3.4.1}$$

Some examples of LARCH(∞) models are now provided. Even if standard LARCH(∞) models simply correspond to the case of real valued X_t and a_j, general LARCH(∞) models include a large variety of models, such as

1. **Bilinear models**, precisely addressed in the forthcoming subsection 3.4.2. They are solution of the equation:

$$X_t = \zeta_t \left(\alpha + \sum_{j=1}^{\infty} \alpha_j X_{t-j} \right) + \beta + \sum_{j=1}^{\infty} \beta_j X_{t-j}$$

   where the variables are real valued and ζ_t is the innovation. This is easy to see that such models take the previous form with $m = 2$ and $d = 1$: write for this $\xi_t = \begin{pmatrix} \zeta_t & 1 \end{pmatrix}$, $a = \begin{pmatrix} \alpha \\ \beta \end{pmatrix}$ and $a_j = \begin{pmatrix} \alpha_j \\ \beta_j \end{pmatrix}$ for $j = 1, 2, \ldots$

2. **GARCH**(p,q) **models**, are defined by

$$\begin{cases} r_t = \sigma_t \varepsilon_t \\ \sigma_t^2 = \sum_{j=1}^{q} \beta_j \sigma_{t-j}^2 + \gamma + \sum_{j=1}^{q} \gamma_j r_{t-j}^2 \end{cases}$$

where $\alpha_0 > 0$, $\alpha_j \geqslant 0$, $\beta_i \geqslant 0$ (and the variables ε are centered at expectation). They may be written as bilinear models: for this set $\alpha_0 = \gamma_0/(1-\sum \beta_i)$ and $\sum \alpha_i z^i = \sum \gamma_i z^i/(1-\sum \beta_i z^i)$ (see Giraitis *et al.*(2006) [93]).

3. **ARCH**(∞) **processes**, given by equations,

$$\begin{cases} r_t = \sigma_t \varepsilon_t \\ \sigma_t^2 = \beta_0 + \sum_{j=1}^{\infty} \beta_j \varepsilon_{t-j}^2 \end{cases}$$

They may be written as bilinear models: set $\xi_t = \begin{pmatrix} \varepsilon_t & 1 \end{pmatrix}$, $a = \begin{pmatrix} \kappa\beta_0 \\ \lambda_1\beta_0 \end{pmatrix}$, $a_j = \begin{pmatrix} \kappa\beta_j \\ \lambda_1\beta_j \end{pmatrix}$ with $\lambda_1 = \mathbb{E}(\varepsilon_0^2)$, $\kappa^2 = \text{Var}\,(\varepsilon_0^2)$.

4. **Models with vector valued innovations**

$$\begin{cases} X_t = \zeta_t^1 \left(\alpha^1 + \sum_{j=1}^{\infty} \alpha_j^1 X_{t-j} \right) + \mu_t^1 \left(\beta^1 + \sum_{j=1}^{\infty} \beta_j^1 X_{t-j} \right) + \gamma^1 + \sum_{j=1}^{\infty} \gamma_j^1 X_{t-j} \\ Y_t = \zeta_t^2 \left(\alpha^2 + \sum_{j=1}^{\infty} \alpha_j^2 Y_{t-j} \right) + \mu_t^2 \left(\beta^2 + \sum_{j=1}^{\infty} \beta_j^2 Y_{t-j} \right) + \gamma^2 + \sum_{j=1}^{\infty} \gamma_j^2 Y_{t-j} \end{cases}$$

may clearly be written as LARCH(∞) models with now $m = d = 2$.

3.4.1 Chaotic expansion of LARCH(∞) models

We provide sufficient conditions for the following chaotic expansion

$$X_t = \xi_t \left(a + \sum_{k=1}^{\infty} \sum_{j_1,\ldots,j_k \geqslant 1} a_{j_1} \xi_{t-j_1} a_{j_2} \ldots a_{j_k} \xi_{t-j_1-\cdots-j_k} a \right). \tag{3.4.2}$$

For a $k \times l$ matrix M, and $\|\cdot\|$ a norm on $\mathbb{R}^k$, let as usual

$$\|M\| = \sup_{x \in \mathbb{R}^l, \|x\| \leq 1} \|Mx\|.$$

For a random matrix M we set $\|M\|_p^p = \mathbb{E}(\|M\|^p)$.

Theorem 3.1. *Assume that either* $\sum_{j\geq 1}\|a_j\|^p\mathbb{E}\|\xi_0\|^p < 1$ *for some* $p \leq 1$*, or* $\sum_{j\geq 1}\|a_j\|\left(\mathbb{E}\|\xi_0\|^p\right)^{\frac{1}{p}} < 1$ *for* $p > 1$*. Then one stationary of solution of eqn. (3.4.1) in* $\mathbb{L}^p$ *is given by (3.4.2).*

In this section we set

$$A(x) = \sum_{j\geq x}\|a_j\|, \quad A = A(1), \quad \text{and} \quad \lambda_p = A\|\xi_0\|_p. \tag{3.4.3}$$

Proof. We first prove that expression (3.4.2) is well defined. Set

$$S = \sum_{k=1}^{\infty}\sum_{j_1,\ldots,j_k\geq 1}\|a_{j_1}\xi_{t-j_1}\cdots a_{j_k}\xi_{t-j_1-\cdots-j_k}\|$$

Clearly

$$S \leq \sum_{k=1}^{\infty}\sum_{j_1,\ldots,j_k\geqslant 1}\|a_{j_1}\|\cdots\|a_{j_k}\|\|\xi_{t-j_1}\|\cdots\|\xi_{t-j_1-\cdots-j_k}\|$$

Using that the sequence $(\xi_n)_{n\in\mathbb{Z}}$ is i.i.d., we obtain for $p \geq 1$,

$$\begin{aligned}\|S\|_p &\leq \sum_{k=1}^{\infty}\sum_{j_1,\ldots,j_k\geqslant 1}\|a_{j_1}\|\cdots\|a_{j_k}\|\|\xi_{t-j_1}\|_p\cdots\|\xi_{t-j_1-\cdots-j_k}\|_p \\ &\leq \sum_{k=1}^{\infty}\left(\|\xi_0\|_pA\right)^k.\end{aligned}$$

If $\lambda_p < 1$, the last series is convergent and S belongs to $\mathbb{L}^p$. If $p < 1$ we conclude as for $p = 1$, using the bound

$$\left(\sum_{k=1}^{\infty}\|a_{j_1}\xi_{t-j_1}\cdots a_{j_k}\xi_{t-j_1-\cdots-j_k}\|\right)^p \leqslant \sum_{k=1}^{\infty}\|a_{j_1}\xi_{t-j_1}\cdots a_{j_k}\xi_{t-j_1-\cdots-j_k}\|^p$$

Now we prove that the expression (3.4.2) is a solution of eqn. (3.4.1),

$$
\begin{aligned}
X_t &= \xi_t\Big(a + \sum_{\substack{k \geq 1,\\ j_1,\ldots,j_k \geqslant 1}} a_{j_1}\xi_{t-j_1}\cdots a_{j_k}\xi_{t-j_1-\cdots-j_k}a\Big)\\
&= \xi_t\left(a + \sum_{j_1\geqslant 1} a_{j_1}\xi_{t-j_1}a + \sum_{k\geq 2, j_1\geqslant 1} a_{j_1}\xi_{t-j_1}\sum_{j_2,\ldots,j_k\geqslant 1} a_{j_2}\xi_{t-j_1-j_2}\cdots a_{j_k}\xi_{t-j_1-j_2-\cdots-j_k}a\right)\\
&= \xi_t\Big(a + \sum_{j_1\geqslant 1} a_{j_1}\xi_{t-j_1}\big(a + \sum_{\substack{k \geq 2,\\ j_2,\ldots,j_k \geqslant 1}} a_{j_2}\xi_{(t-j_1)-j_2}\cdots a_{j_k}\xi_{(t-j_1)-j_2-\cdots-j_k}a\big)\Big)\\
&= \xi_t\Big(a + \sum_{j=1}^{\infty} a_j X_{t-j}\Big). \qquad \square
\end{aligned}
$$

Theorem 3.2. *Assume that $p \geqslant 1$ in the assumption of theorem 3.1, and assume that $\varphi = \sum_j \|a_j\|\|\xi_0\|_p < 1$. If a stationary solution $(Y_t)_{t\in\mathbb{Z}}$ to equation (3.4.1) exists (a.s.), if Y_t is independent of the σ-algebra generated by $\{\xi_s; s > t\}$, for each $t \in \mathbb{Z}$, then this solution is also in L^p and it is (a.s.) equal to the previous solution $(X_t)_{t\in\mathbb{Z}}$ defined by equation (3.4.2).*

Proof. Step 1. We first prove that $\|Y_0\|_p < \infty$. From the equation (3.4.1), from the stationarity of $\{Y_t\}_{t\in\mathbb{Z}}$ and from the independence assumption, we derive that

$$\|Y_0\|_p \leqslant \|\xi_0\|_p\left(\|a\| + \sum_{j=1}^{\infty}\|a_j\|\|Y_0\|_p\right).$$

Hence, the first point in the theorem follows from $\|Y_0\|_p \leqslant \dfrac{\|\xi_0\|_p\|a\|}{1-\varphi} < \infty$.

Step 2. As in Giraitis *et al.* (2000) [92] we write $Y_t = \xi_t\left(a + \sum_{j\geqslant 1} a_j Y_{t-j}\right) = X_t^m + S_t^m$ with

$$
\begin{aligned}
X_t^m &= \xi_t\left(a + \sum_{k=1}^{m}\sum_{j_1,\cdots,j_k\geqslant 1} a_{j_1}\xi_{t-j_1}\cdots a_{j_k}\xi_{t-j_1\cdots-j_k}a\right),\\
S_t^m &= \xi_t\left(\sum_{j_1,\cdots,j_{m+1}\geqslant 1} a_{j_1}\xi_{t-j_1}\cdots a_{j_m}\xi_{t-j_1\cdots-j_m}a_{j_m+1}Y_{t-j_1\cdots-j_m}\right).
\end{aligned}
$$

We have

$$\|S_t^m\|_p \leqslant \|\xi\|_p \sum_{j_1,\cdots,j_{m+1}\geqslant 1}\|a_{j_1}\|\cdots\|a_{j_{m+1}}\|\|\xi\|_p^m\|Y_0\|_p = \|Y_0\|_p\varphi^{m+1}.$$

We recall the additive decomposition of the chaotic expansion X_t in equation (3.4.2) as a finite expansion plus a negligible remainder that can be controlled $X_t = X_t^m + R_t^m$ where

$$R_t^m = \xi_t \left(\sum_{k>m} \sum_{j_1,\cdots,j_k \geqslant 1} a_{j_1}\xi_{t-j_1} \cdots a_{j_k}\xi_{t-j_1\cdots-j_k} a \right),$$

satisfies

$$\|R_t^m\|_p \leqslant \|a\| \|\xi_0\|_p \sum_{k>m} \varphi^k \leqslant \|a\| \|\xi_0\|_p \frac{\varphi^m}{1-\varphi} \to 0.$$

Then, the difference between those two solutions is controlled as a function of m with $X_t - Y_t = R_t^m - S_t^m$, hence

$$\begin{aligned} \|X_t - Y_t\|_p &\leqslant \|R_t^m\|_p + \|S_t^m\|_p \\ &\leqslant \frac{\varphi^m}{1-\varphi}\|a\|\|\xi_0\|_p + \|Y_0\|_p \varphi^m \\ &\leqslant 2\frac{\varphi^m}{1-\varphi}\|a\|\|\xi_0\|_p \end{aligned}$$

thus, $Y_t = X_t$ a.s. □

Dependence properties. To our knowledge, there is no study of the weak dependence properties of ARCH or GARCH type models with an infinite number of coefficients. Mixing seems difficult to prove for such models excepted in the Markov case ($a_j = 0$ for large enough j), because this case is included in the Markov section for which we already mention that additional assumptions are needed to derive mixing. This section refers to Doukhan, Teyssière and Winant (2005) [75], who improve on the special case of bilinear models precisely considered in Doukhan, Madré and Rosenbaum (2006) [69]. We use the notations given in (3.4.3).

Theorem 3.3. *The solution (3.4.2) of eqn. (3.4.1) is $\theta-$weakly dependent with*

$$\theta(r) \leq 2\mathbb{E}\|\xi_0\| \left(\mathbb{E}\|\xi_0\| \sum_{k=1}^{t-1} k\lambda^{k-1} A\left(\frac{t}{k}\right) + \frac{\lambda^t}{1-\lambda} \right) \|a\| \quad \textit{for any } t \leq r.$$

Proof. Consider $f : (\mathbb{R}^d)^u \to \mathbb{R}$ with $\|f\|_\infty < \infty$ and $g : (\mathbb{R}^d)^v \to \mathbb{R}$ with $\mathrm{Lip}\,(g) < \infty$, that is $|g(x) - g(y)| \leq \mathrm{Lip}(g)(\|x_1 - y_1\| + \cdots + \|x_u - y_u\|)$. Let $i_1 < \cdots < i_u$, $j_1 < \cdots < j_v$, such that $j_1 - i_u \geqslant r$. To bound weak dependence coefficients we use an approximation of the vector $\mathbf{v} = \left(X_{j_1}, \ldots, X_{j_v}\right)$ by $\hat{\mathbf{v}}$ in such a way that, for each index $j \in \{j_1, \ldots, j_k\}$ and $s \le r$, the random variable $\hat{X}_j$ is independent of X_{j-s}. More precisely, let

$$\hat{X}_t = \xi_t \left(a + \sum_{k=1}^{\infty} \sum_{j_1+\cdots+j_k<s} a_{j_1}\xi_{t-j_1} \cdots a_{j_k}\xi_{t-j_1-\cdots-j_k} a \right).$$

Now, let $f(\mathbf{u}) = f(X_{i_1}, \ldots, X_{i_u})$ and $g(\mathbf{v}) = g(X_{j_1}, \ldots, X_{j_v})$. We have that

$$\begin{aligned}
|\mathrm{Cov}(f(\mathbf{u}), g(\mathbf{v}))| &\leqslant |\mathbb{E}(f(\mathbf{u})(g(\mathbf{v}) - g(\hat{\mathbf{v}})) - \mathbb{E}(f(\mathbf{u}))\mathbb{E}(g(\mathbf{v}) - g(\hat{\mathbf{v}}))| \\
&\leqslant 2\|f\|_\infty \mathbb{E}\,|g(\mathbf{v}) - g(\hat{\mathbf{v}})| \\
&\leqslant 2\|f\|_\infty \mathrm{Lip}\,(g) \sum_{k=1}^{v} \mathbb{E}\|X_{j_k} - \hat{X}_{j_k}\| \\
&\leqslant 2v\|f\|_\infty \mathrm{Lip}\,(g)\mathbb{E}\|X_0 - \hat{X}_0\|.
\end{aligned}$$

Hence $\theta(r) \le 2\mathbb{E}\|X_0 - \hat{X}_0\|$ for any $s \le r$, which implies the bound of the theorem. This bound is made explicit for simple decay rates. More precisely

$$\theta(t) \leqslant \begin{cases} Kt^{-b}, & \text{under Riemanian decay } A(x) \leqslant Cx^{-b} \\ K(q \vee \lambda)^{\sqrt{t}}, & \text{under geometric decay } A(x) \leqslant Cq^x \end{cases}$$

Further approximations. In order to simulate and also to better understand their behaviour, it is an important feature to see how far those models are from simple processes. Weak dependence was proved with independent approximations

$$\hat{X}_t = \xi_t \left(a + \sum_{k=1}^{\infty} \sum_{j_1+\cdots+j_k<s} a_{j_1}\xi_{t-j_1} \cdots a_{j_k}\xi_{t-j_1-\cdots-j_k} a \right)$$

of the LARCH models; we precise this approximation through coupling arguments and we also prove below proximity to the Markov sequence obtained by truncating the series which defines them.

Coupling. The approximation $\hat{X}_t$ of X_t has not the same distribution as X_t. But we are in the case of causal Bernoulli shifts with independent inputs, so that the method of Section 3.1.4 applies. Let $(\xi'_i)_{i\in\mathbb{Z}}$ be an independent copy of $(\xi_i)_{i\in\mathbb{Z}}$, and let $\tilde{\xi}_n = \xi_n$ if $n > 0$ and $\tilde{\xi}_n = \xi'_n$ for $n \le 0$. Finally, let

$$\tilde{X}_t = \tilde{\xi}_t \Big(a + \sum_{k=1}^{\infty} \sum_{j_1,\ldots,j_k} a_{j_1}\tilde{\xi}_{t-j_1} \cdots a_{j_k}\tilde{\xi}_{t-j_1-\cdots-j_k} a \Big).$$

Here $\tilde{X}_t$ has the same distribution as X_t and is independent of the σ-algebra $\mathcal{M}_0 = \sigma(X_i, i \le 0)$. Consequently, if $\tilde{\delta}_{p,n}$ is a non increasing sequence satisfying

(3.2.2), we obtain the upper bounds $\tau_{p,\infty}(n) \leq \tilde{\delta}_{p,n}$. Since for any $s \leq n$, we have that $\|X_n - \tilde{X}_n\|_p \leq 2\|X_0 - \hat{X}_0\|_p$, we obtain that

$$\tau_{p,\infty}(n) \leq \inf_{t \leq r} 2\mathbb{E}\|\xi_0\|_p \left(\mathbb{E}\|\xi_0\|_p \sum_{k=1}^{t-1} k\lambda_p^{k-1} A\left(\frac{t}{k}\right) + \frac{\lambda_p^t}{1-\lambda_p} \right) \|a\|.$$

For $p = 1$, we recover the upper bound of Theorem 3.3. If $d = 1$, we obtain the same upper bounds for $\tilde{\alpha}_k(n)$, $\tilde{\beta}_k(n)$ and $\tilde{\phi}_k(n)$ as in Section 3.1.4, by assuming that the distribution function of X_0 is continuous (similar bounds me bay obtained for $d > 1$ by assuming that each component of X_0 has a continuous distribution function, see Lemma 5.1).

Markov approximation. Consider equation (3.4.1) truncated at rank N,

$$X_t^N = \xi_t \left(a + \sum_{j=1}^{N} a_j X_{t-j}^N \right).$$

The previous solution rewrites as

$$X_t^N = \xi_t \left(a + \sum_{k=1}^{\infty} \sum_{N \geqslant j_1, \ldots, j_k \geqslant 1} a_{j_1} \xi_{t-j_1} \cdots a_{j_k} \xi_{t-j_1-\cdots-j_k} a \right).$$

The corresponding error is then bounded by

$$\mathbb{E}\|X_t - X_t^N\| \leqslant \sum_{k=1}^{\infty} A(N)^k.$$

In the Riemanian decay case, the error is $\sum_{k=1}^{\infty} N^{-bk}$, and in the geometric decay case, the error is $q^N/(1-q^N)$.

3.4.2 Bilinear models

Those models are defined through the recurrence relation

$$X_t = \zeta_t \left(\alpha + \sum_{j=1}^{\infty} \alpha_j X_{t-j} \right) + \beta + \sum_{j=1}^{\infty} \beta_j X_{t-j},$$

the variables here are real valued and ζ_t now denotes the innovation. To see that this is still a $LARCH(\infty)$ model, we set as before

$$\xi_t = \begin{pmatrix} \zeta_t & 1 \end{pmatrix}, \qquad a = \begin{pmatrix} \alpha \\ \beta \end{pmatrix}, \qquad \text{and } a_j = \begin{pmatrix} \alpha_j, \\ \beta_j \end{pmatrix}, \quad \text{for } j \geq 1.$$

One usually takes $\beta = 0$. ARCH(∞) and GARCH(p, q) models are particular cases of the bilinear models. Giraitis and Surgailis (2002) [95] prove that under some restrictions, there is a unique second order stationary solution for these models. This solution has a chaotic expansion. Assume that the power series $A(z) = \sum_{j=1}^{\infty} \alpha_j z^j$ and $B(z) = \sum_{j=1}^{\infty} \beta_j z^j$ exist for $|z| \leq 1$, let

$$G(z) = \frac{1}{1 - B(z)} = \sum_{j=1}^{\infty} g_j z^j \quad \text{and} \quad H(z) = \frac{A(z)}{1 - B(z)} = \sum_{j=1}^{\infty} h_j z^j$$

we will note $\|h\|_p^p = \sum_{j=0}^{\infty} |h_j|^p$. Then:

Proposition 3.3 (Giraitis, Surgailis, 2002 [95]). *If $(\zeta_t)_t$ is i.i.d., and $\|h\|_2 < 1$, then there is a unique second order stationary solution :*

$$X_t = \alpha \sum_{k=1}^{\infty} \sum_{s_k < \ldots < s_1 \leq t} g_{t-s_1} h_{s_1 - s_2} \cdots h_{s_{k-1}-s_k} \, \zeta_{s_1} \cdots \zeta_{s_k} \tag{3.4.4}$$

Lemma 3.6. *Expansion (3.4.2) coincides with the chaotic expansion in proposition 3.3.*

Proof. Assuming that $\beta = 0$, the expansion (3.4.2) writes as:

$$X_t = \zeta_t \alpha + \sum_{k=1}^{\infty} \sum_{s_k < \cdots < s_1 < t} (\zeta_t \alpha_{t-s_1} + \beta_{t-s_1}) \times$$
$$\times \, (\zeta_{s_1} \alpha_{s_1 - s_2} + \beta_{s_1 - s_2}) \cdots (\zeta_{s_{k-1}} \alpha_{s_{k-1}-s_k} + \beta_{s_{k-1}-s_k}) \zeta_{s_k} \alpha$$

or $X_t = \zeta_t \alpha + S_1 + S_2$ with

$$S_1 = \sum_{\substack{k \geq 1 \\ s_k < \cdots < s_1 < t}} \zeta_t \alpha_{t-s_1} (\zeta_{s_1} \alpha_{s_1 - s_2} + \beta_{s_1 - s_2}) \cdots (\zeta_{s_{k-1}} \alpha_{s_{k-1}-s_k} + \beta_{s_{k-1}-s_k}) \zeta_{s_k} \alpha$$

$$S_2 = \sum_{\substack{k \geq 1 \\ s_k < \cdots < s_1 < t}} \beta_{t-s_1} (\zeta_{s_1} \alpha_{s_1 - s_2} + \beta_{s_1 - s_2}) \cdots (\zeta_{s_{k-1}} \alpha_{s_{k-1}-s_k} + \beta_{s_{k-1}-s_k}) \zeta_{s_k} \alpha$$

Under additional assumptions Giraitis and Surgailis prove the expansion:

$$X_t = \alpha \sum_{k=1}^{\infty} \sum_{s_k < \cdots < s_1 \leqslant t} g_{t-s_1} h_{s_1 - s_2} \cdots h_{s_{k-1}-s_k} \zeta_{s_1} \cdots \zeta_{s_k}$$

rewritten as $X_t = \zeta_t\alpha + T_1 + T_2$ with

$$T_1 = \sum_{k=1}^{\infty} \sum_{s_k<\cdots<s_1<t} \zeta_t h_{t-s_1}\zeta_{s_1} h_{s_1-s_2}\zeta_{s_2}\cdots h_{s_{k-1}-s_k}\zeta_{s_k}\alpha$$
$$T_2 = \sum_{k=1}^{\infty} \sum_{s_k<\cdots<s_1<t} g_{t-s_1} h_{s_1-s_2}\zeta_{s_1}\cdots h_{s_{k-1}-s_k}\zeta_{s_k}\alpha$$

The terms in the sum (T_1) take a form:

$$\zeta_t(\alpha_{t-i_1^{(1)}}\beta_{i_1^{(1)}-i_2^{(1)}}\cdots\beta_{i_{p_1}^{(1)}-s_1})\zeta_{s_1}(\alpha_{s_1-i_1^{(2)}}\beta_{i_1^{(2)}-i_2^{(2)}}\cdots\beta_{i_{p_2}^{(2)}-s_2})\cdots \\ \cdots\zeta_{s_{k-1}}(\alpha_{s_{k-1}-i_1^{(k)}}\beta_{i_1^{(k)}-i_2^{(k)}}\cdots\beta_{i_{p_k}^{(k)}-s_k})\zeta_{s_k}\alpha$$

hence between each couple ζ_{s_p}, $\zeta_{s_{p+1}}$ read from the left to the right one founds a term (α_j) followed with several (β_j)'s in such a way that the sum of all indices equals $s_{p+1} - s_p$.
On the one hand, quote that expanding terms in the (S_1) yields a sum of products of such terms proves that (S_1)'s is included in (T_1)'s.
On the other hand, expand the term $\bar{k} = k + p_1 + \cdots + p_k$ $(\bar{s}_{\bar{k}}, \bar{s}_{\bar{k}-1}, \ldots, \bar{s}_1) = (s_k, i_{p_k}^{(k)}, \ldots, i_1^{(k)}, s_{k-1}, \ldots, s_1, i_{p_1}^{(1)}, \ldots, i_2^{(1)}, i_1^{(1)}, t)$ in the sum (S_1) yields the generic term in (T_1) expansion. Hence (S_1) and (T_1)'s expansions coincide.
Analogously $(T_2) = (S_2)$, by quoting that the generic term in (T_2) writes:

$$(\beta_{t-i_1^{(1)}}\beta_{i_1^{(1)}-i_2^{(1)}}\cdots\beta_{i_{p_1}^{(1)}-s_1})\zeta_{s_1}(\alpha_{s_1-i_1^{(2)}}\beta_{i_1^{(2)}-i_2^{(2)}}\cdots\beta_{i_{p_2}^{(2)}-s_2})\cdots \\ \cdots\zeta_{s_{k-1}}(\alpha_{s_{k-1}-i_1^{(k)}}\beta_{i_1^{(k)}-i_2^{(k)}}\cdots\beta_{i_{p_k}^{(k)}-s_k})\zeta_{s_k}\alpha. \ \square$$

Conditional densities of Bilinear models. Another more specific feature of those models is the following result on the existence of conditional densities (see Doukhan *et al.*1995 [69]). This is relevant for subsampling techniques and density estimation. We use the fact that the bilinear equation may be written

$$X_t = \tilde{A}_t\zeta_t + \tilde{B}_t \quad \text{with} \quad \tilde{A}_t = a + \sum_{j=1}^{\infty} a_j X_{t-j}, \quad \tilde{B}_t = \sum_{j=1}^{\infty} b_j X_{t-j}$$

Hence, conditionally to the past σ-algebra (the history of $\{\zeta_s, s < t\}$), the distribution of X_t is as smooth as the one of ζ_t if $\tilde{A}_t \neq 0$. Now if we are interested by higher order marginal distributions we need a bit more work.

Split $\tilde{A}_t$ and $\tilde{B}_t$ into

$$\tilde{A}_t = \sum_{j=1}^{t-1} a_j X_{t-j} + A_t, \qquad A_t = \sum_{j=t}^{\infty} a_j X_{t-j}$$
$$\tilde{B}_t = \sum_{j=1}^{t-1} b_j X_{t-j} + B_t, \qquad B_t = \sum_{j=t}^{\infty} b_j X_{t-j}$$

where A_t and B_t are measurable with respect to the σ algebra of the past $\sigma\{\zeta_s, s \leq 0\}$. This entails, for instance:

$$X_1 = \tilde{A}_1\zeta_1 + \tilde{B}_1$$
$$X_2 = (a_1\zeta_1 + A_2)\zeta_2 + (b_1\zeta_1 + B_2).$$

Thus conditionally to $\tilde{A}_1, \tilde{B}_1, A_2$ and B_2 the previous system is *triangular* and it thus may be solved if $\tilde{A}_1, a_1\zeta_1 + A_2$ do not vanish. The following result extends this observation:

Lemma 3.7 (Conditional densities Doukhan *et al.* [69]). *Assume that the random variables $(\zeta_t)_{t\in\mathbb{Z}}$ and the coefficients a_j are non negative for $j = 1, 2, \ldots$. Also suppose that ζ_i are independent random variables with a density f_{ζ_i} for all $i \in \{1, \ldots, n\}$. Then, conditionally with respect to the past of the process $\sigma\{\zeta_s, s \leq 0\}$, the random vector $(X_1, \ldots, X_n)$ admits the density $f_n(x_1, \ldots, x_n)$ defined by:*

$$f_n(x_1, \ldots, x_n) = \frac{1}{|\alpha_1\alpha_2 \cdots \alpha_n|} f_{\zeta_1}\left(\frac{\beta_1}{\alpha_1}\right) \cdots f_{\zeta_n}\left(\frac{\beta_n}{\alpha_n}\right)$$

with $\beta_j = x_j - b_1x_{j-1} - b_2x_{j-2} - \cdots - b_{j-1}x_1 - B_j$ and $\alpha_j = a_1x_{j-1} + a_2x_{j-2} + \cdots + a_{j-1}x_1 + A_j$ for $1 \leq j \leq n$ (here $\alpha_1 = A_1$).

Corollary 3.1 (Density). *Under the same assumptions as in lemma 3.7, with $a \neq 0$ and $(\zeta_t)_t$ i.i.d. with a density f bounded by M then*
$$f_n(x_1, \ldots, x_n) \leq (M/a)^n \text{ for all } (x_1, \ldots, x_n).$$

Corollary 3.2 (Density of a couple). *Under the same assumptions as in lemma 3.7, and if ζ_t are i.i.d. with density f, then g_i the density of the couple (X_1, X_i) satisfies $\|g_i\|_\infty \leq \|f\|^2/A_1$ for all $i > 1$.*

Remark 3.4. *Asymptotic properties of a standard kernel density estimate relies on such bounds. Indeed an expression of its variance follows jointly from weak dependence properties and such assumptions on the two dimensional distributions.*

Proof of lemma 3.7. We work conditionally to the infinite past from X_0. We write

$$M\begin{pmatrix} X_n \\ X_{n-1} \\ \vdots \\ X_1 \end{pmatrix} = \begin{pmatrix} A_n\zeta_n + B_n \\ A_{n-1}\zeta_{n-1} + B_{n-1} \\ \vdots \\ A_1\zeta_1 + B_1 \end{pmatrix},$$

where

$$M = \begin{pmatrix} 1 & -a_1\zeta_n - b_1 & -a_2\zeta_n - b_2 & \dots & -a_{n-1}\zeta_n - b_{n-1} \\ 0 & 1 & -a_1\zeta_{n-1} - b_1 & \dots & -a_{n-2}\zeta_{n-1} - b_{n-2} \\ \dots & \dots & \dots & \dots & \dots \\ 0 & 0 & 0 & 1 & -a_1\zeta_2 - b_1 \\ 0 & 0 & 0 & 0 & 1 \end{pmatrix}.$$

This may be rewritten as,

$$\begin{aligned} \zeta_1 &= \frac{X_1 - B_1}{A_1} \\ \zeta_2 &= \frac{X_2 - b_1 X_1 - B_2}{a_1 X_1 + A_2} \\ &\vdots \\ \zeta_n &= \frac{X_n - b_1 X_{n-1} - b_2 X_{n-2} - \cdots - b_{n-1} X_1 - B_n}{a_1 X_{n-1} + a_2 X_{n-2} + \cdots + a_{n-1} X_1 + A_n} \end{aligned}$$

where the previous coefficients A_t and B_t are deterministic in this conditional setting. Thus,

$$\mathbb{E}g(X_1, X_2, \dots, X_n) = \int g\left(\phi^{-1}(u_1, \dots, u_n)\right) f_{\zeta_1}(u_1) \cdots f_{\zeta_n}(u_n) du_1 \cdots du_n$$

with f_{ζ_i} the density of ζ_i and with $(\zeta_1, \dots, \zeta_n) = \phi(X_1, \dots, X_n)$. Here, put $(u_1, \dots, u_n) = \phi(x_1, \dots, x_n)$. The function ϕ has a diagonal Jacobian, hence

$$\frac{\partial u_j}{\partial x_j} = \frac{1}{\alpha_j}$$

for $1 \le j \le n$ and the result follows. □

Proof of the corollaries. The first corollary follows by integration from lemma 3.7. We prove the result for the density of the couple (X_1, X_4), and we can prove the general result in the same way. With

$$f_4(x_1, x_2, x_3, x_4) = \frac{1}{|\alpha_1 \cdots \alpha_4|} f\left(\frac{\beta_1}{\alpha_1}\right) \cdots f\left(\frac{\beta_4}{\alpha_4}\right),$$

an integration with respect to x_2 and x_3 implies that

$$g_4(x_1,x_4) = \int f_4(x_1,x_2,x_3,x_4)dx_2dx_3 \le \frac{\|f\|^2}{A_1}\int f\left(\frac{\beta_2}{\alpha_2}\right) f\left(\frac{\beta_3}{\alpha_3}\right)\frac{dx_2\,dx_3}{|\alpha_2\alpha_3|}.$$

Hence with

$$u = \frac{x_2 - b_1x_1 - B_2}{a_1x_1 + A_2}, \qquad v = \frac{x_3 - b_1x_2 - b_2x_1 - B_3}{a_1x_2 + a_2x_1 + A_3},$$

we write $x_2 = u \times (a_1x_1 + A_2) + b_1x_1 + B_2$, $x_3 = v \times (a_1(u \times (a_1x_1 + A_2) + b_1x_1 + B_2) + a_2x_1 + A_3) + b_1(u \times (a_1x_1 + A_2) + b_1x_1 + B_2) + b_2x_1 + B_3$. The Jacobian matrix is diagonal and the absolute value of the Jacobian is equal to $|(a_1x_1 + A_2)(a_1x_2 + a_2x_1 + A_3)|$, and thus

$$g_4(x_1,x_4) \le \frac{\|f\|^2}{A_1}\int f(u)\,f(v)\,du dv \le \frac{\|f\|^2}{A_1}. \quad \square$$

3.5 ζ-dependent models

In this section, we give some classes of ζ-dependent models: associated processes, Gaussian processes and interacting particle systems.

3.5.1 Associated processes

An analogous formula to (1.4.1) proves that associated random variables belong to the class of ζ-dependent models. Several associated models are obtained from nondecreasing transformations of independent variables. For Gaussian vector, Pitt (1982) [146] gave a necessary and sufficient condition to be associated. We discuss Pitt's result in the sequel.

Theorem 3.4. *Let $X = (X_1,\dots,X_n)$ be a Gaussian vector with mean vector 0 and covariance matrix $\Sigma = (\sigma_{i,j} = \mathrm{Cov}(X_i,X_j))_{1\le i,j\le n}$. The condition*

$$\mathrm{Cov}(X_i,X_j) \ge 0, \quad \text{for all} \ \ i,j = 1,\dots,n \tag{3.5.1}$$

is necessary and sufficient for the variables to be associated.

Proof of Theorem 3.4. The method of the proof below is due to Pitt (1982). Assuming Condition (3.5.1), the task is to prove that

$$\mathrm{Cov}(f(X), g(X)) \ge 0,$$

for all nondecreasing functions f and g defined on $\mathbb{R}^n$ (the second implication being trivial).

We suppose, without loss of generality, that Σ is non-singular and that the function f and g are continuously differentiable with bounded derivatives $\frac{\partial f}{\partial x_i}$ and $\frac{\partial g}{\partial x_i}$, for $i = 1, \ldots, n$. Let Z be an independent copy of X. For any $\lambda \in [0,1]$, let $Y(\lambda)$ be the random vector defined by

$$Y(\lambda) = \lambda X + \sqrt{1-\lambda^2} Z.$$

Clearly, for each fixed λ, $Y(\lambda)$ is a Gaussian vector with covariance matrix Σ and

$$\mathrm{Cov}(X_i, Y_j(\lambda)) = \lambda \sigma_{i,j}.$$

Set

$$F(\lambda) = \mathbb{E}\left(f(X) g(Y(\lambda))\right).$$

Clearly

$$\mathrm{Cov}(f(X), g(X)) = F(1) - F(0). \tag{3.5.2}$$

It is sufficient to show that $F'(\lambda)$ exists and $F'(\lambda) \geq 0$ for $0 \leq \lambda \leq 1$. To this end, let ϕ and p be respectively the density of X and the conditional density of $Y(\lambda)$ given $X = x$. We have :

$$\phi(x) = \frac{1}{\sqrt{(2\pi)^n |\Sigma|}} \exp\left(-\frac{1}{2}\sum_{i,j=1}^{n} c_{i,j} x_i x_j\right), \quad \text{with } \Sigma^{-1} = (c_{i,j})_{1 \leq i,j \leq n}$$

and

$$p(\lambda; x, y) = \frac{1}{(1-\lambda^2)^{n/2}} \phi\left(\frac{\lambda x - y}{\sqrt{1-\lambda^2}}\right)$$

Hence

$$F(\lambda) = \int_{\mathbb{R}^n} \phi(x) f(x) g(\lambda; x) dx, \tag{3.5.3}$$

where $g(\lambda; x)$ is defined by

$$g(\lambda; x) = \int_{\mathbb{R}^n} p(\lambda; x, y) g(y) dy,$$

which is equal to

$$g(\lambda; x) = \int_{\mathbb{R}^n} g(\lambda x - y) \phi_\lambda(y) dy, \tag{3.5.4}$$

where

$$\phi_\lambda(x) = \frac{1}{(1-\lambda^2)^{n/2}} \phi\left(\frac{x}{\sqrt{1-\lambda^2}}\right).$$

Now Equation (3.5.4) proves that the partial derivatives $\frac{\partial g(\lambda, x)}{\partial x_i}$ exist and are bounded. Moreover, since g is increasing and λ is positive, we have

$$\frac{\partial g(\lambda, x)}{\partial x_i} \geq 0. \tag{3.5.5}$$

Next an explicit calculation based on the heat equations

$$\frac{\partial \phi}{\partial \sigma_{i,i}} = \frac{1}{2}\frac{\partial^2 \phi}{\partial x_i^2}, \qquad \frac{\partial \phi}{\partial \sigma_{i,j}} = \frac{\partial^2 \phi}{\partial x_i \partial x_j}, \quad \text{for } i \neq j,$$

shows that

$$\begin{aligned} \frac{\partial p}{\partial \lambda} &= \frac{p}{1-\lambda^2}\left(k\lambda - \sum_{i,j} x_i c_{ij}(\lambda x_j - y_j) - \frac{\lambda}{1-\lambda^2}\sum_{i,j}(\lambda x_i - y_i)(\lambda x_j - y_j)\right) \\ &= -\frac{1}{\lambda}\left(\sum_{i,j}\sigma_{i,j}\frac{\partial^2 p}{\partial x_i \partial x_j} - \sum_i x_i \frac{\partial p}{\partial x_i}\right). \end{aligned} \tag{3.5.6}$$

We obtain, combining (3.5.3), (3.5.4) and (3.5.6)

$$F'(\lambda) = -\frac{1}{\lambda}\int_{\mathbb{R}^n} \phi(x) f(x)\left(\sum_{i,j}\sigma_{i,j}\frac{\partial^2 g(\lambda;x)}{\partial x_i \partial x_j} - \sum_i x_i \frac{\partial g(\lambda;x)}{\partial x_i}\right) dx.$$

An integration by parts gives

$$F'(\lambda) = \frac{1}{\lambda}\int_{\mathbb{R}^n} \phi(x)\left(\sum_{i,j}\sigma_{i,j}\frac{\partial f(x)}{\partial x_i}\frac{\partial g(\lambda;x)}{\partial x_j}\right) dx. \tag{3.5.7}$$

We have $\frac{\partial f(x)}{\partial x_i} \geq 0$, since f is increasing. This fact, together with (3.5.5), (3.5.1) and (3.5.7) proves that $F'(\lambda) \geq 0$, for any $\lambda \in [0,1]$. This conclusion together with (3.5.2) proves Theorem 3.4. □

Remark 3.5. *Stable processes have the same linear structure as normal processes since arbitrary linear combinations of stable variables are stable. Lee, Rachev and Samorodnitsky (1990) [118] gave necessary and sufficient conditions for a stable random vector to be associated.*

3.5.2 Gaussian processes

Gaussian processes belong to the class of ζ-dependent models. This property is a consequence of the following lemma.

Lemma 3.8. *Denote $X_C = (X_i)_{i\in C}$ if $C \subset \mathbb{Z}$. Let $(X_n)_{n\in\mathbb{Z}}$ be a Gaussian centered process. Then for all real-valued functions h, k with bounded first partial derivatives, one has*

$$|\mathrm{Cov}(h(X_A), k(X_B))| \leq \sum_{i\in A, j\in B}\left\|\frac{\partial h}{\partial x_i}\right\|_\infty \left\|\frac{\partial k}{\partial x_j}\right\|_\infty |\mathrm{Cov}(X_i, X_j)|. \tag{3.5.8}$$

Remark 3.6. *The proof of Lemma 3.8 is along the paper of Pitt (1982) [146]. For more details, we refer the reader to the proof of Lemma 19 in Doukhan and Louhichi (1999) [67].*

3.5.3 Interacting particle systems

In this subsection, we develop an example of ζ-dependent interacting particle systems (cf. Proposition 3.4 below). Before stating the main result of this paragraph, we briefly recall the basic construction of general interacting particle systems, described in sections I.3 and I.4 of Liggett's book (1985) [122].

Let S be a countable set of sites, W a finite set of states, and $\mathcal{X} = W^S$ the set of configurations, endowed with its product topology, that makes it a compact set. On each site the state evoluate as a Markov chain. But we are interested in the case where the evolution of neighbour sites are linked. We define a Feller process on $\mathcal{X}$ by specifying the local transition rates: to a configuration η and a finite set of sites T is associated a nonnegative measure $c_T(\eta, \cdot)$ on W^T. Loosely speaking, we want the configuration to change on T after an exponential time with parameter

$$c_{T,\eta} = \sum_{\zeta \in W^T} c_T(\eta, \zeta) \ .$$

After that time, the configuration becomes equal to ζ on T, with probability $c_T(\eta, \zeta)/c_{T,\eta}$. Let η^ζ denote the new configuration, which is equal to ζ on T, and to η outside T. The infinitesimal generator should be:

$$\Omega f(\eta) = \sum_{T \subset S} \sum_{\zeta \in W^T} c_T(\eta, \zeta)(f(\eta^\zeta) - f(\eta)) \ . \tag{3.5.9}$$

For Ω to generate a Feller semigroup acting on continuous functions from X into $\mathbb{R}$, some hypotheses have to be imposed on the transition rates $c_T(\eta, \cdot)$. The first condition is that the mapping $\eta \mapsto c_T(\eta, \cdot)$ should be continuous (and thus bounded, since $\mathcal{X}$ is compact). Let us denote by c_T its supremum norm.

$$c_T = \sup_{\eta \in X} c_{T,\eta}.$$

It is the maximal rate of change of a configuration on T. One essential hypothesis is that the maximal rate of change of a configuration at one given site is bounded.

$$B = \sup_{x \in S} \sum_{T \ni x} c_T < \infty. \tag{3.5.10}$$

If f is a continuous function on $\mathcal{X}$, one defines $\Delta_f(x)$ as the degree of dependence of f on x:

$$\Delta_f(x) = \sup\{\, |f(\eta) - f(\zeta)| \,/\, \eta, \zeta \in X \text{ and } \eta(y) = \zeta(y)\ \forall\, y \neq x \,\} \ .$$

Since f is continuous, $\Delta_f(x)$ tends to 0 as x tends to infinity, and f is said to be *smooth* if Δ_f is summable:

$$|||f||| = \sum_{x \in S} \Delta_f(x) < \infty.$$

It can be proved that if f is smooth, then Ωf defined by (3.5.9) is indeed a continuous function on $\mathcal{X}$ and moreover:

$$\|\Omega f\| \leq B|||f|||.$$

We also need to control the dependence of the transition rates on the configuration at other sites. If $y \in S$ is a site, and $T \subset S$ is a finite set of sites, one defines

$$c_T(y) = \sup \left\{ \|c_T(\eta_1, \cdot) - c_T(\eta_2, \cdot)\|_{tv} \,/\, \eta_1(z) = \eta_2(z) \; \forall \, z \neq y \right\},$$

where $\| \cdot \|_{tv}$ is the total variation norm:

$$\|c_T(\eta_1, \cdot) - c_T(\eta_2, \cdot)\|_{tv} = \frac{1}{2} \sum_{\zeta \in W^T} |c_T(\eta_1, \zeta) - c_T(\eta_2, \zeta)| \, .$$

If x and y are two sites such that $x \neq y$, the *influence* of y on x is defined as:

$$\gamma(x, y) = \sum_{T \ni x} c_T(y).$$

We will set $\gamma(x, x) = 0$ for all x. The influences $\gamma(x, y)$ are assumed to be summable:

$$M = \sup_{x \in S} \sum_{y \in S} \gamma(x, y) < \infty. \tag{3.5.11}$$

Under both hypotheses (3.5.10) and (3.5.11), it can be proved that the closure of Ω generates a Feller semigroup $\{S_t \, , \, t \geq 0\}$ (Theorem 3.9 p. 27 of Liggett (1985)). A generic process with semigroup $\{S_t \, , \, t \geq 0\}$ will be denoted by $\{\eta_t \, , \, t \geq 0\}$. The expectations with respect to its distribution, starting from $\eta_0 = \eta$ will be denoted by $\mathbb{E}_\eta$. For each continuous function f, one has:

$$S_t f(\eta) = \mathbb{E}_\eta[f(\eta_t)] = \mathbb{E}[f(\eta_t) \,|\, \eta_0 = \eta].$$

We have now all the ingredients to control the covariance of $f(\eta_s)$ and $g(\eta_t)$ for a finite range interacting particle system when the underlying graph structure has bounded degree. Proposition 3.4 shows that if f and g are mainly located on two finite sets R_1 and R_2, then the covariance of f and g decays exponentially in the distance between R_1 and R_2.

From now on, we assume that the set of sites S is endowed with an undirected graph structure, and we denote by d the natural distance on the graph. We will assume not only that the graph is locally finite, but also that the degree of each vertex is uniformly bounded.

$$\forall x \in S \quad \#\{y \in S \,/\, d(x,y) = 1\} \leq r,$$

where # denotes the cardinality of a finite set. Thus the size of the sphere or ball with center x and radius n is uniformly bounded in x, and increases at most geometrically in n.

$$\#\{y \in S \,/\, d(x,y) = n\} \leq \frac{r}{r-1}(r\text{–}1)^n, \quad \#\{y \in S \,/\, d(x,y) \leq n\} \leq \frac{r}{r-2}(r\text{–}1)^n.$$

Let R be a finite subset of S. We shall use the following upper bounds for the number of vertices at distance n, or at most n from R.

$$\#\{x \in S \,/\, d(x,R) = n\} \leq \#\{y \in S \,/\, d(x,R) \leq n\} \leq 2e^{n\rho}\#R, \tag{3.5.12}$$

with $\rho = \log(r-1)$.
In the case of an amenable graph (e.g. a lattice on $\mathbb{Z}^d$), the ball sizes have a subexponential growth. Therefore, for all $\varepsilon > 0$, there exists c such that:

$$\#\{x \in S \,/\, d(x,R) = n\} \leq \#\{y \in S \,/\, d(x,R) \leq n\} \leq c\, e^{n\varepsilon}.$$

What follows is written in the general case, using (3.5.12). It applies to the amenable case replacing ρ by ε, for any $\varepsilon > 0$.
We are going to deal with smooth functions, depending weakly on coordinates away from a fixed finite set R. Indeed, it is not sufficient to consider functions depending only on coordinates in R, because if f is such a function, then for any $t > 0$, $S_t f$ may depend on all coordinates.

Definition 3.2. *Let f be a function from S into $\mathbb{R}$, and R be a finite subset of S. The function f is said to be* mainly located on R *if there exists two constants α and $\beta > \rho$ such that $\alpha > 0$, $\beta > \rho$ and for all $x \in \mathbb{R}$:*

$$\Delta_f(x) \leq \alpha e^{-\beta d(x,R)}. \tag{3.5.13}$$

Since $\beta > \rho$, the sum $\sum_x \Delta_f(x)$ is finite. Therefore a function mainly located on a finite set is necessarily smooth.
The system we are considering will be supposed to have finite range interactions in the following sense (cf. Definition 4.17 p. 39 of Liggett (1985)).

Definition 3.3. *A particle system defined by the rates $c_T(\eta, \cdot)$ is said to have finite range interactions if there exists $k > 0$ such that if $d(x,y) > k$:*

1. *$c_T = 0$ for all T containing both x and y,*

2. $\gamma(x,y) = 0$.

The first condition imposes that two coordinates cannot simultaneously change if their distance is larger than k. The second one says that the influence of a site on the transition rates of another site cannot be felt beyond distance k. Under these conditions, the following covariance inequality holds.

Proposition 3.4. *Assume (3.5.10) and (3.5.11). Assume moreover that the process is of finite range. Let R_1 and R_2 be two finite subsets of S. Let β be a constant such that $\beta > \rho$. Let f and g be two functions mainly located on R_1 and R_2, in the sense that there exist positive constants κ_f, κ_g such that,*

$$\Delta_f(x) \leq \kappa_f e^{-\beta d(x,R_1)} \quad \textit{and} \quad \Delta_g(x) \leq \kappa_g e^{-\beta d(x,R_2)}.$$

Then for all positive reals s, t,

$$\sup_{\eta \in X} \left| \mathrm{Cov}_\eta(f(\eta_s), g(\eta_t)) \right| \leq C\kappa_f \kappa_g (\#R_1 \wedge \#R_2) e^{D(t+s)} e^{-(\beta-\rho)d(R_1,R_2)}, \quad (3.5.14)$$

where,

$$D = 2Me^{(\beta+\rho)k} \quad \textit{and} \quad C = \frac{2Be^{\beta k}}{D}\left(1 + \frac{e^{\rho k}}{1 - e^{-\beta+\rho}}\right).$$

Proof. We refer the reader to the proof of Proposition 3.3 in Doukhan *et al.* (2005) [64]. □

Remark 3.7. *Shashkin (2005) [176] obtains a similar inequality for random fields indexed by $\mathbb{Z}^d$. For transitive graphs, the covariance inequality stated in Proposition 3.4 was studied by Doukhan et al. (2005) [64] in order to derive a functional central limit theorem for interacting particle systems.*

3.6 Other models

3.6.1 Random AR models

Assume here that

$$X_t = A_t X_{t-1} + \xi_t,$$

where the sequence $\xi_t \in \mathbb{R}^d$ is still i.i.d. but now A_t is assumed to be a stationary sequence of random $d \times d$ matrices. A stationary solution of the previous equation has the formal expansion

$$X_t = \sum_{k=0}^{\infty} A_t \cdots A_{t-k} \xi_{t-k}$$

A first and simple case for convergence of this series is $\mathbb{E}\|A_0\|^p < 1$ for a suitable matrix norm, in the case where the sequence (A_t) is i.i.d. and that $A_t, A_{t-1}, \ldots$ are independent of the inputs (ξ_t). For this we also assume $\mathbb{E}\|\xi_0\|^p < \infty$ for some $p \geq 1$. This condition* also implies convergence of the previous series in $\mathbb{L}^p$.

For $d = 1$, a simple example of this situation is the bilinear model $A_t = a + b\xi_{t-1}$. If now $A_t = \zeta_t + \sum_{j=1}^J b_j \xi_{t-j}$ for a stationary sequence (ζ_t) independent of (ξ_t) the condition

$$\mathbb{E}\left|\zeta_0 + \sum_{j=1}^J b_j \xi_j\right|^{Jp} < 1$$

implies absolute convergence of the previous series in $\mathbb{L}^p$ through Hölder's inequality. The previous relation holds if

$$\|\zeta_0\|_{pJ} + \|\xi_0\|_{pJ} \sum_{j=1}^J \|b_j\| < 1.$$

Those models are also suitable for the previous section related to Markov chains, but a special case of this situation is provided if the sequence (A_t) is stationary and independent of the sequence (ξ_t). In this case the assumption

$$\sum_{k=0}^{\infty} \mathbb{E}\|A_k A_{k-1} \cdots A_0\|^p < \infty$$

implies the convergence of the previous series in $\mathbb{L}^p$ if $\mathbb{E}\|\xi_0\|^p < \infty$.

Extension of such models, solutions of the non Markov equation

$$X_t = \sum_{j \in A} \alpha_t^j X_{t-j} + \zeta_t, \tag{3.6.1}$$

are seen in Doukhan and Truquet (2006) [76] as random fields (†) with infinitely many interactions. If $b = \sum_{j \in A} \|\alpha_0^j\|_p < 1$ the solution of equation (3.6.1) writes a.s. and in $\mathbb{L}^p$,

$$X_t = \zeta_t + \sum_{i=1}^{\infty} \sum_{j_1, \ldots, j_i \in A} \alpha_t^{j_1} \alpha_{t-j_1}^{j_2} \cdots \alpha_{t-j_1-\cdots-j_{i-1}}^{j_i} \zeta_{t-(j_1+\cdots+j_i)}.$$

*Existence of the model in this case also relies on the weaker assumption $\mathbb{E} \log \|A_0\| < 0$; in this case the previous series only converges *a.s.* and dependence conditions are not easy to derived; for this a concentration inequality is needed and a log transformation should be applied to the obtained coefficients.

†Innovations ζ_t are vectors of $\mathbb{R}^k$ and coefficients α_t^j are $k \times k$ matrices, $\|\cdot\|$ is a norm of algebra on this set of matrices and X will be an E valued random field. Let $A \subset \mathbb{Z}^d \setminus \{0\}$, we assume that the i.i.d. random field $\xi = \left((\alpha_t^j)_{i \in A}, \zeta_t\right)_{t \in \mathbb{Z}^d}$ takes now its values in $(M_{k \times k})^A \times E$; here $M_{k \times k}$ denotes the set of $k \times k$ matrices.

3.6.2 Integer valued models

The idea of Galton Watson models conducted Alain Latour (see [116], [117], [65]) to the construction of integer valued extensions of the standard econometric models. As they are discrete valued no mixing condition may usually be expected from such models (see section 1.5) and this is why they fit nicely in the weak dependent frame.

Definition 3.4 (Steutel and van Harn Operator). *Let $(Y_j)_{j\in\mathbb{N}}$ be a sequence of independent and identically distributed (i.i.d.) non-negative integer-valued variables with mean α and variance λ, independent of X, a non-negative integer-valued variable. The Steutel and van Harn operator, $\alpha\circ$ is defined by:*

$$\alpha \circ X = \begin{cases} \sum_{i=1}^{X} Y_i, & \text{if } X \neq 0, \\ 0, & \text{otherwise.} \end{cases}$$

The sequence $(Y_i)_{i\in\mathbb{N}}$ is called a counting sequence. Note that, as indicated in Definition 3.4, the mean of the summands Y_i associated with the operator $\alpha\circ$ is denoted by α. Suppose that $\beta\circ$ is another Steutel and van Harn operator based on a counting sequence $(\widetilde{Y}_i)_{i\in\mathbb{N}}$. The operator $\alpha\circ$ and $\beta\circ$ are said to be independent if, and only if, the counting sequences $(Y_i)_{i\in\mathbb{N}}$ and $(\widetilde{Y}_i)_{i\in\mathbb{N}}$ are mutually independent. One may first think to Poisson distributed variables Y_i with parameter a. The first example, Galton Watson with immigration

$$X_t = a \circ X_{t-1} + \xi_t \tag{3.6.2}$$

was extended in various papers by Alain Latour (see *e.g.* [116] or [117]) for bilinear type extensions (see Doukhan, Latour and Oraichi, 2006 [65]).
We would like to extend the integer-valued model class to give a non-negative integer-valued bilinear process, denoted by INBL(p,q,m,n), similar to the real-valued bilinear process. A time series $(X_t)_{t\in\mathbb{N}}$ is generated by a *bilinear model*, if it satisfies the equation:

$$X_t = \alpha + \sum_{i=1}^{p} a_i X_{t-i} + \sum_{j=1}^{q} c_j \varepsilon_{t-j} + \sum_{k=1}^{m}\sum_{\ell=1}^{n} b_{\ell k}\left(\varepsilon_{t-\ell} X_{t-k}\right) + \varepsilon_t \tag{3.6.3}$$

where $(\varepsilon_t)_{t\in\mathbb{N}}$ is a sequence of i.i.d. random variables, usually but not always with zero mean, and where α, a_i, $i=1,\ldots,p$, c_j, $j=1,\ldots,q$, and $b_{\ell k}$, $k=1,\ldots,m$, $\ell=1,\ldots,n$ are real constants. In (3.6.3), we can "formally" substitute Steutel and van Harn operators to some of the parameters giving an equation of the form

$$X_t = \sum_{i=1}^{p} a_i \circ X_{t-i} + \sum_{j=1}^{q} c_j \circ \varepsilon_{t-j} + \sum_{k=1}^{m}\sum_{\ell=1}^{n} b_{\ell k} \circ \left(\varepsilon_{t-\ell} X_{t-k}\right) + \varepsilon_t \tag{3.6.4}$$

where the operators $a_i\circ$, $i = 1,\ldots,p$, $c_j\circ$, $j = 1,\ldots,q$, and $b_{k\ell}\circ$, $k = 1,\ldots,m$, $\ell = 1,\ldots,n$, are mutually independent and $(\varepsilon_t)_{t\in\mathbb{N}}$ is a sequence of i.i.d. non-negative integer-valued random variables of finite mean μ and finite variance σ^2, independent of the operators. As in Latour *et al.*, we restrict to the first-order bilinear model

$$X_t = a\circ X_{t-1} + b\circ(\varepsilon_{t-1}X_{t-1}) + \varepsilon_t \tag{3.6.5}$$

where the sequence involved in the operator $a\circ$ and $b\circ$ are respectively of mean a and b and variance α and β. Y and $\widetilde{Y}$ denote generic variables used in $a\circ$ and $b\circ$, respectively. If $a + b\cdot\mu < 1$ Doukhan, Latour and Oraichi (2006) [65] prove that this model is strictly stationary in $\mathbb{L}^1$; it is $\theta-$weakly dependent with

$$\theta(r) \le 2(a + b\cdot\mu)^r\mathbb{E}(X_0).$$

If moreover $\|Y\|_p + \|\varepsilon_0\|_p\|\widetilde{Y}\|_p < 1$ this solution belongs to $\mathbb{L}^p$. Moment estimators thus yield $\sqrt{n}-$consistent estimators of the parameters in the previously cited paper. We finally mention that in the case of non negative coefficients such models are also associated sequences.

3.6.3 Random fields

Analogously, one may define some simple stationary random fields. Let T be any group (in an additive notation) with some metric d, then Bernoulli shifts still write

$$X_t = H((\xi_{s-t})_{s\in T})$$

for a function $H : \mathbb{R}^T \to \mathbb{R}$ if $(\xi_t)_{t\in T}$ is stationary, this is also the case of $(X_t)_{t\in T}$. In order to derive dependence properties of such models one better considers i.i.d. innovations and we assume that

$$\mathbb{E}\big|H((\xi_s)_{s\in T}) - H((\xi^{(r)})_{s\in T})\big| \to_{r\to\infty} 0$$

if we set $\xi_t^{(r)} = \xi_t$ for $d(s,0) < r$ and $\xi_t^{(r)} = z$ is a fixed point of ξ's values set. Another option is to use a i.i.d. sequence $\xi' = (\xi_t')_{t\in T}$ independent and with the same distribution as ξ and to set $\xi_t^{(r)} = \xi_t'$ for $d(s,0) \ge r$.
Here again linear random fields as well as Volterra random fields are simple to define. Standard sets T are $\mathbb{Z}^d$ and $(\mathbb{Z}/n\mathbb{Z})^d$. It is less natural to work here with continuous time processes because i.i.d. white noise are discontinuous processes: they are thus less natural to define. A nice example of this situation is given in the next subsection.

LARCH(∞) random fields

Let $(\xi_t)_{t\in\mathbb{Z}^d}$ be a stationary sequence of random $d\times m$-matrices, $(a_j)_{j\in\mathbb{N}^*}$ be a sequence of $m\times d$ matrices, and a be a vector in $\mathbb{R}^m$. A vector valued

LARCH(∞) random field model is a solution of the recurrence equation

$$X_t = \xi_t \left(a + \sum_{j \neq 0} a_j X_{t-j} \right), \quad t \in \mathbb{Z}^d \tag{3.6.6}$$

Such LARCH(∞) models include a large variety of models, as those in § 3.4 but the main point is here that causality in no more assumed in general. The same proof as in Section 3.4 entails the

Proposition 3.5 (Doukhan, Teyssière, Winant, 2006 [75]). *Assume that*

$$\|\xi_0\|_\infty \sum_{j \neq 0} \|a_j\| < 1,$$

then one stationary of solution of eqn. (3.6.6) in $\mathbb{L}^p$ is given as

$$X_t = \xi_t \left(a + \sum_{k=1}^{\infty} \sum_{j_1, \ldots, j_k \neq 0} a_{j_1} \xi_{t-j_1} a_{j_2} \ldots a_{j_k} \xi_{t-j_1-\cdots-j_k} a \right) \tag{3.6.7}$$

In the following of this section we set $A(x) = \sum_{\|j\| \geq x} \|a_j\|$, $A = A(1)$ and $\lambda = A\|\xi_0\|_\infty$ where $\|(j_1, \ldots, j_k)\| = |j_1| + \cdots + |j_k|$.

Approximations. We assume here that the random field $(\xi_t)_{t \in \mathbb{Z}^d}$ is i.i.d.. One first approximates here X_t by a random variable independent of X_0. Set

$$\tilde{X}_t = \xi_t \left(a + \sum_{k=1}^{\infty} \sum_{\|j_1\| + \cdots + \|j_k\| < t} a_{j_1} \xi_{t-j_1} \cdots a_{j_k} \xi_{t-j_1-\cdots-j_k} a \right)$$

Proposition 3.6. *One bound for the error is given by:*

$$\mathbb{E}\|X_t - \tilde{X}_t\| \leqslant \mathbb{E}\|\xi_0\| \left(\mathbb{E}\|\xi_0\| \sum_{k=1}^{t-1} k\lambda^{k-1} A\left(\frac{t}{k}\right) + \frac{\lambda^t}{1-\lambda} \right) \|a\|$$

We now specialize this result. Assume that $b, C > 0$ are constants, then there exists some constant $K > 0$ such that

$$\|X_t - \tilde{X}_t\| \leqslant \begin{cases} \frac{K}{t^b}, & \text{under Riemaniann decay } A(x) \leqslant Cx^{-b} \\ K(q \vee \lambda)^{\sqrt{t}}, & \text{under geometric decay } A(x) \leqslant Cq^x \end{cases}$$

Markov approximation. Consider equation (3.6.6) truncated at rank N, $X_t^N = \xi_t \left(a + \sum_{0<\|j\|\le N} a_j X_{t-j}^N\right)$. The previous solution rewrites as

$$X_t^N = \xi_t \left(a + \sum_{k=1}^{\infty} \sum_{0<\|j_1\|,\dots,\|j_k\|\le N} a_{j_1}\xi_{t-j_1}\cdots a_{j_k}\xi_{t-j_1-\cdots-j_k} a \right).$$

Then $\mathbb{E}\|X_t - X_t^N\| \le \sum_{k=1}^{\infty} A(N)^k$. This error has rate $\sum_{k=1}^{\infty} N^{-bk}$ for Riemanian decays and $q^N/(1-q^N)$ in the geometric case. Moreover:

Theorem 3.5. *The solution (3.6.7) of eqn. (3.6.6) is η−weakly dependent with*

$$\eta(r) = \|\xi_0\|_\infty \left(\|\xi_0\|_\infty \sum_{k=1}^{t-1} k\lambda^{k-1} A\left(\frac{t}{k}\right) + \frac{\lambda^t}{1-\lambda} \right) \|a\| .$$

This bound may be made explicit for the decays considered previously.

Models with infinite memory

We also mention rapidly here truly non linear extensions of LARCH(∞) models which are the chains with infinite memory from Doukhan and Wintenberger (2006) [78] and the random fields with infinite interactions from Doukhan and Truquet (2007) [55], those models are respectively solutions of the equations‡

$$\begin{aligned} X_t &= F(X_{t-1}, X_{t-2}, \dots; \xi_t), \qquad t \in \mathbb{Z}, \\ X_t &= F\left(\left(X_{t-j}\right)_{j\neq 0}; \xi_t\right), \qquad t \in \mathbb{Z}^d, \end{aligned}$$

those models are usual excited by i.i.d. inputs ξ. Even if no explicit chaotic solution seems to be available in general, such models are well defined and $\mathbb{L}^p$ stationary if

$$\|F(x;\xi_0) - F(y;\xi_0)\|_m \le \sum_{j\neq 0} \alpha_j \|x_j - y_j\|, \qquad a = \sum_{j\neq 0} \alpha_j < 1,$$

in the previous inequality§ one should take $m = p$ both for causal random processes and causal random fields (accurately defined in the above mentioned work) and $m = \infty$ else. Moreover the weak dependence coefficients are proved to follow analogous decays with now $\alpha_i = \|a_i\|$ in theorem 3.4.2. More precisely, the respectively τ or η weak dependence coefficients have rates driven by the relation

$$\inf_p \left(a^{\frac{r}{p}} + \sum_{|i|\ge p} \alpha_i \right).$$

‡Here the function $F(x;u)$ is perhaps not defined over all $\mathbb{R}^{\mathbb{N}} \times \mathbb{R}$ or $\mathbb{R}^{\mathbb{Z}^d} \times \mathbb{R}$ but it is enough that it is defined on trajectories of the solution.

§For respectively $j \in \mathbb{N}^*$ and $j \in \mathbb{Z}^d \setminus \{0\}$.

With the previous geometric decays, the bound is the same as in the paragraph related to theorem 3.3 and for Riemanian decays $\sum_{i\geq p} a_i \leq C\, i^{-b}$ a log loss appears and $\tau_{1,\infty}(r) \leq K\log^{b\vee 1} r \cdot r^{-b}$.

3.6.4 Continuous time

In the continuous time setting one better consider a process with independent increments, $(Z_t)_{t\in\mathbb{R}}$, then a simple extension of linear processes is defined through the Wiener integral

$$X_t = \int_{-\infty}^{\infty} f(t-s)dZ_s$$

It is for example simple to define such integrals for a Brownian motion but other possibilities are all the classes of Lévy processes. Among them, the $S\alpha S$-Lévy motion is described in Samorodnitsky and Taqqu (1994) [172].
Analogues of Volterra processes are now multiple stochastic integrals. A complete theory is developed by Major (1981) [126].
More examples are provided in the monograph by Doukhan, Oppenheim and Taqqu (2003) [72].

Chapter 4

Tools for non causal cases

Moment inequalities are the main tools when using non causal weak dependence. A first useful section addresses the weak dependence properties of indicators of processes, useful both for moment inequalities and for the empirical process. After this separate sections address variances of sums, $(2+\delta)$-order moments and higher order moments. They yield both Rosenthal type and Marcinkiewicz-Zygmund inequalities. Finally cumulants sums are also considered as dependence coefficients and they are used in order to derive sharp exponential inequalities. A last section is devoted to prove tightness criteria for empirical processes through suitable moment inequalities.

4.1 Indicators of weakly dependent processes

Define, for positive real number x, the function $g_x : \mathbb{R} \to \mathbb{R}$ by

$$g_x(y) = \mathbf{1}_{x \le y} - \mathbf{1}_{x \le -y}.$$

We are interested along this chapter by $(\mathcal{I}, \Psi)$-dependent sequences, where

$$\mathcal{I} = \left\{ \bigotimes_{i=1}^{u} g_{x_i} \Big/ \ x_i > 0, \ u \in \mathbb{N}^* \right\},$$

and $\Psi(f,g) = c(d_f, d_g)$, for some positive function c defined on $\mathbb{N}^* \times \mathbb{N}^*$, in this case we will simply say that the sequence is $(\mathcal{I}, c)$-dependent. Set

$$\Lambda_0 = \left\{ \bigotimes_{i=1}^{u} f_i \Big/ \ f_i \in \Lambda \cap \mathbb{L}^\infty, \ f_i : \mathbb{R} \to \mathbb{R}, \ i = 1, \ldots, u, \ u \in \mathbb{N}^* \right\}.$$

The following lemma relates η, κ or θ weak dependence to $\mathcal{I}$ weak dependence under additional concentration assumptions.

Lemma 4.1. *Let $(X_n)_{n\in\mathbb{N}}$ be a sequence of r.v's. Suppose that, for some positive real constants C, α, λ*

$$\sup_{x\in\mathbb{R}}\sup_{i\in\mathbb{N}} \mathbb{P}\left(x \le X_i \le x+\lambda\right) \le C\lambda^\alpha. \tag{4.1.1}$$

(i) If the sequence (X_n) is (Λ_0, η)-dependent, then it is $(\mathcal{I}, c)$-dependent with $\epsilon(r) = \eta(r)^{\frac{\alpha}{1+\alpha}}$ and $c(u,v) = 2(8C)^{\frac{1}{1+\alpha}}(u+v)$.

(ii) If the sequence (X_n) is (Λ_0, κ)-dependent, then it is $(\mathcal{I}, c)$-dependent with $\epsilon(r) = \kappa(r)^{\frac{\alpha}{2+\alpha}}$ and $c(u,v) = 2(8C)^{\frac{2}{2+\alpha}}(u+v)^{\frac{2(1+\alpha)}{2+\alpha}}$.

(iii) If the sequence (X_n) is (Λ_0, θ)-dependent, then it is $(\mathcal{I}, c)$-dependent with $\epsilon(r) = \theta(r)^{\frac{\alpha}{1+\alpha}}$ and $c(u,v) = 2(8C)^{\frac{1}{1+\alpha}}(u+v)^{\frac{1}{1+\alpha}}$.

(iv) If the sequence (X_n) is (Λ_0, λ)-dependent (with $\lambda(r) \le 1$), then it is $(\mathcal{I}, c)$-dependent with $c(u,v) = 2\left((8C)^{\frac{1}{1+\alpha}} + (8C)^{\frac{2}{2+\alpha}}\right)(u+v)^{\frac{2(1+\alpha)}{2+\alpha}}$ and $\epsilon(r) = \lambda(r)^{\frac{\alpha}{2+\alpha}}$.

Proof of Lemma 4.1. First recall that for all real numbers $0 \le x_i, y_i \le 1$, $|x_1\cdots x_m - y_1\cdots y_m| \le \sum_{i=1}^m |x_i - y_i|$. Let then $g, f \in \mathcal{I}$, *i.e.*

$$g(y_1,\ldots,y_u) = g_{x_1}(y_1)\cdots g_{x_u}(y_u), \text{ and } f(y_1,\ldots,y_v) = g_{x'_1}(y_1)\cdots g_{x'_v}(y_v)$$

for some $u, v \in \mathbb{N}^*$ and $x_i, x'_j \ge 0$.
For fixed $x > 0$ and $a > 0$ let

$$f_x(y) = \mathbf{1}_{y>x} - \mathbf{1}_{y\le -x} + \left(\frac{y}{a} - \frac{x}{a} + 1\right)\mathbf{1}_{x-a<y<x} + \left(\frac{y}{a} + \frac{x}{a} - 1\right)\mathbf{1}_{-x<y<-x+a}.$$

Then $\text{Lip}(f_x) = a^{-1}$ and $\|f_x\|_\infty = 1$.

Define now h and k respectively by

$$h(y_1,\ldots,y_u) = f_{x_1}(y_1)\cdots f_{x_u}(y_u), \qquad k(y_1,\ldots,y_v) = f_{x'_1}(y_1)\cdots f_{x'_v}(y_v)$$

then $\text{Lip}(h) \le a^{-1}$, $\text{Lip}(k) \le a^{-1}$. Consider $i_1 \le \cdots \le i_u \le i_u + r \le j_1 \le \cdots \le j_v$ and set $\text{Cov}(h,k) := \text{Cov}(h(X_{i_1},\ldots,X_{i_u}), k(X_{j_1},\ldots,X_{j_v}))$.

(i) η-weak dependence $\Rightarrow |\text{Cov}(h,k)| \le (u+v)\eta(r)/a$.

(ii) κ-weak dependence $\Rightarrow |\text{Cov}(h,k)| \le ((u+v)/a)^2\,\kappa(r)$.

(iii) θ-weak dependence $\Rightarrow |\text{Cov}(h,k)| \le v\theta(r)/a$.

Inequality (4.1.1) yields $|\text{Cov}(g,f) - \text{Cov}(h,k)| \le 8Ca^\alpha(u+v)$ and

(i) $|\text{Cov}(g,f)| \le 8Ca^\alpha(u+v) + (u+v)\eta(r)/a$,

(ii) $|\mathrm{Cov}(g,f)| \le 8Ca^\alpha(u+v) + \left(\frac{u+v}{a}\right)^2 \kappa(r)$, or

(iii) $|\mathrm{Cov}(g,f)| \le 8Ca^\alpha(u+v) + \frac{v}{a}\theta(r)$.

The lemma follows by setting respectively

$$(i)\, a = \left(\frac{\eta(r)}{8C}\right)^{1/(1+\alpha)} \quad (ii)\, a = \left(\frac{(u+v)\kappa(r)}{8C}\right)^{1/(2+\alpha)} \quad (iii)\, a = \left(\frac{\theta(r)}{8C(u+v)}\right)^{1/(1+\alpha)}$$

The case of λ dependence is obtained by the summation of both cases *(i)* and *(ii)*. □

4.2 Low order moments inequalities

Our proof for central limit theorems is based on a truncation method. For a truncation level $T \ge 1$ we shall denote $\overline{X}_k = f_T(X_k) - \mathbb{E}f_T(X_k)$ with $f_T(X) = X \vee (-T) \wedge T$. Let us simply remark that $\overline{X}_k$ has moments of any orders because it is bounded. Suppose that $\mu = \mathbb{E}|X|^m$ is finite for some $m > 0$. Furthermore, for any $a \le m$, we control the difference $\mathbb{E}|f_T(X_0) - X_0|^a$ with Markov inequality:

$$\mathbb{E}|f_T(X_0) - X_0|^a \le \mathbb{E}|X_0|^a \mathbf{1}_{\{|X_0| \ge T\}} \le \mu T^{a-m},$$

thus using Jensen inequality yields

$$\|\overline{X}_0 - X_0\|_a \le 2\mu^{\frac{1}{a}} T^{1-\frac{m}{a}}. \tag{4.2.1}$$

Deriving from this truncation, we are now able to control the limiting variance as well as the higher order moments.

4.2.1 Variances

Lemma 4.2 (Variances). *If one of the following conditions holds then the series $\sum_{k\ge0} |\mathrm{Cov}(X_0, X_k)|$ is convergent*

$$\sum_{k=0}^{\infty} \kappa(k) < \infty \tag{4.2.2}$$

$$\sum_{k=0}^{\infty} \lambda(k)^{\frac{m-2}{m-1}} < \infty \tag{4.2.3}$$

Proof. Using the fact that $\overline{X}_0 = g_T(X_0)$ is a function of X_0 with $\mathrm{Lip}\, g_T = 1$, $\|g_T\|_\infty \le 2T$ we derive, for T large enough,

$$|\mathrm{Cov}(\overline{X}_0, \overline{X}_k)| \le \kappa(k), \quad \text{or } \le (2T+1)\lambda(k) \le 4T\lambda(k) \text{ respectively} \tag{4.2.4}$$

In the κ-dependent case, truncation can thus be omitted and

$$|\mathrm{Cov}(X_0, X_k)| \le \kappa(k) \tag{4.2.5}$$

we only consider λ dependence below. Now we develop

$$\mathrm{Cov}(X_0, X_k) = \mathrm{Cov}(\overline{X}_0, \overline{X}_k) + \mathrm{Cov}(X_0 - \overline{X}_0, X_k) + \mathrm{Cov}(\overline{X}_0, X_k - \overline{X}_k)$$

and using a truncation T to be determined we use the two previous bounds (4.2.1) and (4.2.4) with Hölder inequality with the exponents $\frac{1}{a} + \frac{1}{m} = 1$ to derive

$$\begin{aligned} |\mathrm{Cov}(X_0, X_k)| &\le 4T\lambda(k) + 2\|X_0\|_m \|\overline{X}_0 - X_0\|_a \\ &\le 4T\lambda(k) + 4\mu^{1/a+1/m} T^{1-m/a} \\ &\le 4(T\lambda(k) + \mu T^{2-m}). \end{aligned}$$

Note that we used the relation $1 - m/a = 2 - m$. Thus using the truncation such that $T^{m-1} = \frac{\mu}{\lambda(k)}$ yields the bound

$$|\mathrm{Cov}(X_0, X_k)| \le 8\mu^{\frac{1}{m-1}} \lambda(k)^{\frac{m-2}{m-1}}. \ \square \tag{4.2.6}$$

4.2.2 A $(2+\delta)$-order moment bound

Lemma 4.3. *Assume that the stationary and centered process $(X_i)_{i\in\mathbb{Z}}$ satisfies $\mathbb{E}|X_0|^{2+\zeta} < \infty$, and it is either κ-weakly dependent with $\kappa(r) = \mathcal{O}(r^{-\kappa})$ or λ-weakly dependent with $\lambda(r) = \mathcal{O}(r^{-\lambda})$. Then if $\kappa > 2 + \frac{1}{\zeta}$, or $\lambda > 4 + \frac{2}{\zeta}$, then for all $\delta \in]0, A \wedge B \wedge 1[$ (where A and B are constants smaller than ζ and only depending of ζ and respectively κ or λ, see 4.2.10 and 4.2.11), there exist $C > 0$ such that:*

$$\|S_n\|_\Delta \le C\sqrt{n}, \qquad \textit{where} \quad \Delta = 2 + \delta.$$

Remarks.

- The constant C satisfies $C > \left(\frac{5}{2^{\delta/2} - 1}\right)^{1/\Delta} \sum_{k\in\mathbb{Z}} |\mathrm{Cov}(X_0, X_k)|$. Under the conditions of this lemma, Lemma 4.2 entails

$$c \equiv \sum_{k\in\mathbb{Z}} |\mathrm{Cov}(X_0, X_k)| < \infty.$$

- The result is sketched from Bulinski and Sashkin (2005) [33]; notice, however that their condition of dependence is of a causal nature while our is not which explains a loss with respect to the exponents λ and κ. In their $\zeta-$weak dependence setting the best possible value of the exponent is 1 while it is 2 for our non causal dependence.

Proof of lemma 4.3. Analogously to Bulinski and Sashkin (2005) [33], who sketch Ibragimov (1962) [110], we proceed by recurrence on k for $n \le 2^k$ to prove the property:

$$\left\|1+|S_n|\right\|_\Delta \le C\sqrt{n}. \tag{4.2.7}$$

We then assume (4.2.7) for all $n \le 2^{K-1}$. We note $N = 2^K$ and we want to bound $\|1+|S_N|\,\|_\Delta$. We always can share the sum S_N in three blocks, the two first with the same size $n \le 2^{K-1}$ denoted A and B, and the third V placed between the two first and of size $q < n$. We then have $\|1+|S_N|\,\|_\Delta \le \|1+|A|+|B|\,\|_\Delta + \|V\|_\Delta$. The term $\|V\|_\Delta$ is directly bounded with $\|1+|V|\,\|_\Delta$ and the property of recurrence, *i.e.* $C\sqrt{q}$. Writing $q = N^b$ with $b < 1$, then this term is of order strictly smaller than $\sqrt{N}$. For $\|1+|A|+|B|\,\|_\Delta$, we have:

$$\begin{aligned}\mathbb{E}(1+|A|+|B|)^\Delta &\le \mathbb{E}(1+|A|+|B|)^2(1+|A|+|B|)^\delta,\\ &\le \mathbb{E}(1+2|A|+2|B|+(|A|+|B|)^2)(1+|A|+|B|)^\delta.\end{aligned}$$

An expansion yields the terms:

- $\mathbb{E}(1+|A|+|B|)^\delta \le 1+|A|_2^\delta+|B|_2^\delta \le 1+2c^\delta(\sqrt{n})^\delta$,
- $\mathbb{E}|A|(1+|A|+|B|)^\delta \le \mathbb{E}|A|((1+|B|)^\delta+|A|^\delta) \le \mathbb{E}|A|(1+|B|)^\delta+\mathbb{E}|A|^{1+\delta}$. The term $\mathbb{E}|A|^{1+\delta}$ is bounded with $\|A\|_2^{1+\delta}$ and then $c^{1+\delta}(\sqrt{n})^{1+\delta}$. The term $\mathbb{E}|A|(1+|B|)^\delta$ is bounded using Hölder $\|A\|_{1+\delta/2}\|1+|B|\,\|_\Delta^\delta$ and then is at least of order $cC^\delta(\sqrt{n})^{1+\delta}$.
- We have the analogous with B instead of A.
- $\mathbb{E}(|A|+|B|)^2(1+|A|+|B|)^\delta$. For this term, we use an inequality from Bulinski:

$$\begin{aligned}&\mathbb{E}\left((|A|+|B|)^2(1+|A|+|B|)^\delta\right)\\ &\qquad\le \mathbb{E}|A|^\Delta+\mathbb{E}|B|^\Delta+5(\mathbb{E}A^2(1+|B|)^\delta+\mathbb{E}B^2(1+|A|)^\delta).\end{aligned}$$

  The term $\mathbb{E}|A|^\Delta$ is bounded using (4.2.7) by $C^\Delta(\sqrt{n})^\Delta$. The second term is the analogous with B. The third is treated with particular care in the following.

We now want to control $\mathbb{E}A^2(1+|B|)^\delta$ and the analogous with B. For this, we introduce the weak dependence. We then have to truncate the variables. We denote $\overline{X}$ the variable $X \vee (-T) \wedge T$ for a real T that will determined later. We then note by extension $\overline{A}$ and $\overline{B}$ the sums of the truncated variables $\overline{X}_i$. Remarking that $|B|-|\overline{B}| \ge 0$, we have:

$$\mathbb{E}|A|^2(1+|B|)^\delta \le \mathbb{E}A^2(|B|-|\overline{B}|)^\delta+\mathbb{E}(A^2-\overline{A}^2)(1+|\overline{B}|)^\delta+\mathbb{E}\overline{A}^2(1+|\overline{B}|)^\delta.$$

We begin by control $\mathbb{E}A^2(|B|-|\overline{B}|)^\delta$. Set $m=2+\zeta$, then using Hölder inequality with $2/m+1/m'=1$ yields:

$$\mathbb{E}A^2(|B|-|\overline{B}|)^\delta \leq \|A\|_m^2 \|(|B|-\||\overline{B}|)^\delta\|_{m'}$$

$\|A\|_\Delta$ is bounded using (4.2.7) and we remark that:

$$(|B|-|\overline{B}|)^{\delta m'} \leq (|B|-|B|\mathbf{1}_{\{\forall i, |X_i|\leq T\}})^{\delta m'} \leq |B|^{\delta m'}\mathbf{1}_{\{\exists i, |X_i|>T\}} \leq |B|^{\delta m'}\mathbf{1}_{|B|>T}.$$

We then bound $\mathbf{1}_{|B|>T} \leq (|B|/T)^\alpha$ with $\alpha = m-\delta m'$. Then

$$\mathbb{E}||B|-|\overline{B}||^{\delta m'} \leq \mathbb{E}|B|^m T^{\delta m'-m}.$$

Then, by convexity and stationarity, we have $\mathbb{E}|B|^m \leq n^m \mathbb{E}|X_0|^m$. Then:

$$\mathbb{E}A^2(|B|-|\overline{B}|)^\delta \preceq n^{2+m/m'} T^{\delta - m/m'}.$$

Finally, remarking that $m/m'=m-2$, we obtain:

$$\mathbb{E}A^2(|B|-|\overline{B}|)^\delta \preceq n^m T^{\Delta-m}.$$

We obtain the same bound for the second term:

$$\mathbb{E}(A^2-\overline{A}^2)(1+|\overline{B}|)^\delta \preceq n^m T^{\Delta-m}.$$

For the third term, we introduce a covariance term:

$$\mathbb{E}\overline{A}^2(1+|\overline{B}|)^\delta \leq \mathrm{Cov}(\overline{A}^2,(1+|\overline{B}|)^\delta) + \mathbb{E}\overline{A}^2\mathbb{E}(1+|\overline{B}|)^\delta.$$

The last term is bounded with $|A|_2^2|B|_2^\delta \leq c^\Delta\sqrt{n}^\Delta$. The covariance is controlled by the weak-dependence notions:

- in the κ-dependent case: $n^2T\kappa(q)$,
- in the λ-dependent case: $n^3T^2\lambda(q)$.

We then choose either the truncation $T^{m-\delta-1} = n^{m-2}/\kappa(q)$ or $T^{m-\delta} = n^{m-3}/\lambda(q)$. Now the three terms of the decomposition have the same order:

$$\begin{aligned}
\mathbb{E}|A|^2(1+|B|)^\delta &\preceq \left(n^{3m-2\Delta}\kappa(q)^{m-\Delta}\right)^{1/(m-\delta-1)} && \text{under } \kappa\text{-dependence},\\
\mathbb{E}|A|^2(1+|B|)^\delta &\preceq \left(n^{5m-3\Delta}\lambda(q)^{m-\Delta}\right)^{1/(m-\delta)} && \text{under } \lambda\text{-dependence}.
\end{aligned}$$

Set $q=N^b$, we note that $n\leq N/2$ and this term has order $N^{\frac{3m-2\Delta+b\kappa(\Delta-m)}{m-\delta-1}}$ under κ-weak dependence and the order $N^{\frac{5m-3\Delta+b\lambda(\Delta-m)}{m-\delta}}$ under λ-weak dependence. Those terms are thus negligible with respect to $N^{\Delta/2}$ if:

$$\kappa > \frac{3m-2\Delta-\Delta/2(m-\delta-1)}{b(m-\Delta)}, \quad \text{under } \kappa\text{-dependence}, \tag{4.2.8}$$

$$\lambda > \frac{5m-3\Delta-\Delta/2(m-\delta)}{b(m-\Delta)}, \quad \text{under } \lambda\text{-dependence}. \tag{4.2.9}$$

Finally, using this assumption, $b < 1$ and $n \leq N/2$ we derive the bound for some suitable constants $a_1, a_2 > 0$:

$$\mathbb{E}(1 + |S_N|)^\Delta \leq \left(2^{-\delta/2}C^\Delta + 5 \cdot 2^{-\delta/2}c^\Delta + a_1 N^{-a_2}\right)\left(\sqrt{N}\right)^\Delta .$$

Using the relation linking C and c, we conclude that (4.2.7) is also true at the step N if the constant C satisfies $2^{-\delta/2}C^\Delta + 5 \cdot 2^{-\delta/2}c^\Delta + a_1 N^{-a_2} \leq C^\Delta$. Choose $C > \left(\frac{5c^\Delta + a_1 2^{\delta/2}}{2^{\delta/2} - 1}\right)^{1/\Delta}$ with $c = \sum_{k \in \mathbb{Z}} |\mathrm{Cov}(X_0, X_k)|$, then the previous relation holds.

Finally, we use eqns. (4.2.8) and (4.2.9) to find a condition on δ. In the case of κ-weak dependence, we rewrite inequality (4.2.8) as:

$$0 > \delta^2 + \delta(2\kappa - 3 - \zeta) - \kappa\zeta + 2\zeta + 1.$$

It leads to the following condition on λ:

$$\delta < \frac{\sqrt{(2\kappa - 3 - \zeta)^2 + 4(\kappa\zeta - 2\zeta - 1)} + \zeta + 3 - 2\kappa}{2} = A. \tag{4.2.10}$$

We do the same in the case of the λ-weak dependence:

$$\delta < \frac{\sqrt{(2\lambda - 6 - \zeta)^2 + 4(\lambda\zeta - 4\zeta - 2)} + \zeta + 6 - 2\lambda}{2} = B. \tag{4.2.11}$$

Remark: those bounds are always smaller than ζ. □

4.3 Combinatorial moment inequalities

Let $(X_n)_{n \in \mathbb{N}}$ be a sequence of centered r.v.s. Let $S_n = \sum_{i=1}^n X_i$. In this section, we obtain bounds for $|\mathbb{E}(S_n^q)|$, when $q \in \mathbb{N}$ and $q \geq 2$. Our main references are here Doukhan and Portal (1983) [73], Doukhan and Louhichi (1999) [67], and Rio (2000) [161].

We first introduce the following coefficient of weak dependence.

Definition 4.1. *Let (X_n) be a sequence of centered r.v.s. For positive integer r, we define the coefficient of weak dependence as non-decreasing sequences $(C_{r,q})_{q \geq 2}$ such that*

$$\sup |\mathrm{Cov}(X_{t_1} \cdots X_{t_m}, X_{t_{m+1}} \cdots X_{t_q})| =: C_{r,q}, \tag{4.3.1}$$

where the supremum is taken over all $\{t_1, \ldots, t_q\}$ such that $1 \leq t_1 \leq \cdots \leq t_q$ and m, r satisfy $t_{m+1} - t_m = r$.

Below, we provide explicit bounds of $C_{r,q}$ in order to obtain inequalities for moments of the partial sums S_n. We shall assume, that *either* there exist constants $C, M > 0$ such that for any convenient q-tuple $\{t_1, \dots, t_q\}$ as in the definition,

$$C_{r,q} \le CM^q \epsilon(r), \tag{4.3.2}$$

or, denoting by Q_X the quantile function of $|X|$ (inverse of the tail function $t \mapsto \mathbb{P}(|X| > t)$, see (2.2.14)),

$$C_{r,q} \le c(q) \int_0^{\epsilon(r)} Q_{X_{t_1}}(x) \cdots Q_{X_{t_q}}(x) dx, \tag{4.3.3}$$

The bound (4.3.2) holds mainly for bounded sequences. *E.g.* if $\|X_n\|_\infty \le M$ and X is $(\Lambda \cap \mathbb{L}^\infty, \Psi)$-weak dependent, we have:

$$C_{r,q} \le \max_{1 \le m < q} \Psi(j^{\otimes m}, j^{\otimes (q-m)}, m, q-m) M^q \epsilon(r),$$

where $j(x) = x\mathbf{1}_{|x| \le 1} + \mathbf{1}_{x>1} - \mathbf{1}_{x<-1}$. If $\Psi(h, k, u, v) = c(u, v)\mathrm{Lip}\,(h)\mathrm{Lip}\,(k)$, the bound becomes

$$C_{r,q} \le \max_{1 \le m < q} c(m, q-m) M^{q-2} \epsilon(r).$$

The bound (4.3.3) holds for more general r.v.s, using moment or tail assumptions. With Lemma 4.1, we derive that if the concentration property (4.1.1) holds then the η (resp. κ) weak dependence implies $(\mathcal{I}, c)$-weak dependence.

Now relation (4.3.3) follows from the following lemma.

Lemma 4.4. *If the sequence $(X_n)_{n \in \mathbb{N}}$ is $(\mathcal{I}, c)$-weak dependent, then*

$$|\mathrm{Cov}(X_{t_1} \cdots X_{t_m}, X_{t_{m+1}} \cdots X_{t_q})| \le C_q \int_0^{\epsilon(r)} Q_{t_1}(u) \cdots Q_{t_q}(u) du,$$

where $C_q = (\max_{u+v \le q} c(u, v)) \vee 2$. The quantity Q_{t_i} denotes the inverse of the tail function $t \mapsto \mathbb{P}(|X_{t_i}| > t)$ (inverse are defined in eqn. 2.2.14)).

Proof of Lemma 4.4. Let $Y^+ = 0 \vee Y$ and $Y^- = 0 \vee (-Y)$,

$$\begin{aligned} Y^+ &= \int_0^{+\infty} \mathbf{1}_{x \le Y^+} dx = \int_0^{+\infty} \mathbf{1}_{x \le Y} dx, \qquad \text{and} \\ Y^- &= \int_0^{+\infty} \mathbf{1}_{x \le Y^-} dx = \int_0^{+\infty} \mathbf{1}_{x \le -Y} dx. \end{aligned}$$

The inclusion-exclusion formula entails,

$$Y_1 \cdots Y_q = \prod_{i=1}^{q} (Y_i^+ - Y_i^-) = \sum (-1)^{q-r} Y_{i_1}^+ \cdots Y_{i_r}^+ Y_{i_{r+1}}^- \cdots Y_{i_q}^-,$$

where $\sum$ denotes a summation over all the permutations $\{i_1, \ldots, i_q\}$ of $\{1, \ldots, q\}$. Using Fubini's theorem, the preceding integral representation yields

$$\begin{aligned} Y_1 \cdots Y_q &= \sum (-1)^{q-r} \\ &\times \int_{\mathbb{R}_+^d} \mathbf{1}_{x_1 \le Y_{i_1}} \cdots \mathbf{1}_{x_r \le Y_{i_r}} \mathbf{1}_{x_{r+1} \le -Y_{r+1}} \cdots \mathbf{1}_{x_q \le -Y_{i_q}} dx_1 \cdots dx_q \\ &= \int_{\mathbb{R}_+^d} \prod_{i=1}^{q} (\mathbf{1}_{x_i \le Y_i} - \mathbf{1}_{x_i \le -Y_i}) dx_1 \cdots dx_q. \end{aligned}$$

Again Fubini's theorem yields

$$\mathbb{E}(Y_1 \cdots Y_q) = \int_{\mathbb{R}_+^d} \mathbb{E} \prod_{i=1}^{q} (\mathbf{1}_{x_i \le Y_i} - \mathbf{1}_{x_i \le -Y_i})\, dx_1 \cdots dx_q. \tag{4.3.4}$$

Now, eqn. (4.3.4) applied with $Y_i = X_{t_i}$ for $i = 1, \ldots, q$, together with Fubini's theorem implies

$$\operatorname{Cov}\left(X_{t_1} \cdots X_{t_m}, X_{t_{m+1}} \cdots X_{t_q}\right) = \int_{\mathbb{R}_+^d} \operatorname{Cov}\left(\prod_{i=1}^{m} f_i(X_{t_i}), \prod_{i=m+1}^{q} f_i(X_{t_i})\right) dx_1 \cdots dx_q,$$

where $f_i(y) = \mathbf{1}_{x_i \le y} - \mathbf{1}_{x_i \le -y}$. Define

$$B = \left| \operatorname{Cov}\left(\prod_{i=1}^{m} f_i(X_{t_i}), \prod_{i=m+1}^{q} f_i(X_{t_i})\right) \right|. \tag{4.3.5}$$

In the sequel, we give two bounds of the quantity B. The first bound does not use the dependence structure, only that $|f_i(y)| = \mathbf{1}_{x_i \le |y|}$. Thus

$$B \le 2 \inf(\Phi_{X_{t_1}}(x_1), \ldots, \Phi_{X_{t_q}}(x_q)), \tag{4.3.6}$$

with $\Phi_X(x) = \mathbb{P}(|X| \ge x)$. The second bound is deduced from the $(\mathcal{I}, c)$-weak dependence property. In fact, we have (recall that $r = t_{m+1} - t_m$)

$$B \le c(u, v)\epsilon(r). \tag{4.3.7}$$

The bound (4.3.7) together with (4.3.6) yields

$$B \le C_q \inf(\epsilon(r), \Phi_{X_{t_1}}(x_1), \ldots, \Phi_{X_{t_q}}(x_q)).$$

Hence

$$\begin{aligned} &|\operatorname{Cov}(X_{t_1} \cdots X_{t_m}, X_{t_{m+1}} \cdots X_{t_q})| \\ &\le (c_q \vee 2) \int_0^{+\infty} \cdots \int_0^{+\infty} \inf(\epsilon(r), \Phi_{X_{t_1}}(x_1), \ldots, \Phi_{X_{t_q}}(x_q)) dx_1 \cdots dx_q. \end{aligned}$$

The proof of Theorem 1-1 in Rio (1993) [157] can be completely implemented here. We give it for completeness. Let U be an uniform-[0,1] r.v, then

$$\begin{aligned}\epsilon(r) \wedge \min_{1\le j\le q} \Phi_{X_{t_j}}(x_j) &= \mathbb{P}(U \le \epsilon(r), U \le \Phi_{X_{t_1}}(x_1), \dots, U \le \Phi_{X_{t_q}}(x_q)) \\ &= \mathbb{P}(U \le \epsilon(r), x_1 \le Q_{X_{t_1}}(U), \dots, x_q \le Q_{X_{t_q}}(U)).\end{aligned}$$

We obtain, collecting the above results

$$|\mathrm{Cov}(X_{t_1}\cdots X_{t_m}, X_{t_{m+1}}\cdots X_{t_q})| \le C_q \mathbb{E} Q_{X_{t_1}}(U)\cdots Q_{X_{t_q}}(U)\mathbf{1}_{U\le\epsilon(r)}.$$

The lemma is thus proved. □

In order to make possible to use such bounds it will be convenient to express bounds of the quantities

$$s_{a,b,N} = \sum_{r=0}^{N}(r+1)^a \int_0^{\epsilon(r)} Q^b(s)ds \tag{4.3.8}$$

under conditions of summability of the series ϵ, for suitable constants $a \ge 0$ and $N, b > 0$ and a tail function Q of a random variable X such that $\mathbb{E}|X|^{b+\delta} < \infty$ for some $\delta > 0$. Set for convenience $A_r = \sum_{i=0}^{r}(i+1)^a$ for $r \ge 0$ and $= 0$ for $r < 0$, and $B_r = \int_0^{\epsilon(r)} Q^b(s)ds$ for $r \ge 0$ ($= 0$ for $r < 0$), then expression $s_{a,b,N}$ rewrites as follows; Abel transform with Hölder inequality for the conjugate exponents $p = 1 + b/\delta$ and $q = 1 + \delta/b$ implies the succession of inequalities

$$\begin{aligned}s_{a,b,N} &= \sum_{r=0}^{N}(A_r - A_{r-1})B_r \\ &= \sum_{r=0}^{N-1} A_r(B_r - B_{r+1}) + A_N B_N \\ &= \int_0^1 \left(\sum_{r=0}^{N-1} A_r \mathbf{1}_{]\epsilon(r+1),\epsilon(r)]}(s) + A_N \mathbf{1}_{[0,\epsilon(N)]}(s)\right) Q^b(s)ds \\ &\le \Big(\int_0^1 \Big(\sum_{r=0}^{N-1} A_r^p \mathbf{1}_{]\epsilon(r+1),\epsilon(r)]}(s) + A_N^p \mathbf{1}_{[0,\epsilon(N)]}(s)\Big)ds\Big)^{1/p}\Big(\int_0^1 Q^{bq}(s)ds\Big)^{1/q} \\ &\le \Big(\sum_{r=0}^{N-1} A_r^p(\epsilon(r) - \epsilon(r+1)) + A_N^p \epsilon(N)\Big)^{1/p} \left(\mathbb{E}|X|^{bq}\right)^{1/q}\end{aligned}$$

Hence

$$s_{a,b,N} \le \left(\sum_{r=0}^{N}(A_r^p - A_{r-1}^p)\epsilon(r)\right)^{1/p} \left(\mathbb{E}|X|^{b+\delta}\right)^{b/(b+\delta)}$$

Now we note that $r^{a+1}/(a+1) \le A_r \le (r+1)^{a+1}/(a+1)$ so that

$$A_r^p - A_{r-1}^p \le ((r+1)^{p(a+1)} - (r-1)^{p(a+1)})/(a+1)^p \le 2p(r+1)^{p(a+1)-1}/(a+1)^{p-1}$$

hence, setting $c = (2p)^{1/p}(a+1)^{1/p-1}$,

$$s_{a,b,N} \le c \left(\sum_{r=0}^{N} (r+1)^{a+(a+1)b/\delta} \epsilon(r) \right)^{\delta/(b+\delta)} \|X\|_{b+\delta}^b .$$

We summarize this in the following lemma.

Lemma 4.5. *Let $a > 0, b \ge 0$ be arbitrary then there exists a constant $c = c(a,b)$ such that for any real random valued variable X with quantile function Q_X, we have for any $N > 0$*

$$\sum_{r=0}^{N} (r+1)^a \int_0^{\epsilon(r)} Q_X^b(s)ds \le c\Big(\sum_{r=0}^{N} (r+1)^{a+(a+1)b/\delta} \epsilon(r) \Big)^{\delta/(b+\delta)} \|X\|_{b+\delta}^b .$$

4.3.1 Marcinkiewicz-Zygmund type inequalities

Our first result is the following Marcinkiewicz-Zygmund inequality.

Theorem 4.1. *Let $(X_n)_{n\in\mathbb{N}}$ be a sequence of centered r.v.s fulfilling for some fixed $q \in \mathbb{N}$, $q \ge 2$*

$$C_{r,q} = \mathcal{O}(r^{-q/2}), \quad \textit{as} \quad r \to \infty. \tag{4.3.9}$$

Then there exists a positive constant B not depending on n for which

$$|\mathbb{E}(S_n^q)| \le B n^{q/2}. \tag{4.3.10}$$

Proof of Theorem 4.1. For any integer $q \ge 2$, let

$$A_q(n) = \sum_{1 \le t_1 \le \cdots \le t_q \le n} |\mathbb{E}\left(X_{t_1} \cdots X_{t_q} \right)|. \tag{4.3.11}$$

Clearly,

$$|\mathbb{E}(S_n^q)| \le q! A_q(n). \tag{4.3.12}$$

Hence, in order to bound $|\mathbb{E}(S_n^q)|$, it remains to bound $A_q(n)$. For this, we argue as in Doukhan and Portal (1983) [73]. Clearly

$$\begin{aligned} A_q(n) \quad &\le \sum_{1 \le t_1 \le \cdots \le t_q \le n} |\mathbb{E}(X_{t_1} \cdots X_{t_m}) \mathbb{E}(X_{t_{m+1}} \cdots X_{t_q})| \\ &+ \sum_{1 \le t_1 \le \cdots \le t_q \le n} |\mathrm{Cov}(X_{t_1} \cdots X_{t_m}, X_{t_{m+1}} \cdots X_{t_q})|. \end{aligned}$$

The first term on the right hand of the last inequality is bounded by:

$$\sum_{1\le t_1\le\cdots\le t_q\le n} |\mathbb{E}(X_{t_1}\cdots X_{t_m})\mathbb{E}(X_{t_{m+1}}\cdots X_{t_q})| \ \le\ \sum_{m=1}^{q-1} A_m(n)A_{q-m}(n).$$

Hence

$$A_q(n) \le \sum_{m=1}^{q-1} A_m(n)A_{q-m}(n) + V_q(n). \tag{4.3.13}$$

with

$$V_q(n) = \sum_{(t_1,\cdots,t_q)\in G_r} |\mathrm{Cov}(X_{t_1}\cdots X_{t_m}, X_{t_{m+1}}\cdots X_{t_q})|, \tag{4.3.14}$$

where G_r is the set of $\{t_1,\ldots,t_q\}$ fulfilling $1\le t_1\le\cdots\le t_q\le n$ with $r = t_{m+1}-t_m = \max_{1\le i<q}(t_{i+1}-t_i)$.
Our task now is to bound the expression $V_q(n)$ defined by (4.3.14). Clearly

$$V_q(n) \le \sum_{t_1=1}^{n}\sum^{*} |\mathrm{Cov}(X_{t_1}\cdots X_{t_m}, X_{t_{m+1}}\cdots X_{t_q})|,$$

where $\sum^*$ denotes a sum over such a collection $1\le t_1\le\cdots\le t_q\le n$ with fixed t_1, and $r = t_{m+1}-t_m = \max_{1\le i\le q-1}(t_{i+1}-t_i)\in[0,n-1]$. Again

$$\sum^{*} |\mathrm{Cov}(X_{t_1}\cdots X_{t_m}, X_{t_{m+1}}\cdots X_{t_q})| \le \sum_{r=0}^{n-1}\sum^{**} |\mathrm{Cov}(X_{t_1}\cdots X_{t_m}, X_{t_{m+1}}\cdots X_{t_q})|.$$

where $\sum^{**}$ denotes the $(q-2)$-dimensional sums each over

$$\{(t_1,\ldots,t_q)\,/\,t_{i-1}\le t_i\le t_{i-1}+r,\quad i\ne 1, m+1\}.$$

Hence $\sum^{**} 1 = (r+1)^{q-2}$, with $|\mathrm{Cov}(X_{t_1}\cdots X_{t_m}, X_{t_{m+1}}\cdots X_{t_q})| \le C_{r,q}$, we deduce that

$$V_q(n) \le \sum_{t_1=1}^{n}\sum_{r=0}^{n-1}(r+1)^{q-2}C_{r,q}. \tag{4.3.15}$$

We obtain, collecting inequalities (4.3.13) and (4.3.15),

$$A_q(n) \le \sum_{m=1}^{q-1} A_m(n)A_{q-m}(n) + n\sum_{r=0}^{n-1}(r+1)^{q-2}C_{r,q}. \tag{4.3.16}$$

By induction on q, and using the last inequality together with condition (4.3.9), it is easy to check that $A_q(n)\le K_q n^{q/2}$. Theorem 4.1 follows from (4.3.12). □

4.3.2 Rosenthal type inequalities

The following lemma, which is a variant of Lemma 1 page 195 in Billingsley (1968) [20], gives moment inequalities of order $q \in \{2, 4\}$.

Lemma 4.6. *If $(X_n)_{n\in\mathbb{N}}$ is a sequence of centered r.v.s, then*

$$\mathbb{E}(S_n^2) \leq 2n \sum_{r=0}^{n-1} C_{r,2}, \quad \mathbb{E}(S_n^4) \leq 4! \left\{ \left(n \sum_{r=0}^{n-1} C_{r,2} \right)^2 + n \sum_{r=0}^{n-1} (r+1)^2 C_{r,4} \right\}. \tag{4.3.17}$$

Proof of Lemma 4.6. We take respectively $q = 2$ and $q = 4$ in the relation (4.3.16). The obtained formulas, together with (4.3.12) and the fact that $A_1(n) = 0$ for any positive integer n, prove Lemma 4.6. □

The following theorem deals with higher order moments.

Theorem 4.2. *Let q be some fixed integer not less than 2. Assume that the dependence coefficients $C_{r,p}$ associated to the sequence (X_n) satisfy, for every nonnegative integer p, $p \leq q$, and for some positive constants C and M,*

$$C_{r,p} \leq CM^p \epsilon(r). \tag{4.3.18}$$

Then, for any integer $n \geq 2$

$$|\mathbb{E}(S_n^q)| \leq \frac{(2q-2)!}{(q-1)!} M^q \left\{ \left(Cn \sum_{r=0}^{n-1} \epsilon(r) \right)^{q/2} \vee \left(Cn \sum_{r=0}^{n-1} (r+1)^{q-2} \epsilon(r) \right) \right\}. \tag{4.3.19}$$

Proof of Theorem 4.2. The relation (4.3.16) together with Condition (4.3.18) gives,

$$A_q(n) \leq \sum_{m=1}^{q-1} A_m(n) A_{q-m}(n) + CM^q n \sum_{r=0}^{n-1} (r+1)^{q-2} \epsilon(r). \tag{4.3.20}$$

In order to solve the above inductive relation, we need the following lemma.

Lemma 4.7. *Let $(U_q)_{q>0}$ and $(V_q)_{q>0}$ be two sequences of positive real numbers satisfying for some $\gamma \geq 0$, and for all $q \in \mathbb{N}^*$*

$$U_q \leq \sum_{m=1}^{q-1} U_m U_{q-m} + M^q V_q, \tag{4.3.21}$$

with $U_1 = 0 \le V_1$. *Suppose that for every integers* m, q *fulfilling* $2 \le m \le q-1$, *there holds*

$$(V_2^{m/2} \vee V_m)(V_2^{(q-m)/2} \vee V_{q-m}) \le (V_2^{q/2} \vee V_q). \tag{4.3.22}$$

Then, for any integer $q \ge 2$

$$U_q \le \frac{M^q}{q} \binom{2q-2}{q-1} (V_2^{q/2} \vee V_q). \tag{4.3.23}$$

Remark 4.1. Note that a sufficient condition for (4.3.22) to hold is that for all p, q such that $2 \le p \le q-1$

$$V_p^{q-2} \le V_q^{p-2} V_2^{q-p} \tag{4.3.24}$$

Proof of Lemma 4.7. Let $(U_q)_{q>0}$ and $(V_q)_{q>0}$ be two sequences of positive real numbers as defined by Lemma 4.7. We deduce from (4.3.21) and (4.3.22), that the sequence $(\tilde{U}_q)$, defined by $\tilde{U}_q = U_q / M^q (V_2^{q/2} \vee V_q)$ satisfies the relation,

$$\tilde{U}_q \le \sum_{m=1}^{q-1} \tilde{U}_m \tilde{U}_{q-m} + 1, \qquad \tilde{U}_1 = 0.$$

In order to prove (4.3.23), it suffices to show that, for any integer $q \ge 2$,

$$\tilde{U}_q \le d_q := \frac{1}{q} \binom{2q-2}{q-1}, \tag{4.3.25}$$

where d_q is called the q-th number of Catalan. The proof of the last bound is done by induction on q. Clearly (4.3.25) is true for $q = 2$. Suppose now that (4.3.25) is true for every integer m less than $q-1$. The inductive hypothesis yields with $\tilde{U}_1 = 0$:

$$\tilde{U}_q \le \sum_{m=2}^{q-2} d_m d_{q-m} + 1. \tag{4.3.26}$$

The last inequality, together with the identity $d_q = \sum_{m=1}^{q-1} d_m d_{q-m}$ (cf. Comtet (1970) [39], page 64), implies $\tilde{U}_q \le d_q$, proving (4.3.25) and thus Lemma 4.7. □

We continue the proof of Theorem 4.2. We deduce from (4.3.20) that the sequence $(A_q(n))_q$ satisfies (4.3.21) with

$$V_q := V_q(n) = CM^{q-2} n \sum_{r=0}^{n-1} (r+1)^{q-2} \epsilon(r).$$

Hence, to prove Theorem 4.2, it suffices to check condition (4.3.22).

$$(V_2^{m/2} \vee V_m)(V_2^{(q-m)/2} \vee V_{q-m}) \le V_2^{q/2} \vee V_2^{m/2} V_{q-m} \vee V_m V_2^{(q-m)/2} \vee V_m V_{q-m}.$$

To control each of these terms, we use three inequalities. Let p be a positive integer, $2 \le p \le q-1$. We deduce from,

$$\sum_{r=0}^{n-1}(r+1)^{p-2}\epsilon(r) \le \left(\sum_{r=0}^{n-1}\epsilon(r)\right)^{\frac{q-p}{q-2}}\left(\sum_{r=0}^{n-1}(r+1)^{q-2}\epsilon(r)\right)^{\frac{p-2}{q-2}},$$

that, for $2 \le p \le q-1$,

$$V_p \le V_q^{\frac{p-2}{q-2}} V_2^{\frac{q-p}{q-2}}. \tag{4.3.27}$$

Define the discrete r.v. Z by $\mathbb{P}(Z = r+1) = \epsilon(r)/\sum_{i=0}^{n-1}\epsilon_i$. Jensen inequality implies $\|Z\|_{p-2} \le \|Z\|_{q-2}$ if $1 \le p-2 \le q-2$ so that

$$V_p \le V_q^{\frac{p-2}{q-2}}. \tag{4.3.28}$$

For $0 < \alpha < 1$,

$$V_2^{\alpha\frac{q}{2}} V_q^{1-\alpha} \le V_2^{q/2} \vee V_q. \tag{4.3.29}$$

Using (4.3.28), we get

$$V_m V_2^{(q-m)/2} \le V_2^{\frac{(q-m)q}{2(q-2)}} V_q^{\frac{m-2}{q-2}}$$

$$V_2^{m/2} V_{q-m} \le V_2^{\frac{mq}{2(q-2)}} V_q^{\frac{q-m-2}{q-2}}$$

From (4.3.27) we obtain

$$V_m V_{q-m} \le V_2^{\frac{q}{q-2}} V_q^{\frac{q-4}{q-2}}$$

Now (4.3.29) implies that these three bounds are less than $V_2^{q/2} \vee V_q$. □

Theorem 4.3. *If $(X_n)_{n\in\mathbb{N}}$ is a centered and $(\mathcal{I}, c)$-weak dependent sequence, then*

$$\begin{aligned}|\mathbb{E}(S_n^q)| \quad \le \quad & \frac{(2q-2)!}{(q-1)!}\left\{C_q \sum_{i=1}^{n}\int_0^1 \left(\epsilon^{-1}(u)\wedge n\right)^{q-1} Q_i^q(u)du\right. \\ & \left.\vee\left(C_2\sum_{i=1}^{n}\int_0^1\left(\epsilon^{-1}(u)\wedge n\right)Q_i^2(u)du\right)^{q/2}\right\},\end{aligned}$$

where $\epsilon^{-1}(u)$ is the generalized inverse of $\epsilon_{[u]}$ (see eqn. (2.2.14)).

Proof of Theorem 4.3. We begin from relation (4.3.16), and we try to evaluate the coefficient of dependence $C_{r,q}$ for $(\mathcal{I}, \psi)$-weak dependent sequences. For this, we need the forthcoming lemma.

Lemma 4.8. *If the sequence $(X_n)_{n\in\mathbb{N}}$ is $(\mathcal{I}, c)$-weak dependent, then*

$$V_q(n) \le C_q \sum_{i=1}^{n}\int_0^1 \left(\epsilon^{-1}(u)\wedge n\right)^{q-1} Q_i^q(u)du.$$

Proof of Lemma 4.8. We use Lemma 4.4. Arguing exactly as in Rio (1993) [157], we obtain

$$\begin{aligned} V_q(n) &\le C_q \sum_{r=0}^{n-1} \sum_{(i_1,\dots,i_q)\in G_r} \int_0^{\epsilon(r)} Q_{i_1}(u)\cdots Q_{i_q}(u)du \\ &\le \frac{C_q}{q} \sum_{r=0}^{n-1} \sum_{(i_1,\dots,i_q)\in G_r} \int_0^{\epsilon(r)} \sum_{j=1}^{q} Q_{i_j}^q(u)du \\ &\le \frac{C_q}{q} \sum_{r=0}^{n-1} \sum_{(i_1,\dots,i_q)\in \bigcup_{k\le r} G_k} \int_{\epsilon(r+1)}^{\epsilon(r)} Q_{i_j}^q(u)du \end{aligned}$$

Now fixing i_j and noting that the number of completing $(i_1,\dots,i_{j-1},i_{j+1},\dots,i_q)$ to get an sequence in $\bigcup_{k\le r} G_k$ is less than $(r+1)^{q-1}$:

$$\begin{aligned} \sum_{r=0}^{n-1} \sum_{(i_1,\dots,i_q)\in\bigcup_{k\le r}G_k} \int_{\epsilon(r+1)}^{\epsilon(r)} Q_{i_j}^q(u)du &\le \sum_{i_j=0}^{n}\sum_{r=0}^{n-1} \int_{\epsilon(r+1)}^{\epsilon(r)} (r+1)^{q-1} Q_{i_j}^q(u)du \\ &\le \sum_{i=0}^{n} \int_0^1 (\epsilon^{-1}(u)\wedge n)^{q-1} Q_i^q(u)du \end{aligned}$$

Lemma 4.8 is now completely proved. □

We continue the proof of Theorem 4.3. We deduce from Lemma 4.8 and Inequality (4.3.16), that the sequence $(A_q(n))_q$ fulfills relation (4.3.21). So, as in the proof of Theorem 4.2, Theorem 4.3 is proved, if we prove that the sequence $\tilde V_p(n) := (c_p \vee 2)\sum_{i=1}^n \int_0^1 \left(\epsilon^{-1}(u)\wedge n\right)^{p-1} Q_i^p(u)du$ satisfies (4.3.22). We have,

$$\begin{aligned} &\int_0^1 \left(\epsilon^{-1}(u)\wedge n\right)^{p-1} Q_i^p(u)du \\ &\quad\le \left(\int_0^1 \left(\epsilon^{-1}(u)\wedge n\right) Q_i^2(u)du\right)^{\frac{q-p}{q-2}} \left(\int_0^1 \left(\epsilon^{-1}(u)\wedge n\right)^{q-2} Q_i^q(u)du\right)^{\frac{p-2}{q-2}}. \end{aligned}$$

The last bound proves that the sequence $(\tilde V_p(n))_p$ fulfills the convexity inequality (4.3.27), which in turns implies (4.3.22). □

4.3.3 A first exponential inequality

For any positive integers n and $q \ge 2$, we consider the following assumption,

$$M_{q,n} = n\sum_{r=0}^{n-1}(r+1)^{q-2}C_{r,q} \le A_n \frac{q!}{\beta^q} \tag{4.3.30}$$

where β is some positive constant and A_n is a sequence independent of q.
As a consequence of Theorem 4.2, we obtain an exponential inequality.

Corollary 4.1. *Suppose that* (4.3.18) *and* (4.3.30) *hold for some sequence* $A_n \geq 1$ *for any* $n \geq 2$. *Then for any positive real number* x

$$\mathbb{P}\left(|S_n| \geq x\sqrt{A_n}\right) \leq A \exp\left(-B\sqrt{\beta x}\right), \tag{4.3.31}$$

for universal positive constants A *and* B.

Remark 4.2.

- *One may choose the explicit values* $A = e^{4+1/12}\sqrt{8\pi}$, *and* $B = e^{5/2}$.
- *Let us note that condition* (4.3.30) *holds if* $C_{r,q} \leq CM^q e^{-br}$ *for positive constants* C, M, b. *In such a case* A_n *is of order* n. *E.g. this holds if* $\|X_n\|_\infty \leq M$ *and* $\|X_n\|_2 \leq \sigma$ *under* $(\Lambda \cap \mathbb{L}^\infty, \Psi)$*-weak dependence if* $\epsilon(r) = \mathcal{O}(e^{-br})$ *and* $\Psi(h,k,u,v) \leq e^{\delta(u+v)}\mathrm{Lip}\,(h)\mathrm{Lip}\,(k)$ *for some* $\delta \geq 0$. *For this, either compare the series* $\sum_r (r+1)^{q-2}e^{-ar}$ *with integrals or with derivatives of the function* $t \mapsto 1/(1-t) = \sum_i t^i$ *at point* $t = e^{-a}$.
- *The use of combinatorics in those inequalities makes them relatively weak. E.g. Bernstein inequality, valid for independent sequences allows to replace the term* $\sqrt{x}$ *in the previous inequality by* x^2 *under the same assumption* $n\sigma^2 \geq 1$; *in the mixing cases analogue inequalities are also obtained by using coupling arguments (not available here), e.g. the case of absolute regularity is studied in Doukhan (1994) [61].*

Proof of Corollary 4.1. Theorem 4.2 written with $q = 2p$ yields

$$\mathbb{E}(S_n^{2p}) \leq \frac{(2p)!}{2p}\begin{pmatrix} 4p-2 \\ 2p-1 \end{pmatrix}(M_{2p,n} \vee M_{2,n}^p). \tag{4.3.32}$$

Hence inequality (4.3.32) together with condition (4.3.30) implies

$$\begin{aligned}
\mathbb{E}(S_n^{2p}) &\leq \frac{(4p-2)!}{(2p-1)!}\left(\left(\frac{2A_n}{\beta^2}\right)^p \vee A_n\frac{(2p)!}{\beta^{2p}}\right) \\
&\leq \frac{(4p-2)!}{(2p-1)!}(A_n \vee A_n^p)\frac{(2p)!}{\beta^{2p}} \\
&\leq (A_n \vee A_n^p)\frac{(4p)!}{\beta^{2p}}.
\end{aligned}$$

From Stirling formula and from the fact that $A_n \geq 1$ we obtain

$$\begin{aligned}
\mathbb{P}(|S_n| \geq x) &\leq \frac{\mathbb{E}(S_n^{2p})}{x^{2p}} \leq \frac{A_n^p}{x^{2p}\beta^{2p}}e^{1/12-4p}\sqrt{8\pi p}(4p)^{4p} \\
&\leq e^{1/12}\sqrt{8\pi}\left(\frac{16}{x\beta}e^{-7/4}p^2\sqrt{A_n}\right)^{2p}.
\end{aligned}$$

Now setting $h(y) = (C_n y)^{4y}$ with $C_n^2 = \frac{16}{x\beta} e^{-7/4}\sqrt{A_n}$, one obtains

$$\mathbb{P}(|S_n| \geq x) \leq e^{1/12}\sqrt{8\pi} h(p).$$

Define the convex function $g(y) = \log h(y)$. Clearly

$$\inf_{y \in \mathbb{R}^+} g(y) = g\left(\frac{1}{eC_n}\right).$$

Suppose that $eC_n \leq 1$ and let $p_0 = \left[\frac{1}{eC_n}\right]$, then

$$\mathbb{P}(|S_n| \geq x) \leq e^{1/12}\sqrt{8\pi} h(p_0) \leq e^{4+1/12}\sqrt{8\pi} \exp(\frac{-4}{eC_n}).$$

Suppose now that $eC_n \geq 1$, then $1 \leq e^{4+1/12}\sqrt{8\pi} \exp(\frac{-4}{eC_n})$.

In both cases, inequality (4.3.31) holds and Corollary 4.1 is proved. □

Remark. More accurate applications of those inequalities are proposed in Louhichi (2003) [125] and Doukhan & Louhichi (1999) [67]. In particular in Section 3 of [67] conditions for (4.3.18) are checked, providing several other bounds for the coefficients $C_{r,q}$.

4.4 Cumulants

The main objective of this section is to reinterpret the cumulants which classically used expressions to measure the dependence properties of a sequence.

4.4.1 General properties of cumulants

Let $Y = (Y_1, \ldots, Y_k) \in \mathbb{R}^k$ be a random vector, setting $\phi_Y(t) = \mathbb{E}e^{it\cdot Y} = \mathbb{E}\exp\left(i\sum_{j=1}^k t_j Y_j\right)$ for $t = (t_1, \ldots, t_k) \in \mathbb{R}^k$, we write $m_p(Y) = \mathbb{E}Y_1^{p_1} \cdots Y_k^{p_k}$ for $p = (p_1, \ldots, p_k)$ if $\mathbb{E}(|Y_1|^s + \cdots + |Y_1|^s) < \infty$ and $|p| = p_1 + \cdots + p_k = s$. Finally if the previous moment condition holds for some $r \in \mathbb{N}^*$, then the function $\log \phi_Y(t)$ has a Taylor expansion

$$\log \phi_Y(t) = \sum_{|p| \leq r} \frac{i^{|p|}}{p!} \kappa_p(Y) t^p + o(|t|^r), \qquad \text{as } t \to 0$$

for some coefficients $\kappa_p(Y)$ called the cumulant of Y of order $p \in \mathbb{R}^k$ if $|p| \leq s$ where we set $p! = p_1! \cdots p_k!$, $t^p = t_1^{p_1} \cdots t_k^{p_k}$ if $t = (t_1, \ldots, t_k) \in \mathbb{R}^k$ and $p = (p_1, \ldots, p_k)$.

In the case $p = (1, \ldots, 1)$, to which the others may be reduced, we denote $\kappa_{(1,\ldots,1)}(Y) = \kappa(Y)$. Moreover, if $\mu = \{i_1, \ldots, i_u\} \subset \{1, \ldots, k\}$

$$\kappa_\mu(Y) = \kappa(Y_{i_1}, \ldots, Y_{i_u}), \qquad m_\mu(Y) = m(Y_{i_1}, \ldots, Y_{i_u}).$$

Leonov and Shyraev (1959) [119] (see also Rosenblatt, 1985, pages 33-34 [168]) obtained the following expressions

$$\kappa(Y) = \sum_{u=1}^{k} (-1)^{u-1}(u-1)! \sum_{\mu_1,\ldots,\mu_u} \prod_{j=1}^{u} m_{\mu_j}(Y) \tag{4.4.1}$$

$$m(Y) = \sum_{u=1}^{k} \sum_{\mu_1,\ldots,\mu_u} \prod_{j=1}^{u} \kappa_{\mu_j}(Y) \tag{4.4.2}$$

The previous sums are considered for all partitions $\mu_1, \ldots, \mu_u$ of the set $\{1, \ldots, k\}$.

We now recall some notions from Saulis and Statulevicius (1991) [173]. For this we reformulate their notations.

Definition 4.2. *Centered moments of a random vector* $Y = (Y_1, \ldots, Y_k)$ *are defined by setting* $\widehat{\mathbb{E}}\ (Y_1, \ldots, Y_l) = \mathbb{E}Y_1 c(Y_2, \ldots, Y_l)$ *where the centered random variables* $c(Y_2, \ldots, Y_l)$ *are defined recursively, by setting* $c(\xi_1) = \overbrace{\ \xi\ } = \xi_1 - \mathbb{E}\xi_1$,

$$c(\xi_j, \xi_{j-1}, \ldots, \xi_1) = \xi_j\, \overbrace{c(\xi_{j-1}, \ldots, \xi_1)} = \xi_j\, (c(\xi_{j-1}, \ldots, \xi_1) - \mathbb{E}c(\xi_{j-1}, \ldots, \xi_1))$$

We also write $Y_\mu = (Y_j / j \in \mu)$ *as a* $p-$*tuple if* $\mu \subset \{1, \ldots, k\}$.

Quote for comprehension that $\widehat{\mathbb{E}}\ (\xi) = 0, \widehat{\mathbb{E}}\ (\eta, \xi) = \mathrm{Cov}(\eta, \xi)$ and,

$$\widehat{\mathbb{E}}\ (\zeta, \eta, \xi) = \mathbb{E}(\zeta\eta\xi) - \mathbb{E}(\zeta)\mathbb{E}(\eta\xi) \quad \mathbb{E}(\eta)\mathbb{E}(\zeta\xi) - \mathbb{E}(\xi)\mathbb{E}(\zeta\eta)$$

A remarkable result from Saulis and Statulevicius (1991) will be informative

Theorem 4.4 (Saulis, Statulevicius, 1991 [173]).

$$\kappa(Y_1, \ldots, Y_k) = \sum_{u=1}^{k} (-1)^{u-1} \sum_{\mu_1,\ldots,\mu_u} N_u(\mu_1, \ldots, \mu_u) \prod_{j=1}^{u} \widehat{\mathbb{E}}\ Y_{\mu_j}$$

where sums are considered for all partitions $\mu_1, \ldots, \mu_u$ *of the set* $\{1, \ldots, k\}$ *and the integers* $N_u(\mu_1, \ldots, \mu_u) \in \left[0, (u-1)! \wedge \left[\frac{k}{2}\right]!\right]$, *defined for any such partition, satisfy the relations*

- $N(k, u) = \sum_{\mu_1,\ldots,\mu_u} N_u(\mu_1, \ldots, \mu_u) = \sum_{j=1}^{u-1} C_k^j (u-j)^{k-1}$,

- $\sum_{u=1}^{k} N(k,u) = (k-1)!$

Using this representation the following bound will be useful

Lemma 4.9. *Let $Y_1, \ldots, Y_k \in \mathbb{R}$ be centered random variables. For $k \geq 1$, we set $M_k = (k-1)! 2^{k-1} \max_{1 \leq i \leq k} \mathbb{E}|Y_i|^k$, then*

$$\begin{aligned} |\kappa(Y_1, \ldots, Y_k)| &\leq M_k, &(4.4.3)\\ M_k M_l &\leq M_{k+l}, \text{ if } k, l \geq 2. &(4.4.4)\end{aligned}$$

Mention that a consequence of this lemma will be used in the following:

$$\prod_{i=1}^{u} |\kappa_p(Y_1, \ldots, Y_{p_u})| \leq M_{p_1+\cdots+p_u} \qquad (4.4.5)$$

We shall use this inequality for components $Y_i = X_{k_i}^{(a_i)}$ of a stationary sequence of $\mathbb{R}^D$−valued random variable hence $\max_{i \geq 1} \mathbb{E}|Y_i|^p \leq \max_{1 \leq j \leq D} \mathbb{E}|X_0^{(j)}|^p$ and we may set

$$M_p = (p-1)! 2^{p-1} \max_{1 \leq j \leq D} \mathbb{E}|X_0^{(j)}|^p. \qquad (4.4.6)$$

Proof of lemma 4.9. The second point in this lemma follows from the elementary inequality $a!\, b! \leq (a+b)!$ and the first one is a consequence of theorem 4.4 and of the following lemma

Lemma 4.10. *For any $j, p \geq 1$ and any real random variables $\xi_0, \xi_1, \xi_2, \ldots$ with identical distribution,*

$$\|c(\xi_j, \xi_{j-1}, \ldots, \xi_1)\|_p \leq 2^j \max_{1 \leq i \leq j} \|\xi_i\|_{pj}^j, \qquad \text{with } \|\xi\|_q = \mathbb{E}^{1/q}|\xi|^q.$$

Proof of lemma 4.10. For simplicity we shall omit suprema replacing $\max_{i \leq j} \|\xi_i\|_p$ by $\|\xi_1\|_p$. First of all, Hölder inequality implies

$$\|c(\xi_1)\|_p \leq \|\xi_1\|_p + |\mathbb{E}\xi_1| \leq 2\|\xi_1\|_p,$$

We now use recursion; setting $Z_j = c(\xi_j, \xi_{j-1}, \ldots, \xi_1)$ yields $Z_j = \xi_j(Z_{j-1} - \mathbb{E}Z_{j-1})$ hence Minkowski and Hölder inequalities entail

$$\begin{aligned} \|Z_j\|_p &\leq \|\xi_j Z_{j-1}\|_p + \|\xi_0\|_p |\mathbb{E}Z_{j-1}| \\ &\leq \|\xi_0\|_{pj} \|Z_{j-1}\|_q + \|\xi_0\|_{pj} \|Z_{j-1}\|_p \\ &\leq 2^j \|\xi_0\|_{pl} \|\xi_0\|_{q(j-1)}^{j-1} \end{aligned}$$

where $\dfrac{1}{q} + \dfrac{1}{p \cdot l} = 1$; from $p \geq 1$ we infer $q(j-1) \leq pj$ to conclude. □

Proof of lemma 4.9. Here again, we omit *suprema* and we replace $\max_{j \leq J} \|Y_j\|_p$ by $\|Y_0\|_p$. With lemma 4.10 we deduce that $|\widehat{\mathbb{E}}\ Y_\mu| \leq 2^{l-1} \|Y_0\|_l^l$ with $l = \#\mu$. Indeed, write $Z = c(Y_2, \dots, Y_l)$, then with $\frac{1}{q} + \frac{1}{l} = 1$ we get

$$\left| \widehat{\mathbb{E}}\ (Y_1, \dots, Y_l) \right| = |\mathbb{E} Y_1 Z| \leq \|Y_0\|_l \|Z\|_q \leq 2^{l-1} \|Y_0\|_l^l$$

since $q(l-1) = l$. Hence theorem 4.4 implies,

$$\begin{aligned} |\kappa(Y)| &\leq \sum_{u=1}^{k} \sum_{\mu_1, \dots, \mu_u} N_u(\mu_1, \dots, \mu_u) \prod_{i=1}^{u} 2^{\#\mu_i - 1} \|Y_0\|_{\#\mu_i}^{\#\mu_i} \\ &\leq \sum_{u=1}^{k} 2^{k-u} N(k,u) \|Y_0\|_k^k \\ &\leq 2^{k-1} \|Y_0\|_k^k \sum_{u=1}^{k} N(k,u) \\ &= 2^{k-1} (k-1)! \|Y_0\|_k^k. \qquad \square \end{aligned}$$

The following lemmas are essentially proved in Doukhan & León (1989) [66] for real valued sequences $(X_n)_{n \in \mathbb{Z}}$. Let now $(X_n)_{n \in \mathbb{Z}}$ denote a vector valued and stationary sequence (with values in $\mathbb{R}^D$), we define*, extending Doukhan and Louhichi (1999)'s coefficients,

$$c_{X,q}(r) = \sup_{\substack{1 \leq l < q \\ 1 \leq a_1, \dots, a_q \leq D \\ t_1 \leq \cdots \leq t_q \\ t_{l+1} - t_l \geq r}} \left| \operatorname{Cov}\left(X_{t_1}^{(a_1)} \cdots X_{t_l}^{(a_l)}, X_{t_{l+1}}^{(a_{l+1})} \cdots X_{t_q}^{(u_q)} \right) \right| \qquad (4.4.7)$$

We also define the following decreasing coefficients, for further convenience,

$$c^\star_{X,q}(r) = \max_{1 \leq l \leq q} c_{X,l}(r) \mu_{q-l}, \qquad \text{with} \quad \mu_t = \max_{1 \leq d \leq D} \mathbb{E}|X_0|^t. \qquad (4.4.8)$$

In order to state the following results we set, for $1 \leq a_1, \dots, a_q \leq D$,

$$\kappa(q)^{(a_1, \dots, a_q)}(t_2, \dots, t_q) = \kappa_{(1, \dots, 1)}(X_0^{(a_1)}, X_{t_2}^{(a_2)}, \dots, X_{t_q}^{(a_q)})$$

The following decomposition lemma will be very useful. It explain how cumulants behave naturally as covariances. Precisely, it proves that a cumulant

*For $D = 1$ this coefficient was already defined in definition 4.1 as $C_{r,q}$ but the present notation recalls also the underlying process.

$\kappa_Q(X_{k_1},\ldots,X_{k_Q})$ is small if $k_{l+1}-k_l$ is large, $k_1 \le \cdots \le k_Q$, and the process is weakly dependent. This is a natural extension of one essential property of the cumulants which states that such a cumulant vanishes if the index set can be partitioned into two strict subsets such that the vector random variables determined this way are independent.

Definition 4.3. *Let $t=(t_1,\ldots,t_p)$ be any $p-$tuple in $\mathbb{Z}^p$ such that $t_1 \le \cdots \le t_p$, we denote $r(t)=\max_{1\le l<p}(t_{l+1}-t_l)$, the maximal lag within the succession $(t_1,\ldots,t_p)$.*
We now introduce another dependence coefficient

$$\kappa_p(r) = \max_{\substack{t_1 \le \cdots \le t_p \\ r(t_1,\ldots,t_p) \ge r}} \max_{1\le a_1,\ldots,a_p\le D} \left|\kappa_p\left(X_{t_1}^{(a_1)},\ldots,X_{t_p}^{(a_p)}\right)\right| \tag{4.4.9}$$

Lemma 4.11. *Let $(X_n)_{n\in\mathbb{Z}}$ be a stationary process, centered at expectation with finite moments of any order. Then if $Q \ge 2$ we have, using the notation in lemma 4.9,*

$$\kappa_{X,Q}(r) \le c_{X,Q}(r) + \sum_{s=2}^{Q-2} M_{Q-s}\left[\frac{Q}{2}\right]^{Q-s+1} \kappa_{X,s}(r)$$

Proof of lemma 4.11. We denote $X_\eta^{(a)}=\prod_{i\in\eta} X_i^{(a_i)}$ for any $p-$tuples $\eta\in\mathbb{Z}^p$ and $a=(a_1,\ldots,a_p)\in\{1,\ldots,D\}^p$ (this way admits repetitions of the succession η). We assume that $k_1\le\cdots\le k_Q$ satisfy $k_{l+1}-k_l=r=\max_{1\le s<p}(k_{s+1}-k_s)\ge 0$, then if $\mu=\{\mu_1,\ldots,\mu_u\}$ ranges over all the partitions of $\{1,\ldots,Q\}$ there is some μ_i (which we denote ν_μ) satisfies $\nu_\mu^-=[1,l]\cap\nu_\mu\ne\emptyset$ and $\nu_\mu^+=[l+1,Q]\cap\nu_\mu\ne\emptyset$. Using formula (4.4.2), we obtain, with $\eta=\{1,\ldots,l\}$,

$$\kappa(X_{k_1}^{(a_1)},\ldots,X_{k_Q}^{(a_Q)}) = \mathrm{Cov}(X_{\eta(k)}^{(a)},X_{\overline{\eta}(k)}^{(a)}) - \sum_u\sum_{\{\mu\}} \kappa_{\nu_\mu(k)} K_{\mu,k}, \tag{4.4.10}$$

where $K_{\mu,k} = \prod_{\mu_i\ne\nu_u} \kappa_{\mu_i(k)}$ and where the previous sum extends to partitions $\mu=\{\mu_1,\ldots,\mu_u\}$ of $\{1,\ldots,Q\}$ such that there exists some $1\le i\le u$ with $\mu_i\cap\nu\ne\emptyset$ and $\mu_i\cap\overline{\nu}\ne\emptyset$. For simplicity the previous formulas omit the reference to indices $(a_1,\ldots,a_Q)$ which is implicit. We use the simple but essential remark that

$$r(\nu_\mu(k)) \ge r(k) \text{ to derive } |\kappa_{\nu_\mu(k)}| \le \kappa_{X,\#\nu_\mu}(r).$$

We now use lemma 4.9 to deduce $|M_\mu| \le M_{Q-\#\mu_\nu}$ as in eqn. (4.4.5). This yields the bound

$$\begin{aligned}
\left|\kappa\left(X_{k_1}^{(a_1)},\ldots,X_{k_Q}^{(a_Q)}\right)\right| &\le C_{X,Q}(r)+\sum_{u=2}^{[Q/2]}(u-1)!\sum_{\mu_1,\ldots,\mu_u} M_{Q-\#\nu_\mu}|\kappa_{\nu_\mu(k)}(X^{(a)})| \\
&\le C_{X,Q}(r)+\sum_{u=2}^{[Q/2]}(u-1)!\sum_{s=2}^{Q-2} M_{Q-s}\kappa_{X,s}(r)\sum_{\substack{\mu_1,\ldots,\mu_u\\ \#\nu_\mu=s}} 1 \\
&\le C_{X,Q}(r)+\sum_{u=2}^{[Q/2]}(u-1)!\sum_{s=2}^{Q-2}(u-1)^{Q-s}M_{Q-s}\kappa_{X,s}(r) \\
&\le C_{X,Q}(r)+\sum_{s=2}^{Q-2}\frac{1}{Q-s+1}\left[\frac{Q}{2}\right]^{Q-s+1} M_{Q-s}\kappa_{X,s}(r)
\end{aligned}$$

since the inequality $\sum_{u=1}^{U}(u-1)^p \le \frac{1}{p+1}U^{p+1}$ follows from a comparison between a sum and an integral. □

Rewrite now the lemma 4.11 as

$$\kappa_{X,Q}(r)\le c_{X,Q}(r)+\sum_{s=2}^{Q-2} B_{Q,s}\kappa_{X,s}(r)$$

thus the following formulas follow

$$\begin{aligned}
\kappa_{X,2}(r) &\le c_{X,2}(r), \\
\kappa_{X,3}(r) &\le c_{X,3}(r), \\
\kappa_{X,4}(r) &< c_{X,4}(r)+B_{4,2}\kappa_{X,2}(r) \\
&\le c_{X,4}(r)+B_{4,2}c_{X,2}(r), \\
\kappa_{X,5}(r) &\le c_{X,5}(r)+B_{5,3}\kappa_{X,3}(r)+B_{5,2}\kappa_{X,2}(r) \\
&\le c_{X,5}(r)+B_{5,3}c_{X,3}(r)+B_{5,2}c_{X,2}(r), \\
\kappa_{X,6}(r) &\le c_{X,6}(r)+B_{6,4}\kappa_{X,4}(r)+B_{6,3}\kappa_{X,3}(r)+B_{6,2}\kappa_{X,2}(r) \\
&\le c_{X,6}(r)+B_{6,4}\left(c_{X,4}(r)+B_{4,2}c_{X,2}(r)\right)+B_{6,3}c_{X,3}(r)+B_{6,2}c_{X,2}(r) \\
&\le c_{X,6}(r)+B_{6,4}c_{X,4}(r)+B_{6,3}c_{X,3}(r)+(B_{6,2}+B_{6,4}B_{4,2})c_{X,2}(r).
\end{aligned}$$

A main corollary of lemma 4.11 is the following, it is proved by induction.

Corollary 4.2. *For any $Q\ge 2$, there exists a constant $A_Q\ge 0$ only depending on Q, such that*

$$\kappa_{X,Q}(r)\le A_Q c^*_{X,Q}(r).$$

Remark 4.3. *This corollary explains an equivalence between the coefficients $c_{X,Q}(r)$ and the $\kappa_Q(r)$ which may also be preferred as a dependence coefficient. A way to derive sharp bounds for the constants involved is, using theorem 4.4, to decompose the corresponding sum of centered moments in two terms, the first of them involving the maximal covariance of a product.*

Section 12.3.2 will provide multivariate extensions devoted to spectral analysis.

For completeness sake remark that formula (4.4.10) implies with $B_{Q,Q} = 1$ that $c_{X,Q}(r) \leq \sum_{s=2}^{Q} B_{Q,s}\kappa_{X,s}(r)$. Hence there exists some constant $\widetilde{A}_Q$ such that

$$c_{X,Q}(r) \leq \widetilde{A}_Q \kappa^*_{X,Q}(r), \qquad \kappa^*_{X,Q}(r) = \max_{2 \leq l \leq Q} \kappa^*_{X,l}(r)\mu_{Q-l}$$

Finally, we have proved that constants $a_Q, A_Q > 0$ satisfy

$$a_Q c^*_{X,Q}(r) \leq \kappa^*_{X,Q}(r) \leq A_Q c^*_{X,Q}(r)$$

Hence, for fixed Q those inequalities are equivalent.

The previous formula (4.4.10) implies that the cumulant

$$\kappa(X_{k_1}^{(a_1)}, \ldots, X_{k_Q}^{(a_Q)}) = \sum_{\alpha,\beta} K_{\alpha,\beta,k} \mathrm{Cov}(X_{\alpha(k)}^{(a)}, X_{\beta(k)}^{(a)})$$

writes as the linear combination of such covariances with $\alpha \subset \{1, \ldots, l\}$ and $\beta \subset \{l+1, \ldots, Q\}$ where the coefficients $K_{\alpha,\beta,k}$ are some polynomials of the cumulants. For this one replaces the $Q-$tuple $(X_{k_1}^{(a_1)}, \ldots, X_{k_Q}^{(a_Q)})$ by $(X_i^{(a)})_{i \in \nu_\mu(k)}$ for each partition μ, in formula (4.4.10) and use recursion.

Such representation is useful namely if one precisely knows such covariances; let us mention the cases of Gaussian and associated processes for which additional informations are provided.

The main attraction for cumulants with respect to covariance of products is that if a sample $(X_{k_1}, \ldots, X_{k_q})$ is provided, the behaviour of the cumulant is that of $c_{X,q}(r(k))$ with appears as suprema over a position l of the maximal lag in the sequence k. This means an automatic search of this maximal lag $r(k)$ may be performed with the help of cumulants.

Examples. The constants A_Q, which are not explicit, may be determined for some low orders. A careful analysis of the previous proof allows the sharp bounds[†]

$$\begin{aligned}
\kappa_{X,2}(r) &= c_{X,2}(r)\\
\kappa_{X,3}(r) &= c_{X,3}(r)\\
\kappa_{X,4}(r) &\le c_{X,4}(r)+3\mu_2 c_{X,2}(r)\\
\kappa_{X,5}(r) &\le c_{X,5}(r)+10\mu_2 c_{X,3}(r)+10\mu_3 c_{X,2}(r)\\
\kappa_{X,6}(r) &\le c_{X,6}(r)+15\mu_2 c_{X,4}(r)+20\mu_3 c_{X,3}(r))+150\mu_4 c_{X,2}(r)
\end{aligned}$$

Our main application of those inequalities is the

Lemma 4.12. *Set*

$$\kappa_Q = \sum_{k_2=0}^{\infty}\cdots\sum_{k_Q=0}^{\infty}\max_{1\le a_1,\dots,a_Q\le D}\left|\kappa\left(X_0^{(a_1)},X_{k_2}^{(a_2)},\dots,X_{k_Q}^{(a_Q)}\right)\right|. \qquad (4.4.11)$$

We use notation (4.4.8). For each $Q\ge 2$, there exists a constant B_Q such that

$$\kappa_Q \le B_Q\sum_{r=0}^{\infty}(r+1)^{Q-2}C^*_{X,Q}(r).$$

Proof of lemma 4.12. For this we only decompose the sums

$$\begin{aligned}
\kappa_Q &\le (Q-1)!\sum_{k_2\le\cdots\le k_Q}\max_{1\le a_1,\dots,a_Q\le D}\left|\kappa\left(X_0^{(a_1)},X_{k_2}^{(a_2)},\dots,X_{k_Q}^{(a_Q)}\right)\right|\\
&\equiv (Q-1)!\,\widetilde{\kappa}_Q
\end{aligned}$$

by considering the partition of the indice set

$$E=\{k=(k_2,\dots,k_Q)\subset\mathbb{N}^{Q-1}/k_2\le\cdots\le k_Q\}$$

into $E_r=\{k\in E/\ r(k)=r\}$ for $r\ge 0$, then

$$\widetilde{\kappa}_Q=\sum_{r=0}^{\infty}\sum_{k\in E_r}\max_{1\le a_1,\dots,a_Q\le D}\left|\kappa\left(X_0^{(a_1)},X_{k_2}^{(a_2)},\dots,X_{k_Q}^{(a_Q)}\right)\right|$$

The previous lemma implies that there exists some constant $A_Q>0$ such that

$$\sum_{k\in E_r}\max_{1\le a_1,\dots,a_Q\le D}\left|\kappa\left(X_0^{(a_1)},X_{k_2}^{(a_2)},\dots,X_{k_Q}^{(a_Q)}\right)\right|\le A_Q\#E_rC^*_{X,Q}(r)$$

and the simple bound $\#E_r\le(Q-1)(r+1)^{Q-2}$ concludes the proof. □

[†]To bound higher order cumulants we shall prefer lemma 4.11 rough bounds to those involved combinatorics coefficients.

Lemma 4.13. *Let us assume now that $D = 1$ (the process is real valued and we omit the super-indices a_j). If the series (4.4.11) is finite for each $Q \le p$ we set $q = [p/2]$ ($q = p/2$ if p is even and $q = (p-1)/2$ if p is odd) then*

$$\left|\mathbb{E}\left(\sum_{j=1}^{n} X_j\right)^p\right| \le \sum_{u=1}^{q} n^u \gamma_u, \qquad \textit{where} \tag{4.4.12}$$

$$\gamma_u = \sum_{v=1}^{2q} \sum_{p_1+\cdots+p_u=p} \frac{p!}{p_1!\cdots p_u!} \kappa_{p_1}\cdots\kappa_{p_u}$$

Proof. As in Doukhan and Louhichi (1999) [67], we bound

$$\begin{aligned} |\mathbb{E}(X_1+\cdots+X_n)^p| &= \left|\sum_{1\le k_1,\ldots,k_p\le n} \mathbb{E}X_{k_1}\cdots X_{k_p}\right| \\ &\le A_{p,n} \equiv \sum_{1\le k_1,\ldots,k_p\le n} \left|\mathbb{E}X_{k_1}\cdots X_{k_p}\right| \end{aligned}$$

Let now $\mu = \{i_1,\ldots,i_v\} \subset \{1,\ldots,p\}$ and $k = (k_1,\ldots,k_p)$, we set for convenience

$$\mu(k) = (k_{i_1},\ldots,k_{i_v}) \in \mathbb{N}^v \tag{4.4.13}$$

In order to count the terms with their order of multiplicity, it is indeed not suitable to define the previous item as a set and cumulants or moments are defined equivalently in this case. Hence, as in Doukhan and León (1989) [66], we compute, using formula (4.4.2) and using all the partitions $\mu_1,\ldots,\mu_u$ of $\{1,\ldots,p\}$ with exactly $1 \le u \le p$ elements

$$\begin{aligned} A_{p,n} &= \sum_{1\le k_1,\ldots,k_p\le n} \sum_{u=1}^{p} \sum_{\mu_1,\ldots,\mu_u} \prod_{j=1}^{u} \kappa_{\mu_j(k)}(X) \\ &= \sum_{u=1}^{p} \sum_{\mu_1,\ldots,\mu_u} \sum_{1\le k_1,\ldots,k_p\le n} \prod_{j=1}^{u} \kappa_{\mu_j(k)}(X) \\ &= \sum_{r=1}^{p} \sum_{p_1+\cdots+p_r=p} \frac{p!}{p_1!\cdots p_r!} \times \\ &\qquad \times \prod_{u=1}^{r} \sum_{1\le k_1,\ldots,k_{p_u}\le n} \kappa_{p_u}(X_{k_1},\ldots,X_{k_{p_u}}) \qquad (4.4.14) \\ |A_{p,n}| &\le \sum_{u=1}^{q} n^u \sum_{p_1+\cdots+p_u=p} \frac{p!}{p_1!\cdots p_u!} \prod_{j=1}^{u} \kappa_{p_j} \qquad (4.4.15) \end{aligned}$$

Identity (4.4.14) follows from a simple change of variables and takes in account that the number of partitions of $\{1, \ldots, p\}$ into u subsets with respective cardinalities $p_1, \ldots, p_u$ is simply the corresponding multinomial coefficient. Remark that, for $\lambda \in \mathbb{N}$, and taking X's stationarity in account, we obtain

$$\sum_{1 \le k_1, \ldots, k_\lambda \le n} |\kappa_{p_u}(X_{k_1}, \ldots, X_{k_\lambda})| \le n \kappa_\lambda$$

Using this remark and the fact that cumulants of order 1 vanish and the only non-zero terms are those for which $p_1, \ldots, p_u \ge 2$ and thus $2u \le p$, hence $u \le q$ we finally deduce inequality (4.4.15). □

Remark 4.4. *If $\kappa_s \le C^s$ for $s \le p$ and for a constant $C > 0$, the bound (4.4.15) rewrites as $C^p \sum_{u=1}^{q} u^p n^u$ by using the multinomial identity.*

4.4.2 A second exponential inequality

This section is based on Doukhan and Neumann (2005), [71]. In this section we will be concerned with probability and moment inequalities for

$$S_n = X_1 + \cdots + X_n,$$

where $X_1, \ldots, X_n$ are zero mean random variables which fulfill appropriate weak dependence conditions. We denote by σ_n^2 the variance of S_n. Result are here stated without proof and we defer the reader to [71].
The first result is a Bernstein-type inequality which generalizes and improves previous inequalities of this chapter.

Theorem 4.5. *Suppose that $X_1, \ldots, X_n$ are real-valued random variables defined on a probability space $(\Omega, \mathcal{A}, \mathbb{P})$ with $\mathbb{E}X_i = 0$ and $\mathbb{P}(|X_i| \le M) = 1$, for all $i = 1, \ldots, n$ and some $M < \infty$. Let $\Psi : \mathbb{N}^2 \to \mathbb{N}$ be one of the following functions: (a) $\Psi(u, v) = 2v$, (b) $\Psi(u, v) = u + v$, (c) $\Psi(u, v) = uv$, or (d) $\Psi(u, v) = \alpha(u + v) + (1 - \alpha)uv$, for some $\alpha \in (0, 1)$.*
We assume that there exist constants $K, L_1, L_2 < \infty$, $\mu \ge 0$, and a nonincreasing sequence of real coefficients $(\rho(n))_{n \ge 0}$ such that, for all u-tuples $(s_1, \ldots, s_u)$ and all v-tuples $(t_1, \ldots, t_v)$ with $1 \le s_1 \le \cdots \le s_u \le t_1 \le \cdots \le t_v \le n$ the following inequality is fulfilled:

$$|\mathrm{Cov}\,(X_{s_1} \cdots X_{s_u}, X_{t_1} \cdots X_{t_v})| \le K^2 M^{u+v-2} \Psi(u, v) \rho(t_1 - s_u), \qquad (4.4.16)$$

where

$$\sum_{s=0}^{\infty} (s + 1)^k \rho(s) \le L_1 L_2^k (k!)^\mu, \qquad \forall k \ge 0. \qquad (4.4.17)$$

Then

$$\mathbb{P}\left(S_n \geq t\right) \leq \exp\left(-\frac{t^2/2}{A_n + B_n^{1/(\mu+2)} t^{(2\mu+3)/(\mu+2)}}\right), \tag{4.4.18}$$

where A_n *can be chosen as any number greater than or equal to* σ_n^2 *and*

$$B_n = 2(K \vee M) L_2 \left(\left(\frac{2^{4+\mu} n K^2 L_1}{A_n}\right) \vee 1\right).$$

Remark 4.5. *(i) Inequality (4.4.18) resembles the classical Bernstein inequality for independent random variables. Asymptotically,* σ_n^2 *is usually of order* $\mathcal{O}(n)$ *and* A_n *can be chosen equal to* σ_n^2 *while* B_n *is usually* $\mathcal{O}(1)$ *and hence negligible. In cases where* σ_n^2 *is very small or where the knowledge of the value of* A_n *is required for some statistical procedure, it might, however, be better to choose* A_n *larger than* σ_n^2*. It follows from (4.4.16) and (4.4.17) that a rough bound for* σ_n^2 *is given by*

$$\sigma_n^2 \leq 2nK^2 \Psi(1,1) L_1. \tag{4.4.19}$$

Hence, taking $A_n = 2nK^2\Psi(1,1)L_1$ *we obtain from (4.4.18) that*

$$P\left(\sum_{i=1}^{n} X_i \geq t\right) \leq \exp\left(-\frac{t^2}{C_1 n + C_2 t^{(2\mu+3)/(\mu+2)}}\right), \tag{4.4.20}$$

where $C_1 = 4K^2\Psi(1,1)L_1$ *and* $C_2 = 2B_n^{1/(\mu+2)}$ *with* B_n *such that* $B_n = 2(K \vee M)\, L_2 \left((2^{3+\mu}/\Psi(1,1)) \vee 1\right)$*. Inequality (4.4.20) is then more of Hoeffding-type.*
(ii) In the causal case, we obtain in Theorem 5.2 of Chapter 5 a Bennett-type inequality for τ*-dependent random variables. This also implies a Bernstein-type inequality, however, with different constants. In particular, the leading term in the denominator of the exponent differs from* σ_n^2*. This is a consequence of the method of proof which consists of replacing weakly dependent blocks of random variables by independent ones according to some coupling device (an analogue argument is used in [61] for the case of absolute regularity).*
(iii) Condition (4.4.16) in conjunction with (4.4.17) may be interpreted as a weak dependence condition in the sense that the covariances on the left-hand side tend to zero as the time gap between the two blocks of observations increases. Note that the supremum of expression (4.4.16) for all u-tuples $(s_1, \ldots, s_u)$ *and all v-tuples* $(t_1, \ldots, t_v)$ *with* $1 \leq s_1 \leq \cdots \leq s_u \leq t_1 \leq \cdots \leq t_v < \infty$ *such that* $t_1 - s_u = r$ *is denoted by* $C_{u+v,r}$ *in (4.3.1). Conditions (4.4.16) and (4.4.17) are typically fulfilled for truncated versions of random variables from many time series models; see also Proposition 4.1 below. The constant K in (4.4.16) is included to possibly take advantage of a sparsity of data as it appears, for example, in nonparametric curve estimation.*

(iv) For unbounded random variables, the coefficients $C_{p,r}$ may still be bounded by an explicit function of the index p under a weak dependence assumption; see Lemma 4.14 below. For example, assume that $\mathbb{E}\exp(A|X_t|^a) \le L$ holds for some constants $A, a > 0$, $L < \infty$. Since the inequality $u^p \le p!e^u$ $(p \in \mathbb{N}, u \ge 0)$ implies that

$$u^m = (Aa)^{-m/a}(Aau^a)^{m/a} \le (Aa)^{-m/a}(m!)^{1/a}e^{Au^a} \quad \forall m \in \mathbb{N}$$

we obtain that $\mathbb{E}|X_t|^m \le L(m!)^{1/a}(Aa)^{-m/a}$ holds for all $m \in \mathbb{N}$. Lemma 4.14 below provides then appropriate estimates for $C_{p,r}$.

Note that the variance of the sum does not explicitly show up in the Rosenthal-type inequality given in Theorem 4.2. Using the formula of Leonov and Shiryaev (1959) [119], we are able to obtain a more precise inequality which resembles the Rosenthal inequality in the independent case (see Rosenthal (1970) [170] and and Theorem 2.12 in Hall and Heyde (1980) [100] in the case of martingales).

Theorem 4.6. *Suppose that $X_1, \dots, X_n$ are real-valued random variables on a probability space $(\Omega, \mathcal{A}, \mathbb{P})$ with zero mean and let p be a positive integer. We assume that there exist finite constants K, M, and a nonincreasing sequence of real coefficients $(\rho(n))_{n\ge 0}$ such that, for all u-tuples $(s_1, \dots, s_u)$ and all v-tuples $(t_1, \dots, t_v)$ with $1 \le s_1 \le \dots \le s_u \le t_1 \le \dots \le t_v \le n$ and $u + v \le p$, condition (4.4.16) is fulfilled. Furthermore, we assume that*

$$\mathbb{E}|X_i|^{p-2} \le M^{p-2}.$$

Then, with $Z \sim \mathcal{N}(0,1)$,

$$|\mathbb{E}S_n^p - \sigma_n^p \mathbb{E}Z^p| \le B_{p,n} \sum_{1 \le u < p/2} A_{u,p} K^{2u}(M \vee K)^{p-2u} n^u,$$

where $B_{p,n} = (p!)^2 2^p \max_{2\le k\le p}\{\rho_{k,n}^{p/k}\}$, $\rho_{k,n} = \sum_{s=0}^{n-1}(s+1)^{k-2}\rho(s)$ and

$$A_{u,p} = \frac{1}{u!} \sum_{k_1+\dots+k_u=p, k_i\ge 2, \forall i} \frac{p!}{k_1! \cdots k_u!}.$$

Remark 4.6. *For even p, the above result implies that*

$$\mathbb{E}S_n^p \le (p-1)(p-3)\cdots 1\, \sigma_n^p + B_{p,n} \sum_{1\le u<p/2} A_{u,p} K^{2u}(M \vee K)^{p-2u} n^u,$$

which resembles the classical Rosenthal inequality from the independent case. If $\sup_n B_{p,n} < \infty$ and $\sigma_n^2 \asymp n$, then the first term on the right-hand side is asymptotically dominating, as $n \to \infty$. This term is equal to the p-th moment of a Gaussian random variable with mean 0 and variance σ_n^2.

Remark 4.7. *The inequality from Theorem 4.6 is well suited for proving a central limit theorem via the method of moments. Assume first that the random variables* X_i *are uniformly bounded, centred and satisfy condition (4.4.16) with* $\lim_{s\to\infty}\rho(s)/s^p = 0$, *for all* $p > 0$. *Then*

$$\lim_{n\to\infty}\frac{\sigma_n^2}{n} = \sigma^2 = \sum_{k=-\infty}^{\infty}\mathbb{E}(X_0X_k)$$

is a convergent series, and thus the method of moments implies a central limit theorem,

$$\frac{1}{\sqrt{n}}S_n \xrightarrow{d}_{n\to\infty} \sigma Z.$$

4.4.3 From weak dependence to the exponential bound

A large class of weak dependent sequences satisfies the assumption (4.4.16) of Theorem 4.5. If $\delta(x,y) = |x-y|$, we write $\Lambda^{(1)}$ instead of $\Lambda^{(1)}(\delta)$.

(i) Assume that $(X_t)_{t\in\mathbb{Z}}$ is an $\mathbb{R}^d$-valued and stationary process which is $(\Lambda^{(1)},\Psi)$-weakly dependent. Then for any Lipschitz-continuous function $F:\mathbb{R}^d\to\mathbb{R}$ with $\|F\|_\infty = M < \infty$ and $\mathrm{Lip}\,F \le 1$, the process $Y_t = F(X_t)$ is real valued, stationary, and $\|Y_t\|_\infty \le M$. Moreover, it is also $(\Lambda^{(1)},\Psi)$-weakly dependent.

(ii) In the more general case when $\mathrm{Lip}\,F$ possibly exceeds 1 (e.g., if the function F depends on the sample size in a statistical context), then weak dependence still holds where only $\Psi(a,b,u,v)$ has to be replaced by $\psi_Y(a,b,u,v) = \Psi(a\mathrm{Lip}\,F, b\mathrm{Lip}\,F,u,v)$. For the special cases of η, κ and λ weak dependence conditions, one may re-formulate this as $(Y_t)_{t\in\mathbb{Z}}$ is still an η, κ or λ weakly dependent sequence but now we have to respectively consider the coefficients

$$\begin{aligned}\eta_Y(r) &= \mathrm{Lip}\,F\cdot\eta(r), \qquad \kappa_Y(r) = \mathrm{Lip}^{\,2}F\cdot\kappa(r),\\ \lambda_Y(r) &= \max\left\{\mathrm{Lip}\,F, \mathrm{Lip}^{\,2}F\right\}\lambda(r).\end{aligned}$$

Now we relate the conditions of weak dependence to condition (4.4.16). Suppose that $(X_t)_{t\in\mathbb{Z}}$ is a stationary sequence of real-valued random variables with $\|X_t\|_\infty \le M$ which satisfies a weak dependence condition. To see the connection to (4.4.16), we consider the functions g_1 and g_2 with $g_1(x_1,\dots,x_u) = \prod_{i=1}^u f(x_i/M)$ and $g_2(x_1,\dots,x_v) = \prod_{i=1}^v f(x_i/M)$, where $f(u) = u\vee(-1)\wedge 1$. These functions satisfy $\mathrm{Lip}\,g_i \le 1/M$ and $\|g_i\|_\infty \le 1$. The covariance in the definition of weak dependence can be expressed as in equation (4.4.16), up to a factor M^{u+v} since $g_1(X_{i_1},\dots,X_{i_u}) = X_{i_1}\cdots X_{i_u}/M^u$ and $g_2(X_{i_1},\dots,X_{i_v}) = X_{i_1}\cdots X_{i_v}/M^v$.

Proposition 4.1. *Assume that the real valued sequence $(X_t)_{t\in\mathbb{Z}}$ is $(\Lambda^{(1)},\Psi)$-weakly dependent and that $\|X_t\|_\infty \le M$. Then*

$$\left|\mathrm{Cov}(X_{s_1}\cdots X_{s_u}, X_{t_1}\cdots X_{t_v})\right| \le M^{u+v}\Psi(M^{-1},M^{-1},u,v)\epsilon(t_1-s_u) \quad (4.4.21)$$

Moreover, if $\epsilon(r)=e^{-ar}$, for some $a>0$, then we may choose in inequality (4.4.17) $\mu=1$ and $L_1=L_2=1/(1-e^{-a})$. If $\epsilon(r)=e^{-ar^b}$, for some $a>0$, $b\in(0,1)$, then we may choose $\mu=1/b$ and L_1, L_2 appropriately as only depending on a and b.

Remark 4.8. *(i) Notice that Proposition 4.1 implies that $(\Lambda^{(1)},\Psi)$-dependence implies (4.4.16) with*

(a) $\Psi(u,v)=2v$, $K^2=M$ and $\epsilon(r)=\theta(r)/2$, under θ-dependence,

(b) $\Psi(u,v)=u+v$, $K^2=M$ and $\epsilon(r)=\eta(r)$, under η-dependence,

(c) $\Psi(u,v)=uv$, $K=1$ and $\epsilon(r)=\kappa(r)$, under κ-dependence,

(d) $\Psi(u,v)=(u+v+uv)/2$, $K^2=M\vee 1$ and $\epsilon(r)=2\lambda(r)$, under λ-dependence.

(ii) Now if the vector valued process $(X_t)_{t\in\mathbb{Z}}$ is an η, κ or λ-weakly dependent sequence, for any Lipschitz function $F:\mathbb{R}^d\to\mathbb{R}$ such that $\|F\|_\infty=M<\infty$, then the process $Y_t=F(X_t)$ is real valued and the relation (4.4.16) holds with

(a) $\Psi(u,v)=2v$, $K^2=M\,\mathrm{Lip}\,F$ and $\epsilon(r)=\theta(r)/2$, under θ-dependence,

(b) $\Psi(u,v)=u+v$, $K^2=M\,\mathrm{Lip}\,F$ and $\epsilon(r)=\eta(r)$, under η-dependence,

(c) $\Psi(u,v)=uv$, $K=\mathrm{Lip}\,F$ and $\epsilon(r)=\kappa(r)$, under κ-dependence,

(d) $\Psi(u,v)=(u+v+uv)/2$, $K^2=(M\vee 1)(\mathrm{Lip}^2 F\vee \mathrm{Lip}\,F)$ and $\epsilon(r)=2\lambda(r)$, under λ-dependence.

Those bounds allow to use the Bernstein type inequality in Theorem 4.5 for sums of functions of weakly dependent sequences.

A last problem in this setting is to determine sharp bounds for the coefficients $C_{p,r}$. This is even possible when the variables X_i are unbounded and is stated in the following lemma.

Lemma 4.14. *Assume that the real valued sequence $(X_t)_{t\in\mathbb{Z}}$ is η, κ or λ-weakly dependent and that $\mathbb{E}|X_i|^m\le M_m$, for any $m>p$. Then, according to the type of the weak dependence condition:*

$$C_{p,r} \le 2^{p+3}p^2 M_m^{\frac{p-1}{m-1}}\eta(r)^{1-\frac{p-1}{m-1}}, \quad (4.4.22)$$

$$\le 2^{p+3}p^4 M_m^{\frac{p-2}{m-2}}\kappa(r)^{1-\frac{p-2}{m-2}}, \quad (4.4.23)$$

$$\le 2^{p+3}p^4 M_m^{\frac{p-1}{m-1}}\lambda(r)^{1-\frac{p-1}{m-1}}. \quad (4.4.24)$$

Remark 4.9. *This lemma is the essential tool to provide a version of Theorem 4.6 which yields both a Rosenthal-type moment inequality and a rate of convergence for moments in the central limit theorem. We also note that this result does not involve the assumption that the random variables are a.s. bounded. In fact even the use of Theorem 4.5 does not really require such a boundedness; see Remark 4.5-(iv).*

4.5 Tightness criteria

Following Andrews and Pollard (1994) [5], we give in this section a general criterion based on Rosenthal type inequalities and on chaining arguments. Let $d \in \mathbb{N}^*$ and let $(X_i)_{i\in\mathbb{Z}}$ be a stationary sequence with values in $\mathbb{R}^d$. Let $\mathcal{F}$ be a class of functions from $\mathbb{R}^d$ to $\mathbb{R}$. We define the empirical process $\{Z_n(f)\,,\ f \in \mathcal{F}\}$ by

$$Z_n(f) := \sqrt{n}(P_n(f) - P(f))\ ,$$

with P the common law of $(X_i)_{i\in\mathbb{Z}}$ and, for $f \in \mathcal{F}$,

$$P_n(f) := \frac{1}{n}\sum_{i=1}^{n} f(X_i)\,, \quad P(f) = \int_{\mathbb{R}^d} f(x)P(dx)\ .$$

We study the process $\{Z_n(f)\,,\ f \in \mathcal{F}\}$ on the space $\ell^\infty(\mathbb{R}^d)$ of bounded functions from $\mathbb{R}^d$ to $\mathbb{R}$ equipped with the uniform norm $\|\cdot\|_\infty$. For more details on tightness on the non separable space $\ell^\infty(\mathbb{R}^d)$, we refer to van der Vaart and Wellner (1996) [183]. In particular, we shall not discuss any measurability problems which can be handled by using the outer probability. The process $\{Z_n(f)\,,\ f \in \mathcal{F}\}$ is tight on $(\ell_\infty(\mathbb{R}^d), \|.\|_\infty)$ as soon as there exists a semi-metric ρ such that $(\mathcal{F}, \rho)$ is totally bounded for which

$$\lim_{\delta\to 0}\limsup_{n\to\infty}\mathbb{P}\left(\sup_{f,g\in\mathcal{F},\,\rho(f,g)\le\delta}|Z_n(f) - Z_n(g)| > \varepsilon\right) = 0, \quad \forall\varepsilon > 0.$$

We recall the following definition of bracketing number.

Definition 4.4. *Let Q be a finite measure on a measurable space $\mathcal{X}$. For any measurable function f from $\mathcal{X}$ to $\mathbb{R}$, let $\|f\|_{Q,r} = Q(|f|^r)^{1/r}$. If $\|f\|_{Q,r}$ is finite, one says that f belongs to L^r_Q. Let $\mathcal{F}$ be some subset of L^r_Q. The number of brackets $N_{Q,r}(\varepsilon, \mathcal{F})$ is the smallest integer N for which there exist some functions $f_1^- \le f_1, \ldots, f_N^- \le f_N$ in $\mathcal{F}$ such that: for any integer $1 \le i \le N$ we have $\|f_i - f_i^-\|_{Q,r} \le \varepsilon$, and for any function f in $\mathcal{F}$ there exists an integer $1 \le i \le N$ such that $f_i^- \le f \le f_i$.*

Proposition 4.2. *Let $(X_i)_{i\geq 1}$ be a sequence of identically distributed random variables with values in a measurable space $\mathcal{X}$, with common distribution P. Let P_n be the empirical measure $P_n = n^{-1}\sum_{i=1}^n \delta_{X_i}$, and let Z_n be the normalized empirical process $Z_n = \sqrt{n}(P_n - P)$. Let Q be any finite measure on $\mathcal{X}$ such that $Q - P$ is a positive measure. Let $\mathcal{F}$ be a class of functions from $\mathcal{X}$ to $\mathbb{R}$ and $\mathcal{G} = \{f - l/(f,l) \in \mathcal{F}\times\mathcal{F}\}$. Assume that there exist $r \geq 2$, $p \geq 1$ and $q > 2$ such that for any function g of $\mathcal{G}$, we have*

$$\|Z_n(g)\|_p \leq C(\|g\|_{Q,1}^{1/r} + n^{1/q-1/2}), \tag{4.5.1}$$

where the constant C does not depend on g nor n. If moreover

$$\int_0^1 x^{(1-r)/r}(\mathcal{N}_{Q,1}(x,\mathcal{F}))^{1/p}dx < \infty \quad \text{and} \quad \lim_{x\to 0} x^{p(q-2)/q}\mathcal{N}_{Q,1}(x,\mathcal{F}) = 0\,,$$

then

$$\lim_{\delta\to 0}\limsup_{n\to\infty}\mathbb{E}\Big(\sup_{g\in\mathcal{G},\|g\|_{Q,1}\leq\delta}|Z_n(g)|^p\Big) = 0\,. \tag{4.5.2}$$

Proof of Proposition 4.2. It follows the line of Andrews and Pollard (1994) [5] and Louhichi (2000) [124]. It is based on the following inequality: given N real-valued random variables, we have

$$\|\max_{1\leq i\leq N}|Z_i|\,\|_p \leq N^{1/p}\max_{1\leq i\leq N}\|Z_i\|_p\,. \tag{4.5.3}$$

For any positive integer k, denote by $\mathcal{N}_k = \mathcal{N}_{Q,1}(2^{-k},\mathcal{F})$ and by $\mathcal{F}_k$ a family of functions $f_1^{k,-} \leq f_1^k, \ldots, f_{\mathcal{N}_k}^{k,-} \leq f_{\mathcal{N}_k}^k$ in $\mathcal{F}$ such that $\|f_i^k - f_i^{k,-}\|_{Q,1} \leq 2^{-k}$, and for any f in $\mathcal{F}$, there exists an integer $1 \leq i \leq \mathcal{N}_k$ such that $f_i^{k,-} \leq f \leq f_i^k$.

First step. We shall construct a sequence $h_{k(n)}(f)$ belonging to $\mathcal{F}_{k(n)}$ such that

$$\lim_{n\to\infty}\Big\|\sup_{f\in\mathcal{F}}|Z_n(f) - Z_n(h_{k(n)}(f))|\Big\|_p = 0\,. \tag{4.5.4}$$

For any f in $\mathcal{F}$, there exist two functions g_k^- and g_k^+ in $\mathcal{F}_k$ such that $g_k^- \leq f \leq g_k^+$ and $\|g_k^+ - g_k^-\|_{Q,1} \leq 2^{-k}$. Since $Q - P$ is a positive measure, we have

$$\begin{aligned} Z_n(f) - Z_n(g_k^-) &\leq Z_n(g_k^+) - Z_n(g_k^-) + \frac{1}{\sqrt{n}}\sum_{i=1}^n \mathbb{E}((g_k^+ - f)(X_i)) \\ &\leq |Z_n(g_k^+) - Z_n(g_k^-)| + \sqrt{n}2^{-k}\,. \end{aligned}$$

Since $g_k^- \leq f$, we also have that $Z_n(g_k^-) - Z_n(f) \leq \sqrt{n}2^{-k}$, which enables us to conclude that $|Z_n(f) - Z_n(g_k^-)| \leq |Z_n(g_k^+) - Z_n(g_k^-)| + \sqrt{n}2^{-k}$. Consequently

$$\sup_{f\in\mathcal{F}}|Z_n(f) - Z_n(g_k^-)| \leq \max_{1\leq i\leq\mathcal{N}_k}|Z_n(f_i^k) - Z_n(f_i^{k,-})| + \sqrt{n}2^{-k}\,. \tag{4.5.5}$$

Combining (4.5.3) and (4.5.5), we obtain that

$$\Big\| \sup_{f\in\mathcal{F}} |Z_n(f) - Z_n(g_k^-)| \Big\|_p \le \mathcal{N}_k^{1/p} \max_{1\le i\le \mathcal{N}_k} \|Z_n(f_i^k) - Z_n(f_i^{k,-})\|_p + \sqrt{n}2^{-k}\,. \tag{4.5.6}$$

Starting from (4.5.6) and applying the inequality (4.5.1),we obtain

$$\Big\| \sup_{f\in\mathcal{F}} |Z_n(f) - Z_n(g_k^-)| \Big\|_p \le C(\mathcal{N}_k^{1/p} 2^{-k/r} + \mathcal{N}_k^{1/p} n^{1/q-1/2}) + \sqrt{n}2^{-k}\,. \tag{4.5.7}$$

From the integrability condition on $\mathcal{N}_{Q,1}(x,\mathcal{F})$, and since the function $x \mapsto x^{(1-r)/r}\mathcal{N}(x,\mathcal{F})^{1/p}$ is non increasing, we infer that $\mathcal{N}_k^{1/p}2^{-k/r}$ tends to 0 as k tends to infinity. Take $k(n)$ such that $2^{k(n)} = \sqrt{n}/a_n$ for some sequence a_n decreasing to zero. Then $\sqrt{n}2^{-k(n)}$ tends to 0 as n tends to infinity. Il remains to control the second term on right hand in (4.5.7). By definition of $\mathcal{N}_{k(n)}$, we have that

$$\mathcal{N}_{k(n)} n^{p(1/q-1/2)} = \mathcal{N}_{Q,1}\Big(\frac{a_n}{\sqrt{n}}, \mathcal{F}\Big)\Big(\frac{1}{\sqrt{n}}\Big)^{p(q-2)/q}\,. \tag{4.5.8}$$

Since $x^{p(q-2)/p}\mathcal{N}_{Q,1}(x,\mathcal{F})$ tends to 0 as x tends to zero, we can find a sequence a_n such that the right hand term in (4.5.8) converges to 0. The function $h_{k(n)}(f) = g_{k(n)}^-$ satisfies (4.5.4).

Second step. We shall prove that for any $\epsilon > 0$ and n large enough, there exists a function $h_m(f)$ in $\mathcal{F}_m$ such that

$$\Big\| \sup_{f\in\mathcal{F}} |Z_n(h_m(f)) - Z_n(h_{k(n)}(f))| \Big\|_p \le \epsilon\,. \tag{4.5.9}$$

Given h in $\mathcal{F}_k$, choose a function $T_{k-1}(h)$ in $\mathcal{F}_{k-1}$ such that $\|h - T_{k-1}(h)\|_{Q,1} \le 2^{-k+1}$. Denote by $\pi_{k,k} = Id$ and for $l < k$, $\pi_{l,k}(h) = T_l \circ \cdots \circ T_{k-1}(h)$. We consider the function $h_m(f) = \pi_{m,k(n)}(h_{k(n)}(f))$. We have that

$$\Big\| \sup_{f\in\mathcal{F}} |Z_n(h_m) - Z_n(h_{k(n)})| \Big\|_p \le \sum_{l=m+1}^{k(n)} \Big\| \sup_{f\in\mathcal{F}} |Z_n(\pi_{l,k(n)}(h_{k(n)}) - Z_n(\pi_{l-1,k(n)}(h_{k(n)})| \Big\|_p\,. \tag{4.5.10}$$

Clearly

$$\Big\| \sup_{f\in\mathcal{F}} |Z_n(\pi_{l,k(n)}(h_{k(n)}) - Z_n(\pi_{l-1,k(n)}(h_{k(n)})| \Big\|_p \le \Big\| \max_{f\in\mathcal{F}_l} |Z_n(f) - Z_n(T_{l-1}(f))| \Big\|_p\,.$$

Applying the inequality (4.5.1) to (4.5.10) we obtain

$$\Big\| \sup_{f\in\mathcal{F}} |Z_n(h_m) - Z_n(h_{k(n)})| \Big\|_p \le C \sum_{l=m+1}^{k(n)} (2^{1/r}\mathcal{N}_l^{1/p} 2^{-l/r} + \mathcal{N}_l^{1/p} n^{1/q-1/2})$$

Clearly

$$\sum_{l=m+1}^{\infty} \mathcal{N}_l^{1/p} 2^{-l/r} \leq 2 \int_0^{2^{-m-1}} x^{(1-r)/r} (\mathcal{N}_{Q,1}(x, \mathcal{F}))^{1/p} dx,$$

which by assumption is as small as we wish. To control the second term, write

$$n^{1/q-1/2} \sum_{l=m+1}^{k(n)} \mathcal{N}_l^{1/p} \leq n^{1/q-1/2} \sum_{l=0}^{k(n)} 2^l \mathcal{N}_l^{1/p} 2^{-l}$$
$$\leq 2n^{1/q-1/2} \int_{2^{-k(n)}}^{1} \frac{1}{x} (\mathcal{N}_{Q,1}(x, \mathcal{F}))^{1/p} dx.$$

It is easy to see that if $x^{p(q-2)/q} \mathcal{N}_{Q,1}(x, \mathcal{F})$ tends to 0 as x tends to 0, then

$$\lim_{x \to 0} x^{(q-2)/q} \int_x^1 \frac{1}{y} (\mathcal{N}_{Q,1}(y, \mathcal{F}))^{1/p} dy = 0\,.$$

Consequently, we can choose the decreasing sequence a_n such that

$$\lim_{n \to \infty} \Big(\frac{1}{\sqrt{n}} \Big)^{\frac{q-2}{q}} \int_{a_n n^{-1/2}}^{1} \frac{1}{x} (\mathcal{N}_{Q,1}(x, \mathcal{F}))^{1/p} dx = 0.$$

The function $h_m(f) = \pi_{m,k(n)}(h_{k(n)}(f))$ satisfies (4.5.9).

Third step. From steps 1 and 2, we infer that for any $\epsilon > 0$ and n large enough, there exists $h_m(f)$ in $\mathcal{F}_m$ such that

$$\Big\| \sup_{f \in \mathcal{F}} |Z_n(f) - Z_n(h_m(f))| \Big\|_p \leq 2\epsilon\,.$$

Using the same argument as in Andrews and Pollard (1994) [5] (see the paragraph "Comparison of pairs" page 124), we obtain that, for any f and g in $\mathcal{F}$,

$$\Big\| \sup_{\|f-g\|_{Q,1} \leq \delta} |Z_n(f) - Z_n(g)| \Big\|_p \leq 8\epsilon + \mathcal{N}_m^{2/r} \sup_{\|f-g\|_{Q,1} \leq \delta} \|Z_n(f) - Z_n(g)\|_p\,.$$

We conclude the proof by noting that

$$\limsup_{\delta \to 0} \limsup_{n \to \infty} \Big\| \sup_{g \in \mathcal{G}, \|g\|_{Q,1} \leq \delta} |Z_n(g)| \Big\|_p \leq 8\epsilon\,. \quad \square$$

Chapter 5

Tools for causal cases

The purpose of this chapter is to give several technical tools useful to derive limit theorems in the causal cases. The first section gives comparison results between different causal coefficients. The second section deals with covariance inequalities for the coefficients γ_1, $\tilde{\beta}$ and $\tilde{\phi}$ already defined in Chapter 2. Section 3 discusses a coupling result for τ_1-dependent random variables. This coupling result is generalized to the case of variables with values in any Polish space. Section 4 gives various inequalities mainly Bennett, Fuk-Nagaev, Burkholder and Rosenthal type inequalities for different dependent sequences. Finally Section 5 gives a maximal inequality as an extension of Doob's inequality for martingales.

5.1 Comparison results

Weak dependence conditions may be compared as in the case of mixing conditions.

Lemma 5.1. *Let $(\Omega, \mathcal{A}, \mathbb{P})$ be a probability space. Let $(X_i)_{i\in\mathbb{Z}}$ be a sequence of real valued random variables. We have the following comparison results:*

1. $\forall\, r \in \mathbb{N}^*$, $\forall\, k \geq 0$,

$$\tilde{\alpha}_r(k) \leq \tilde{\beta}_r(k) \leq \tilde{\phi}_r(k).$$

Assume now that $X_i \in \mathbb{L}^1(\mathbb{P})$ for all $i \in \mathbb{Z}$. If Y is a real valued random variable Q_Y is the generalized inverse of the tail function, $t \mapsto \mathbb{P}(|Y| > t)$, see (2.2.14).

2. *Let $Q_X \geq \sup_{k\in\mathbb{Z}} Q_{X_k}$. Then, $\forall\, k \geq 0$,*

$$\tau_{1,1}(k) \leq 2\int_0^{\tilde{\alpha}_1(k)} Q_X(u)du. \tag{5.1.1}$$

Assume now that the sequence $(X_i)_{i\in\mathbb{Z}}$ is valued in $\mathbb{R}^d$ for some $d \in \mathbb{N}^*$ and that each X_i belongs to $\mathbb{L}^1(\mathbb{P})$. On $\mathbb{R}^d$, we put the distance $d_1(x,y) = \sum_{p=1}^d |x^{(p)} - y^{(p)}|$, where $x^{(p)}$ (resp. $y^{(p)}$) denotes the pth coordinate of x (resp. y). Assume that for all $i \in \mathbb{Z}$, each component of X_i has a continuous distribution function, and let ω be the supremum of the modulus of continuity, that is

$$\omega(x) = \sup_{i\in\mathbb{Z}} \max_{1\le k\le d} \sup_{|y-z|\le x} |F_{X_i^{(k)}}(y) - F_{X_i^{(k)}}(z)|,$$

where for all $i \in \mathbb{Z}$, $X_i = (X_i^{(1)}, \ldots, X_i^{(d)})$. Define $g(x) = x\omega(x)$. Then we have

3. *$\forall r \in \mathbb{N}^*$, $\forall k \ge 0$,*

$$\begin{aligned} \tilde{\beta}_r(k) &\le \frac{2\,r\,\tau_{1,r}(k)}{g^{-1}\left(\frac{\tau_{1,r}(k)}{d}\right)}. \\ \tilde{\phi}_r(k) &\le \frac{2\,r\,\tau_{\infty,r}(k)}{g^{-1}\left(\frac{\tau_{\infty,r}(k)}{d}\right)}, \end{aligned}$$

   where g^{-1} denotes the generalized inverse of g defined in (2.2.14).

4. *Assume now that there exists some positive constant K such that for all $i \in \mathbb{Z}$, each component of X_i has a density bounded by $K > 0$. Then, for all $r \in \mathbb{N}^*$ and $k \ge 0$,*

$$\tilde{\alpha}_r(k) \le 4\,r\sqrt{K\,d\,\theta_{1,r}(k)}.$$

Remark 5.1. *If each marginal distribution satisfies a concentration condition $|F_{X_i^{(k)}}(y) - F_{X_i^{(k)}}(z)| \le K|y-z|^a$ with $a \le 1, K > 0$ then Item 3. yields*

$$\begin{aligned} \tilde{\beta}_r(k) &\le 2\,r\,\tau_{1,r}(k)^{\frac{a}{1+a}}(Kd)^{\frac{1}{1+a}}, \\ \tilde{\phi}_r(k) &\le 2\,r\,\tau_{\infty,r}(k)^{\frac{a}{1+a}}(Kd)^{\frac{1}{1+a}}. \end{aligned}$$

If e.g. for all $i \in \mathbb{Z}$, each component of X_i has a density bounded by $K > 0$ then those relations write more simply with $a = 1$ as in Item 4.

In order to prove Lemma 5.1, we introduce some general comments related with the notation (2.2.14). Let $(\Omega, \mathcal{A}, \mathbb{P})$ be a probability space, $\mathcal{M}$ a σ-algebra of $\mathcal{A}$ and X a real-valued random variable. Let $F_{\mathcal{M}}(t,\omega) = P_{X|\mathcal{M}}(]-\infty, t], \omega)$ be a conditional distribution function of X given $\mathcal{M}$. For any ω, $F_{\mathcal{M}}(\cdot,\omega)$ is a distribution function, and for any t, $F_{\mathcal{M}}(t,\cdot)$ is a $\mathcal{M}$-measurable random variable. Hence for any ω, define the generalized inverse $F_{\mathcal{M}}^{-1}(u,\omega)$ as in (2.2.14). Now, from the equality $\{\omega/t \ge F_{\mathcal{M}}^{-1}(u,\omega)\} = \{\omega/F_{\mathcal{M}}(t,\omega) \ge u\}$, we infer that $F_{\mathcal{M}}^{-1}(u,\cdot)$ is $\mathcal{M}$-measurable. In the same way, $\{(t,\omega)/F_{\mathcal{M}}(t,\omega) \geq u\} = \{(t,\omega)/t \geq F_{\mathcal{M}}^{-1}(u,\omega)\}$, which implies that the mapping $(t,\omega) \mapsto F_{\mathcal{M}}(t,\omega)$ is measurable with respect to $\mathcal{B}(\mathbb{R}) \otimes \mathcal{M}$. The same arguments imply that the mapping $(u,\omega) \mapsto F_{\mathcal{M}}^{-1}(u,\omega)$ is measurable with respect to $\mathcal{B}([0,1]) \otimes \mathcal{M}$. Denote by $F_{\mathcal{M}}(t)$ (resp. $F_{\mathcal{M}}^{-1}(u)$) the random variable $F_{\mathcal{M}}(t,\cdot)$ (resp. $F_{\mathcal{M}}^{-1}(u,\cdot)$), and let $F_{\mathcal{M}}(t-0) = \sup_{s<t} F_{\mathcal{M}}(s)$.

Proof of Lemma 5.1.
• Item 1. follows from the definition of $\tilde{\alpha}(\mathcal{M},X)$, $\tilde{\beta}(\mathcal{M},X)$ and $\tilde{\phi}(\mathcal{M},X)$.
• Let us now prove Item 2. We first prove that if $\mathcal{M}$ is a σ-algebra of $\mathcal{A}$, and if X is a real valued random variable in $\mathbb{L}^1(\mathbb{P})$, then

$$\tau(\mathcal{M},X) \leq 2\int_0^{\tilde{\alpha}(\mathcal{M},X)} Q_X(u)du\,. \tag{5.1.2}$$

The proof follows from arguments in Peligrad (2002) [140]. Denote $X^+ = \sup(X,0)$ and $X^- = \sup(-X,0)$. Let F denote the distribution function of X. Let $F_{\mathcal{M}}(t,\omega)$ be the conditional distribution of X given $\mathcal{M}$. Assume that there exists a random variable δ uniformly distributed over $[0,1]$, independent of the σ-algebra generated by X and $\mathcal{M}$. As δ is uniformly distributed over $[0,1]$ and independent of the σ-algebra generated by X and $\mathcal{M}$,

$$U = F_{\mathcal{M}}(X-0) + \delta(F_{\mathcal{M}}(X) - F_{\mathcal{M}}(X-0))$$

is uniformly distributed over $[0,1]$ conditionally to $\mathcal{M}$. So U is independent of $\mathcal{M}$ and is uniformly distributed over $[0,1]$ (see Major (1978) [126] and also Rio (2000) [161]). Hence,

$$X^* = F^{-1}(U) \tag{5.1.3}$$

is independent of $\mathcal{M}$ and distributed as X. Moreover,

$$F_{\mathcal{M}}^{-1}(U) = X, \qquad \mathbb{P}\text{-almost surely.}$$

It yields

$$\|X - X^*\|_1 = \mathbb{E}\Big(\int_0^1 |F_{\mathcal{M}}^{-1}(u) - F^{-1}(u)|du\Big)\,. \tag{5.1.4}$$

We start with equality (5.1.4).

$$\begin{aligned}
\|X - X^*\|_1 &= \mathbb{E}\Big(\textstyle\int_0^1 |F_{\mathcal{M}}^{-1}(u) - F^{-1}(u)|du\Big)\\
&= \mathbb{E}\Big(\textstyle\int_0^\infty |F_{\mathcal{M}}(u) - F(u)|du\Big)\\
&= \mathbb{E}\Big(\textstyle\int_0^\infty |\mathbb{P}(X^+ > u) - \mathbb{P}(X^+ > u|\mathcal{M})|du\Big)\\
&\quad +\mathbb{E}\Big(\textstyle\int_0^\infty |\mathbb{P}(X^- > u) - \mathbb{P}(X^- > u|\mathcal{M})|du\Big)
\end{aligned}$$

Now, by definition of $\tilde{\alpha}$, we have the inequalities

$$\begin{aligned}
\mathbb{E}|\mathbb{P}(X^+ > u) - \mathbb{P}(X^+ > u|\mathcal{M})| &\leq \tilde{\alpha}(\mathcal{M}, X^+) \wedge 2\mathbb{P}(X^+ > u) \\
\mathbb{E}|\mathbb{P}(X^- > u) - \mathbb{P}(X^- > u|\mathcal{M})| &\leq \tilde{\alpha}(\mathcal{M}, X^-) \wedge 2\mathbb{P}(X^- > u)
\end{aligned}$$

It is clear that $\sup(\tilde{\alpha}(\mathcal{M}, X^+), \tilde{\alpha}(\mathcal{M}, X^-)) \leq \tilde{\alpha}(\mathcal{M}, X)$. Define $H_X(x) = \mathbb{P}(|X| > x)$. Next, using the inequality $a \wedge b + c \wedge d \leq (a + c) \wedge (b + d)$, we obtain that

$$\mathbb{E}|X - X^*| \leq 2\int_0^\infty \tilde{\alpha}(\mathcal{M}, X) \wedge H_X(u)du \leq 2\int_0^\infty \int_0^{\tilde{\alpha}(\mathcal{M},X)} \mathbf{1}_{t<H_X(u)}dt\, du\,.$$

Then, since $\mathbb{P}(|X| > u) > t$ if and only if $u < Q_X(t)$ and applying Fubini Theorem, we get Inequality (5.1.2).
Let us now prove (5.1.1). For $i \in \mathbb{Z}$, let $\mathcal{M}_i = \sigma(X_j, j \leq i)$. For $i + k \leq j$, we infer from Inequality (5.1.2) that

$$\tau_1(\mathcal{M}_i, X_j) \leq 2\int_0^{\tilde{\alpha}_1(k)} Q_{X_j}(u)du \leq 2\int_0^{\tilde{\alpha}_1(k)} Q_X(u)du,$$

and the result follows from the definition of $\tau_{1,1}(k)$.

• For Item 3. we will write the proof for $r = 1, 2$, the generalization to r points being straightforward. We will make use of the following proposition:

Proposition 5.1. *Let $(\Omega, \mathcal{A}, \mathbb{P})$ be a probability space, $X = (X_1, \ldots, X_d)$ and $Y = (Y_1, \ldots, Y_d)$ two random variables with values in $\mathbb{R}^d$, and $\mathcal{M}$ a σ-algebra of $\mathcal{A}$. If (X^*, Y^*) is distributed as (X, Y) and independent of $\mathcal{M}$ then, assuming that each component X_k and Y_k has a continuous distribution function F_{X_k} and F_{Y_k}, we get for any $x_1, \ldots, x_d, y_1, \ldots, y_d$ in $[0, 1]$,*

$$\tilde{\beta}(\mathcal{M}, X) \leq \Big\| \sum_{k=1}^d x_k + \mathbb{P}\big(|F_{X_k}(X_k^*) - F_{X_k}(X_k)| > x_k \big| \mathcal{M}\big)\Big\|_1, \qquad (5.1.5)$$

$$\begin{aligned}
\tilde{\beta}(\mathcal{M}, X, Y) \leq &\Big\| \sum_{k=1}^d x_k + \mathbb{P}\big(|F_{X_k}(X_k^*) - F_{X_k}(X_k)| > x_k \big| \mathcal{M}\big)\Big\|_1 \\
&+ \Big\| \sum_{k=1}^d y_k + \mathbb{P}\big(|F_{Y_k}(Y_k^*) - F_{Y_k}(Y_k)| > y_k \big| \mathcal{M}\big)\Big\|_1. \qquad (5.1.6)
\end{aligned}$$

We prove Proposition 5.1 at the end of this section and we continue the proof of Item 3. Starting from (5.1.5) with $x_1 = \cdots = x_k = \omega(x)$ and applying Markov's inequality, we infer that

$$\tilde{\beta}(\mathcal{M}, X) \leq d\omega(x) + \frac{1}{x}\Big\|\mathbb{E}\Big(\sum_{k=1}^d |X_k - X_k^*| \Big| \mathcal{M}\Big)\Big\|_1.$$

Now, from Proposition 6 in Rüschendorf (1985) [171] (see also Equality (5.3.5) of Lemma 5.3), one can choose X^* such that

$$\Big\|\mathbb{E}\Big(\sum_{k=1}^{d}|X_k - X_k^*|\Big|\, \mathcal{M}\Big)\Big\|_1 = \tau_1(\mathcal{M}, X).$$

Hence

$$\tilde{\beta}(\mathcal{M}, X) \leq d\omega(x) + \frac{\tau_1(\mathcal{M}, X)}{x},$$

and Item 3. follows for $r = 1$ by noting that $d\,x\omega(x) = \tau_1(\mathcal{M}, X)$ for $x = g^{-1}\left(\frac{\tau_1(\mathcal{M},X)}{d}\right)$. Starting now from (5.1.6), Item 3. may be proved in the same way for $r = 2$. Proposition 5.1 is still true when replacing $\tilde{\beta}$ by $\tilde{\phi}$ and $\|\cdot\|_1$ by $\|\cdot\|_\infty$ (see Proposition 7 in Dedecker & Prieur, 2006 [49]). The proof for $\tilde{\phi}$ follows then the same lines and is therefore omitted here.

• It remains now to prove Item 4. We write the proof for $r = 1$ and $d = 1$ and then generalize to any dimension $d \geq 1$ and any $r \geq 1$.

Case $r = 1, d = 1$. For $i \in \mathbb{Z}$, let $\mathcal{M}_i = \sigma(X_j, j \leq i)$. Let $i + k \leq j_1$. We know from Definition 2.5 of Chapter 2 that

$$\tilde{\alpha}(\mathcal{M}_i, X_{j_1}) = \sup_{t\in\mathbb{R}} \|L_{X_{j_1}|\mathcal{M}_i}(t) - L_{X_{j_1}}(t)\|_1.$$

Define $\varphi_t(x) = \mathbf{1}_{x\leq t} - \mathbb{P}(X_{j_1} \leq t)$. We want to smooth the function φ_t which is not Lipschitz. Let $\varepsilon > 0$. We consider the following Lipschitz function φ_t^ε smoothing φ_t:

$$\varphi_t^\varepsilon(x) = \begin{cases} \varphi_t(x), & x \in]-\infty, t] \cup [t+\varepsilon, +\infty[, \\ \varphi_t(x) + \frac{t-x+\varepsilon}{\varepsilon}\mathbf{1}_{t<x<t+\varepsilon}, & t < x < t + \varepsilon. \end{cases}$$

We then have $||\varphi_t^\varepsilon||_\infty \leq 1$ and $\text{Lip}\,(\varphi_t^\varepsilon) \leq \frac{1}{\varepsilon}$. Hence,

$$\begin{aligned} &\|L_{X_{j_1}|\mathcal{M}_i}(t) - L_{X_{j_1}}(t)\|_1 = \mathbb{E}\,|\mathbb{E}(\varphi_t(X_{j_1})\,|\mathcal{M}_i)| \\ &\leq \|\varphi_t - \varphi_t^\varepsilon\|_\infty \mathbb{P}(X_{j_1} \in]t, t+\varepsilon[) + \mathbb{E}\,|\mathbb{E}(\varphi_t^\varepsilon(X_{j_1}) - \mathbb{E}\varphi_t^\varepsilon(X_{j_1})\,|\mathcal{M}_i)| \\ &+ |\mathbb{E}\,(\varphi_t^\varepsilon(X_{j_1}) - \varphi_t(X_{j_1}))| \\ &\leq 2\,K\,\varepsilon + \tfrac{1}{\varepsilon}\,\theta_{1,1}(k) + 2K\varepsilon. \end{aligned}$$

Then, if we take $\varepsilon = \sqrt{\dfrac{\theta_{1,1}(k)}{4K}}$ and if we consider the supremum in t, we get

$$\tilde{\alpha}_1(k) \leq 4\sqrt{K\,\theta_{1,1}(k)}\,.$$

Case $r \geq 1$, $d \geq 1$. The proof follows essentially the same lines. Let $i+k \leq j_1 < \cdots < j_r$. Let $X = (X_{j_1}, \ldots, X_{j_r})$ in $\left(\mathbb{R}^d\right)^r$. We know from Definition 2.5 of Chapter 2 that

$$\tilde{\alpha}(\mathcal{M}_i, X) = \sup_{t \in (\mathbb{R}^d)^r} \|L_{X|\mathcal{M}_i}(t) - L_X(t)\|_1.$$

For $t = (t_1, \ldots, t_r) \in (\mathbb{R}^d)^r$ and $x = (x_1, \ldots, x_r) \in (\mathbb{R}^d)^r$, define

$$\varphi_t(x) = \prod_{i=1}^{r} (\mathbf{1}_{x_i \leq t_i} - \mathbb{P}(X_{j_i} \leq t)) = \prod_{i=1}^{r} \varphi_{t_i}(x_i).$$

We want to smooth the function φ_t which is not Lipschitz. Let $\varepsilon > 0$. We first smooth each of the functions φ_{t_i}. In the following, if x is in $\mathbb{R}^d$, $x^{(p)}$ denotes its pth coordinate. We consider the following Lipschitz function $\varphi_{t_i}^{\varepsilon}$ smoothing φ_{t_i}:

$\varphi_{t_i}^{\varepsilon}$ is equal to φ_{t_i} on $]-\infty, t_i] \cup [t_i+\varepsilon, +\infty[$, and for $x_i \notin]-\infty, t_i] \cup [t_i+\varepsilon, +\infty[$,

$$\varphi_{t_i}^{\varepsilon}(x_i) = \prod_{j=1}^{d} \left(\mathbf{1}_{x_i^{(j)} \leq t_i^{(j)}} + \frac{t_i^{(j)} - x_i^{(j)} + \varepsilon}{\varepsilon} \mathbf{1}_{t_i^{(j)} < x_i^{(j)} < t_i^{(j)} + \varepsilon} \right) - \mathbb{P}(X_{j_i} \leq t_i).$$

We then have $||\varphi_{t_i}^{\varepsilon}||_\infty \leq 1$ and $\mathrm{Lip}\,(\varphi_{t_i}^{\varepsilon}) \leq \frac{1}{\varepsilon}$, where the distance used on $\mathbb{R}^d$ is $d_1(x,y) = \sum_{j=1}^d |x^{(j)} - y^{(j)}|$. We then smooth $\varphi_t(x)$ by $\varphi_t^{\varepsilon}(x) = \prod_{i=1}^r \varphi_{t_i}^{\varepsilon}(x_i)$. We have $||\varphi_t^{\varepsilon}||_\infty \leq 1$ and $\mathrm{Lip}\,(\varphi_t^{\varepsilon}) \leq \frac{1}{\varepsilon}$. Let $X = (X_{j_1}, X_{j_2})$. We get

$$\begin{aligned} \|L_{X|\mathcal{M}_i}(t) - L_X(t)\|_1 &= \mathbb{E}\,|\mathbb{E}(\varphi_t(X) - \mathbb{E}\varphi_t(X)\,|\mathcal{M}_i)| \\ &\leq 2rKd\varepsilon + \tfrac{1}{\varepsilon} r\theta_{1,r}(k) + 2\,rKd\varepsilon. \end{aligned}$$

We conclude by taking the supremum in $t \in (\mathbb{R}^d)^r$ and $\varepsilon = \sqrt{\frac{\theta_{1,r}(k)}{4\,K\,d}}$. □

Proof of Proposition 5.1. Let Z be a random variable with values in $\mathbb{R}^m$ and let $f : \mathbb{R}^m \to \mathbb{R}$ such that

$$|f(z_1, \ldots, z_i, \ldots, z_m) - f(z_1, \ldots, z_i', \ldots, z_m)| \leq |\mathbf{1}_{z_i \leq a_i} - \mathbf{1}_{z_i' \leq a_i}|$$

for some real numbers $a_1, \ldots, a_m$. Let $\mathcal{U}$ be a σ-algebra and let Z^* be a random variable distributed as Z and independent of $\mathcal{U}$. Then

$$\begin{aligned} |f(Z) - f(Z^*)| &= \left| \sum_{k=1}^{m} f(Z_1, \ldots Z_k, Z_{k+1}^*, \ldots Z_m^*) - f(Z_1, \ldots Z_{k-1}, Z_k^*, \ldots Z_m^*) \right| \\ &\leq \sum_{k=1}^{m} |\mathbf{1}_{Z_k \leq a_k} - \mathbf{1}_{Z_k^* \leq a_k}|. \end{aligned}$$

Hence

$$|\mathbb{E}(f(Z)|\mathcal{U}) - \mathbb{E}(f(Z))| \le \mathbb{E}(|f(Z) - f(Z^*)||\mathcal{U}) \le \sum_{k=1}^{m} \mathbb{E}\big(|\mathbf{1}_{Z_k \le a_k} - \mathbf{1}_{Z_k^* \le a_k}|\,\big|\,\mathcal{U}\big)\,. \tag{5.1.7}$$

Let $t \in \mathbb{R}^d$. We first apply (5.1.7) to $Z = X$, $Z^* = X^*, \mathcal{U} = \mathcal{M}$, and $f(z) = \mathbf{1}_{z \le t}$ with $a_1 = t_1, \ldots, a_d = t_d$. Since $F_{X_k}^{-1}(F_{X_k}(X_k)) = X_k$ almost surely, we obtain

$$\begin{aligned} |\mathbb{E}(\mathbf{1}_{X \le t}|\mathcal{M}) - \mathbb{P}(X \le t)| &\le \sum_{k=1}^{d} \mathbb{E}\big(|\mathbf{1}_{X_k \le t_k} - \mathbf{1}_{X_k^* \le t_k}|\,\big|\,\mathcal{M}\big) \qquad (5.1.8)\\ &\le \sum_{k=1}^{d} \mathbb{E}\big(|\mathbf{1}_{F_{X_k}(X_k) \le F_{X_k}(t_k)} - \mathbf{1}_{F_{X_k}(X_k^*) \le F_{X_k}(t_k)}|\,\big|\,\mathcal{M}\big). \end{aligned}$$

Define $T_k = F_{X_k}(X_k)$, $T_k^* = F_{X_k}(X_k^*)$. We have

$$\begin{aligned} &\mathbb{E}\big(|\mathbf{1}_{T_k \le F_{X_k}(t_k)} - \mathbf{1}_{T_k^* \le F_{X_k}(t_k)}|\,\big|\,\mathcal{M}\big)\\ &\quad \le \max\Big(F_{T_k|\mathcal{M}}(F_{X_k}(t_k)) - F_{X_k}(t_k), 1 - F_{T_k|\mathcal{M}}(F_{X_k}(t_k)) - (1 - F_{X_k}(t_k))\Big). \end{aligned}$$

Now, for any y in $[0, 1]$,

$$\begin{aligned} F_{T_k|\mathcal{M}}(y) &= \int \mathbf{1}_{v+u-v \le y} \mathbb{P}_{T_k, T_k^*|\mathcal{M}}(du, dv)\\ &\le \int \mathbf{1}_{v \le y + x_k} \mathbb{P}_{T_k^*|\mathcal{M}}(dv) + \int \mathbf{1}_{v-u > x_k} \mathbb{P}_{T_k, T_k^*|\mathcal{M}}(du, dv)\\ &\le y + x_k + \int \mathbf{1}_{v-u > x_k} \mathbb{P}_{T_k, T_k^*|\mathcal{M}}(du, dv)\,. \end{aligned}$$

In the same way,

$$1 - F_{T_k|\mathcal{M}}(y) \le 1 - (y - x_k) + \int \mathbf{1}_{u-v > x_k} \mathbb{P}_{T_k, T_k^*|\mathcal{M}}(du, dv)\,.$$

Consequently, taking $y = F_{X_k}(t_k)$,

$$\mathbb{E}\big(|\mathbf{1}_{T_k \le F_{X_k}(t_k)} - \mathbf{1}_{T_k^* \le F_{X_k}(t_k)}|\,\big|\,\mathcal{M}\big) \le x_k + \int \mathbf{1}_{|u-v| > x_k} \mathbb{P}_{T_k, T_k^*|\mathcal{M}}(du, dv), \tag{5.1.9}$$

and Inequality (5.1.5) follows from (5.1.9) and (5.1.8) by taking the supremum in t and the expectation. Let s, $t \in \mathbb{R}^d$. In the same way, applying (5.1.7) to $Z = (Z^{(1)}, Z^{(2)}) = (X, Y)$, $Z^* = (X^*, Y^*)$, $\mathcal{U} = \mathcal{M}$ and

$$f(z^{(1)}, z^{(2)}) = (\mathbf{1}_{z^{(1)} \le s} - F_X(s))(\mathbf{1}_{z^{(2)} \le t} - F_Y(t))\,,$$

we obtain that

$$|\mathbb{E}((\mathbf{1}_{X\le s} - F_X(s))(\mathbf{1}_{Y\le t} - F_Y(t))|\,\mathcal{M}) - \mathbb{E}((\mathbf{1}_{X\le s} - F_X(s))(\mathbf{1}_{Y\le t} - F_Y(t)))|$$
$$\le \sum_{k=1}^{d} \mathbb{E}\big(|\mathbf{1}_{X_k\le s_k} - \mathbf{1}_{X_k^*\le s_k}|\,\big|\,\mathcal{M}\big) + \sum_{k=1}^{d} \mathbb{E}\big(|\mathbf{1}_{Y_k\le t_k} - \mathbf{1}_{Y_k^*\le t_k}|\,\big|\,\mathcal{M}\big),$$

and we conclude the proof of (5.1.6) by using the same arguments as for (5.1.5). □

5.2 Covariance inequalities

In this section we present some covariance inequalities for the coefficients γ_1, $\tilde{\beta}$ and $\tilde{\phi}$. We begin with the weakest of those coefficients.

5.2.1 A covariance inequality for γ_1

Definition 5.1. *Let X, Y be real valued random variables. Denote by*
- Q_X the generalized inverse of the tail function $H_X : x \mapsto \mathbb{P}(|X| > x)$.
- G_X the inverse of $x \mapsto \int_0^x Q_X(u)du$.
- $H_{X,Y}$ the generalized inverse of $x \mapsto \mathbb{E}(|X|\mathbf{1}_{|Y|>x})$.

Proposition 5.2. *Let $(\Omega, \mathcal{A}, \mathbb{P})$ be a probability space and $\mathcal{M}$ be a σ-algebra of $\mathcal{A}$. Let X be an integrable random variable and Y be an $\mathcal{M}$-measurable random variable such that $|XY|$ is integrable. The following inequalities hold*

$$|\mathbb{E}(YX)| \le \int_0^{\|\mathbb{E}(X|\mathcal{M})\|_1} H_{X,Y}(u)du \le \int_0^{\|\mathbb{E}(X|\mathcal{M})\|_1} Q_Y \circ G_X(u)du\,. \qquad (5.2.1)$$

If furthermore Y is integrable, then

$$|\mathrm{Cov}(Y, X)| \le \int_0^{\gamma_1(\mathcal{M},X)} Q_Y \circ G_{X-\mathbb{E}(X)}(u)du \le 2\int_0^{\gamma_1(\mathcal{M},X)/2} Q_Y \circ G_X(u)du\,. \quad (5.2.2)$$

Combining Proposition 5.2 and the comparison results (2.2.13), (2.2.18) and Item 2. of Lemma 5.1, we easily derive covariance inequalities for θ_1, τ_1 and $\tilde{\alpha}$.

Proof of Proposition 5.2. We start from the inequality

$$|\mathbb{E}(YX)| \le \mathbb{E}(|Y\mathbb{E}(X|\mathcal{M})|) = \int_0^\infty \mathbb{E}\big(|\mathbb{E}(X|\mathcal{M})|\mathbf{1}_{|Y|>t}\big)dt.$$

Clearly we have that $\mathbb{E}\big(|\mathbb{E}(X|\mathcal{M})|\mathbf{1}_{|Y|>t}\big) \le \|\mathbb{E}(X|\mathcal{M})\|_1 \wedge \mathbb{E}(|X|\mathbf{1}_{|Y|>t})$. Hence

$$|\mathbb{E}(YX)| \le \int_0^\infty \Big(\int_0^{\|\mathbb{E}(X|\mathcal{M})\|_1} \mathbf{1}_{u<\mathbb{E}(|X|\mathbf{1}_{|Y|>t})}du\Big)dt \le \int_0^{\|\mathbb{E}(X|\mathcal{M})\|_1} \Big(\int_0^\infty \mathbf{1}_{t<H_{X,Y}(u)}dt\Big)du,$$

and the first inequality in (5.2.1) is proved. In order to prove the second one we use Fréchet's inequality (1957) [88]:

$$\mathbb{E}(|X|\mathbf{1}_{|Y|>t}) \leq \int_0^{\mathbb{P}(|Y|>t)} Q_X(u)du. \tag{5.2.3}$$

We infer from (5.2.3) that $H_{X,Y}(u) \leq Q_Y \circ G_X(u)$, which yields the second inequality in (5.2.1).

We now prove (5.2.2). The first inequality in (5.2.2) follows directly from (5.2.1). To prove the second one, note that $Q_{X-\mathbb{E}(X)} \leq Q_X + \|X\|_1$ and consequently

$$\int_0^x Q_{X-\mathbb{E}(X)}(u)du \leq \int_0^x Q_X(u)du + x\|X\|_1 . \tag{5.2.4}$$

Set $R(x) = \int_0^x Q_X(u)du - x\|X\|_1$. Clearly, R' is non-increasing over $]0,1]$, $R'(\epsilon) \geq 0$ for ϵ small enough and $R'(1) \leq 0$.

We infer that R is first non-decreasing and next non-increasing, and that for any $x \in [0,1]$, $R(x) \geq \min(R(0), R(1))$. Since $\int_0^1 Q_X(u)du = \|X\|_1$, we have that $R(1) = R(0) = 0$ and we infer from (5.2.4) that

$$\int_0^x Q_{X-\mathbb{E}(X)}(u)du \leq \int_0^x Q_X(u)du + x\|X\|_1 \leq 2\int_0^x Q_X(u)du .$$

This implies $G_{X-\mathbb{E}(X)}(u) \geq G_X(u/2)$ which concludes the proof of (5.2.2). □
Combining Proposition 5.2 and the comparison results (2.2.13), (2.2.18) and Item 2. of Lemma 5.1, we easily derive covariance inequalities for θ_1, τ_1 and $\tilde{\alpha}$.

Corollary 5.1. *Let $(\Omega, \mathcal{A}, \mathbb{P})$ be a probability space. Let X be a real valued integrable random variable, and $\mathcal{M}$ be a $\sigma-$algebra of $\mathcal{A}$. We have that*

$$|\mathrm{Cov}(Y,X)| \leq 2\int_0^{\tilde{\alpha}(\mathcal{M},X)} Q_Y(u)Q_X(u)du . \tag{5.2.5}$$

Proof of Corollary 5.1. To prove (5.2.5), put $z = G_X(u)$ in the second integral of (5.2.2), and we use the results of comparison (2.2.13), (2.2.18) and Item 2. of Lemma 5.1. □

5.2.2 A covariance inequality for $\tilde{\beta}$ and $\tilde{\phi}$

Proposition 5.3. *Let X and Y be two real-valued random variables on the probability space $(\Omega, \mathcal{A}, \mathbb{P})$. Let $F_{X|Y} : t \mapsto \mathbb{P}_{X|Y}(]-\infty, t])$ be a distribution function of X given Y and let F_X be the distribution function of X. Define the* *random variable* $b(\sigma(Y), X) = \sup_{x\in\mathbb{R}} |F_{X|Y}(x) - F_X(x)|$. *For any conjugate exponents* p *and* q, *we have the inequalities*

$$|\mathrm{Cov}(Y,X)| \leq 2\{\mathbb{E}(|X|^p b(\sigma(X),Y))\}^{\frac{1}{p}}\{\mathbb{E}(|Y|^q b(\sigma(Y),X))\}^{\frac{1}{q}} \quad (5.2.6)$$

$$\leq 2\tilde{\phi}(\sigma(X),Y)^{\frac{1}{p}}\tilde{\phi}(\sigma(Y),X)^{\frac{1}{q}}\|X\|_p\|Y\|_q\,. \quad (5.2.7)$$

Remark 5.2. *Inequality (5.2.6) is a weak version of that of Delyon (1990) [57] (see also Viennet (1997) [185], Lemma 4.1) in which appear two random variables* $b_1(\sigma(Y),\sigma(X))$ *and* $b_2(\sigma(X),\sigma(Y))$ *each having mean* $\beta(\sigma(Y),\sigma(X))$. *Inequality (5.2.7) is a weak version of that of Peligrad (1983) [139], where the dependence coefficients are* $\phi(\sigma(Y),\sigma(X))$ *and* $\phi(\sigma(X),\sigma(Y))$.

Proof of Proposition 5.3. We start from the equality

$$\mathrm{Cov}(Y,X) = \int_0^\infty\int_0^\infty \mathrm{Cov}(\mathbf{1}_{X>x} - \mathbf{1}_{X<-x}, \mathbf{1}_{Y>y} - \mathbf{1}_{Y<-y})\,dx\,dy\,. \quad (5.2.8)$$

Since $\|b(\sigma(Y),X)\|_\infty = \tilde{\phi}(\sigma(Y),X)$, we only need to prove (5.2.6). Here, note that the value of $b(\sigma(Y),X)$ does not change if we replace $F_{X|Y}(x)$ and $F_X(x)$ by $P_{X|Y}(]-\infty,x[)$ and $P_X(]-\infty,x[)$ respectively. Consequently, the following inequalities hold:

$$\big|\mathbb{E}(\mathbf{1}_{X>x}\mathbf{1}_{Y>y} - \mathbb{P}(X>x)\mathbb{P}(Y>y))\big| \leq \mathbb{E}\big(\mathbf{1}_{Y>y}b(\sigma(Y),X)\big)\wedge\mathbb{E}\big(\mathbf{1}_{X>x}b(\sigma(X),Y)\big)$$

$$\big|\mathbb{E}(\mathbf{1}_{X<-x}\mathbf{1}_{Y>y} - \mathbb{P}(X<-x)\mathbb{P}(Y>y))\big| \leq \mathbb{E}\big(\mathbf{1}_{Y>y}b(\sigma(Y),X)\big)\wedge\mathbb{E}\big(\mathbf{1}_{X<-x}b(\sigma(X),Y)\big)$$

$$\big|\mathbb{E}(\mathbf{1}_{X>x}\mathbf{1}_{Y<-y} - \mathbb{P}(X>x)\mathbb{P}(Y<-y))\big| \leq \mathbb{E}\big(\mathbf{1}_{Y<-y}b(\sigma(Y),X)\big)\wedge\mathbb{E}\big(\mathbf{1}_{X>x}b(\sigma(X),Y)\big)$$

$$\big|\mathbb{E}(\mathbf{1}_{X<-x}\mathbf{1}_{Y<-y} - \mathbb{P}(X<-x)\mathbb{P}(Y<-y))\big| \leq \mathbb{E}\big(\mathbf{1}_{Y<-y}b(\sigma(Y),X)\big)\wedge\mathbb{E}\big(\mathbf{1}_{X<-x}b(\sigma(X),Y)\big).$$

Since $a_1\wedge b_1 + a_1\wedge b_2 + a_2\wedge b_1 + a_2\wedge b_2 \leq 2(a_1+a_2)\wedge(b_1+b_2)$, we infer from (5.2.8) that

$$|\mathrm{Cov}(Y,X)| \leq 2\int_0^\infty\int_0^\infty \mathbb{E}(\mathbf{1}_{|X|>x}b(\sigma(X),Y))\wedge\mathbb{E}(\mathbf{1}_{|Y|>y}b(\sigma(Y),X))\,dx\,dy\,. \quad (5.2.9)$$

Let $\beta_1 = \tilde{\beta}(\sigma(X), Y)$ and $\beta_2 = \tilde{\beta}(\sigma(Y), X)$. Note β_1 (or β_2) is 0 if and only if X is independent of Y, and then (5.2.6) is true in that case. Otherwise, let $\mathbb{P}_{\beta_1}$, $\mathbb{P}_{\beta_2}$ be the probabilities with density $b(\sigma(X), Y)/\beta_1$ and $b(\sigma(Y), X)/\beta_2$ with respect to $\mathbb{P}$. The inequality (5.2.9) writes

$$|\mathrm{Cov}(Y, X)| \leq 2 \int_0^\infty \int_0^\infty \beta_1 \mathbb{P}_{\beta_1}(|X| > x) \wedge \beta_2 \mathbb{P}_{\beta_2}(|Y| > y)\, dx\, dy\,. \tag{5.2.10}$$

Let $Q_{\beta_1,X}$ et $Q_{\beta_2,Y}$ be the generalized inverse of $x \mapsto \mathbb{P}_{\beta_1}(|X| > x)$ and $y \mapsto \mathbb{P}_{\beta_2}(|Y| > y)$. Starting from (5.2.10), we have successively

$$\begin{aligned} |\mathrm{Cov}(Y, X)| &\leq 2 \int_0^\infty \int_0^\infty \int_0^{\beta_1 \wedge \beta_2} \mathbf{1}_{u < \beta_1 \mathbb{P}_{\beta_1}(|X| > x)} \mathbf{1}_{u < \beta_2 \mathbb{P}_{\beta_2}(|Y| > y)}\, du\, dx\, dy \\ &\leq 2 \int_0^\infty \int_0^\infty \int_0^{\beta_1 \wedge \beta_2} \mathbf{1}_{Q_{\beta_1,X}(u/\beta_1) > x} \mathbf{1}_{Q_{\beta_2,Y}(u/\beta_2) > y}\, du\, dx\, dy \\ &\leq 2 \int_0^{\beta_1 \wedge \beta_2} Q_{\beta_1,X}(u/\beta_1) Q_{\beta_2,Y}(u/\beta_2)\, du\,. \end{aligned}$$

Applying Hölder's inequality, and setting $s = u/\beta_1$ and $t = u/\beta_2$, we obtain

$$|\mathrm{Cov}(Y, X)| \leq 2 \left(\int_0^1 \beta_1 Q^p_{\beta_1,X}(s)\, ds \right)^{1/p} \left(\int_0^1 \beta_2 Q^q_{\beta_2,Y}(t)\, dt \right)^{1/q}.$$

To complete the proof of (5.2.6), note that, by definition of $Q_{\beta_1,X}$,

$$\int_0^1 \beta_1 Q^p_{\beta_1,X}(s)\, ds = \mathbb{E}(|X|^p b(\sigma(X), Y))\,. \quad \square$$

Corollary 5.2. *Let f_1, f_2, g_1, g_2 be four increasing functions, and let $f = f_1 - f_2$ et $g = g_1 - g_2$. For any random variable Z, let $\Delta_p(Z) = \inf_{a \in \mathbb{R}} \|Z - a\|_p$ and $\Delta_{p,\sigma(X),Y}(Z) = \inf_{a \in \mathbb{R}} (\mathbb{E}(|Z - a|^p b(\sigma(X), Y)))^{1/p}$. For any conjugate exponents p and q, we have the inequalities*

$$\begin{aligned} |\mathrm{Cov}(g(Y), f(X))| \leq 2 \big\{ &\Delta_{p,\sigma(X),Y}(f_1(X)) + \Delta_{p,\sigma(X),Y}(f_2(X)) \big\} \\ &\times \big\{ \Delta_{q,\sigma(Y),X}(g_1(Y)) + \Delta_{q,\sigma(Y),X}(g_2(Y)) \big\}\,, \end{aligned}$$

$$\begin{aligned} |\mathrm{Cov}(g(Y), f(X))| \leq\ & 2\phi(\sigma(X), Y)^{\frac{1}{p}} \phi(\sigma(Y), X)^{\frac{1}{q}} \\ &\times \big\{ \Delta_p(f_1(X)) + \Delta_p(f_2(X)) \big\} \big\{ \Delta_q(g_1(Y)) + \Delta_q(g_2(Y)) \big\}\,. \end{aligned}$$

In particular, if μ is a signed measure with total variation $\|\mu\|$ and $f(x) = \mu(]-\infty, x])$, we have

$$|\mathrm{Cov}(Y, f(X))| \leq \|\mu\| \mathbb{E}(|Y| b(\sigma(Y), X)) \leq \tilde{\phi}(\sigma(Y), X) \|\mu\|\, \|Y\|_1\,. \tag{5.2.11}$$

Proof of Corollary 5.2. For the two first inequalities, note that, for any a_1, a_2, b_1, b_2,

$$\begin{aligned}|\mathrm{Cov}(g(Y), f(X))| \leq\ & |\mathrm{Cov}(g_1(Y) - b_1, f_1(X) - a_1)| \\ & + |\mathrm{Cov}(g_1(Y) - b_1, f_2(X) - a_2)| \\ & + |\mathrm{Cov}(g_2(Y) - b_2, f_1(X) - a_1)| \\ & + |\mathrm{Cov}(g_2(Y) - b_2, f_2(X) - a_2)| \,.\end{aligned}$$

the functions $f_1 - a_1, f_2 - a_2, g_1 - b_1, g_2 - b_2$ being nondecreasing, we infer that $b(\sigma(f_i(X)), g_j(Y) - b_j) \leq \mathbb{E}(b(\sigma(X), Y)|\sigma(f_i(X)))$ almost surely. Now, apply (5.2.6) and (5.2.7), and take the infimum over a_1, b_1, a_2, b_2.
To show (5.2.11), we take $q = 1$ and $p = \infty$. Let $\mu = \mu_+ - \mu_-$ be the Jordan decomposition of μ.
We have $f(x) = f_1(x) - f_2(x)$, with $f_1(x) = \mu_+(]-\infty, x])$ and $f_2(x) = \mu_-(]-\infty, x])$. To conclude, apply the preceding inequalities and note that $\|\mu\| = \mu_+(\mathbb{R}) + \mu_-(\mathbb{R})$ and $\Delta_{1,\sigma(Y),X}(Y) \leq \mathbb{E}(|Y|b(\sigma(Y), X))$, $2\Delta_\infty(f_1(X)) \leq \mu_+(\mathbb{R})$, and $2\Delta_\infty(g_1(X)) \leq \mu_-(\mathbb{R})$. □

5.3 Coupling

There exist several methods to obtain limit theorems for sequences of dependent random variables. One of the most popular and useful is the coupling of the initial sequence with an independent one. The main result in Section 5.3.1 is a coupling result (Dedecker and Prieur (2004) [45]) allowing to replace a sequence of τ_1-dependent random variables by an independent one, having the same marginals. Moreover, a variable in the newly constructed sequence is independent of the past of the initial one and it is close, for the $\mathbb{L}_1$ norm, to the variable having the same rank. The price to pay to replace the initial sequence by an independent one depends on the τ_1-dependence properties of the initial sequence.
Various approaches to coupling have been developed by different authors. We refer to a recent paper by Merlevède and Peligrad (2002) [130] for a survey on the various coupling methods and their applications. The approach used in this chapter lies on the quantile transform of Major (1978) [126]. It has been used for strongly mixing sequences by Rio (1995) [159] and Peligrad (2002) [140] to obtain a coupling result in $\mathbb{L}^1$.
In Section 5.3.2, the coupling result of Section 5.3.1 is generalized to the case of variables with values in any Polish space $\mathcal{X}$.

5.3.1 A coupling result for real valued random variables

The main result of this section is a coupling result for the coefficient τ_1, which appears to be the appropriate coefficient for coupling in $\mathbb{L}^1$.
We first recall the nice coupling properties known for the usual mixing coefficients. Berbee (1979) [16] and Goldstein (1979) [96] proved: if Ω is rich enough, there exists a random variable X^* distributed as X and independent of $\mathcal{M}$ such that $\mathbb{P}(X \neq X^*) = \beta(\mathcal{M}, \sigma(X))$. For the mixing coefficient $\alpha(\mathcal{M}, \sigma(X))$, Bradley (1983) [29] proved the following result: if Ω is rich enough, then for each $1 \leq p \leq \infty$ and each $\lambda < \|X\|_p$, there exists X^* distributed as X and independent of $\mathcal{M}$ such that

$$\mathbb{P}(|X - X^*| \geq \lambda) \leq 18 \left(\frac{\|X\|_p}{\lambda}\right)^{p/(2p+1)} (\alpha(\mathcal{M}, \sigma(X)))^{2p/(2p+1)}.$$

For the weaker coefficient $\tilde{\alpha}(\mathcal{M}, X)$, Rio (1995, 2000) [159, 160] obtained the following upper bound, which is not directly comparable to Bradley's: if X belongs to $[a, b]$ and if Ω is rich enough, there exists X^* independent of $\mathcal{M}$ and distributed as X such that $\|X - X^*\|_1 \leq (b - a)\tilde{\alpha}(\mathcal{M}, X)$. This result has then been extended by Peligrad (2002) [140] to the case of unbounded variables.
Recall that the random variable X^* appearing in the results by Rio (1995, 2000) [159, 160] and Peligrad (2002) [140] is based on Major's quantile transformation (1978) [126]. X^* has the following remarkable property: $\|X - X^*\|_1$ is the infimum of $\|X - Y\|_1$ where Y is independent of $\mathcal{M}$ and distributed as X. Starting from the exact expression of X^*, Dedecker and Prieur (2004) [45] proved that $\tau_1(\mathcal{M}, X)$ is the appropriate coefficient for the coupling in $\mathbb{L}^1$ (see Lemma 5.2 below).

Lemma 5.2. *Let $(\Omega, \mathcal{A}, \mathbb{P})$ be a probability space, X an integrable real-valued random variable, and $\mathcal{M}$ a σ-algebra of $\mathcal{A}$. Assume that there exists a random variable δ uniformly distributed over $[0, 1]$, independent of the σ-algebra generated by X and $\mathcal{M}$. Then there exists a random variable X^*, measurable with respect to $\mathcal{M} \vee \sigma(X) \vee \sigma(\delta)$, independent of $\mathcal{M}$ and distributed as X, such that*

$$\|X - X^*\|_1 = \tau_1(\mathcal{M}, X). \tag{5.3.1}$$

This coupling result is a useful tool to obtain suitable inequalities, to prove various limit theorems and to obtain upper bounds for $\tilde{\beta}(\mathcal{M}, X)$ (see Dedecker and Prieur (2004) [48]).

Remark 5.3. *From Berbee's lemma and Lemma 5.2 above, we see that both $\beta(\mathcal{M}, \sigma(X))$ and $\tau_1(\mathcal{M}, X)$ have a property of optimality: they are equal to the infimum of $\mathbb{E}(d_0(X, Y))$ where Y is independent of $\mathcal{M}$ and distributed as X, for the distances $d_0(x, y) = \mathbf{1}_{x \neq y}$ and $d_0(x, y) = |x - y|$ respectively.*

Proof of Lemma 5.2. We first construct X^* using the conditional quantile transform of Major (1978), see (5.1.3) in the proof of Lemma 5.1. The choice of this transform is due to its property to minimize the distance between X and X^* in $\mathbb{L}^1(\mathbb{R})$.
We recall equality (5.1.4):

$$\|X - X^*\|_1 = \mathbb{E}\Big(\int_0^1 |F_{\mathcal{M}}^{-1}(u) - F^{-1}(u)|du\Big)\,. \tag{5.3.2}$$

For two distribution functions F and G, denote by $M(F,G)$ the set of all probability measures on $\mathbb{R} \times \mathbb{R}$ with marginals F and G. Define

$$d(F,G) = \inf\Big\{\int |x-y|\mu(dx,dy)\Big/ \mu \in M(F,G)\Big\}\,,$$

and recall that (see Dudley (1989) [80], Section 11.8, Problems 1 and 2 page 333)

$$d(F,G) = \int_{\mathbb{R}} |F(t) - G(t)|dt = \int_0^1 |F^{-1}(u) - G^{-1}(u)|du\,. \tag{5.3.3}$$

On the other hand, Kantorovich and Rubinstein (Theorem 11.8.2 in Dudley (1989) [80]) have proved that

$$d(F,G) = \sup\Big\{\Big|\int f dF - \int f dG\Big| \Big/ f \in \Lambda^{(1)}(\mathbb{R})\Big\}. \tag{5.3.4}$$

Combining (5.1.4), (5.3.3) and (5.3.4), we have that

$$\|X - X^*\|_1 = \mathbb{E}\Big(\sup\Big\{\Big|\int f dF_{\mathcal{M}} - \int f dF\Big| \Big/ f \in \Lambda^{(1)}(\mathbb{R})\Big\}\Big)\,,$$

and the proof of Lemma 5.2 is complete. □

5.3.2 Coupling in higher dimension

We show that the coupling properties of τ_1 described in the previous section remain valid when $\mathcal{X}$ is any Polish space. This is due to a conditional version of the Kantorovich Rubinstein Theorem (Theorem 5.1).

Lemma 5.3. *Let $(\Omega, \mathcal{A}, \mathbb{P})$ be a probability space, $\mathcal{M}$ a σ-algebra of $\mathcal{A}$ and X a random variable with values in a Polish space $(\mathcal{X}, d)$. Assume that $\int d(x, x_0)$ $\mathbb{P}_X(dx)$ is finite for some (and therefore any) $x_0 \in \mathcal{X}$. Assume that there exists a random variable δ uniformly distributed over $[0,1]$, independent of the σ-algebra generated by X and $\mathcal{M}$. Then there exists a random variable X^*, measurable with respect to $\mathcal{M} \vee \sigma(X) \vee \sigma(\delta)$, independent of $\mathcal{M}$ and distributed as X, such that*

$$\tau_p(\mathcal{M}, X) = \|\mathbb{E}(d(X, X^*)|M)\|_p. \tag{5.3.5}$$

We first need some notations. We denote $\mathcal{P}(\Omega)$ the set of probability measures on a space Ω and define

$$\mathcal{Y}(\Omega, \mathcal{A}, \mathbb{P}; \mathcal{X}) = \left\{\mu \in \mathcal{P}(\Omega \times \mathcal{X}), \mathcal{A} \otimes \mathcal{B}_{\mathcal{X}} \Big/ \forall A \in \mathcal{A},\, \mu(A \times \mathcal{X}) = \mathbb{P}(A)\right\}.$$

Recall that every $\mu \in \mathcal{Y}(\Omega, \mathcal{A}, \mathbb{P}; \mathcal{X})$ is *disintegrable*, that is, there exists a (unique, up to $\mathbb{P}$-a.s. equality) $\mathcal{A}_\mu^*$-measurable mapping $\omega \mapsto \mu_\omega$, $\Omega \to \mathcal{P}(\mathcal{X})$, such that

$$\mu(f) = \int_\Omega \int_{\mathcal{X}} f(\omega, x)\, d\mu_\omega(x)\, d\mathbb{P}(\omega)$$

for every measurable $f : \Omega \times \mathcal{X} \to [0, +\infty]$ (see Valadier (1973) [184]). Moreover, the mapping $\omega \mapsto \mu_\omega$ can be chosen $\mathcal{A}$-measurable. If $\mathcal{X}$ is endowed with the distance d, let denote

$$\mathcal{Y}^{d,1}(\Omega, \mathcal{A}, \mathbb{P}; \mathcal{X}) = \left\{\mu \in \mathcal{Y} \Big/ \int_{\Omega \times \mathcal{X}} d(x, x_0)\, d\mu(\omega, x) < \infty\right\},$$

where x_0 is some fixed element of $\mathcal{X}$ (this definition is independent of the choice of x_0). For any $\mu, \nu \in \mathcal{Y}$, let $\underline{D}(\mu, \nu)$ be the set of probability laws π on $\Omega \times \mathcal{X} \times \mathcal{X}$ such that $\pi(\cdot \times \cdot \times \mathcal{X}) = \mu$ and $\pi(\cdot \times \mathcal{X} \times \cdot) = \nu$. We now define the parametrized versions of $\Delta_{\mathrm{KR}}^{(d)}$ and $\Delta_{\mathrm{L}}^{(d)}$. Set, for $\mu, \nu \in \mathcal{Y}^{d,1}$,

$$\underline{\Delta}_{\mathrm{KR}}^{(d)}(\mu, \nu) = \inf_{\pi \in \underline{D}(\mu, \nu)} \int_{\Omega \times \mathcal{X} \times \mathcal{X}} d(x, y)\, d\pi(\omega, x, y).$$

Let also $\underline{\Lambda^{(1)}}$ denote the set of measurable integrands $f : \Omega \times \mathcal{X} \to \mathbb{R}$ such that $f(\omega, \cdot) \in \Lambda^{(1)} \cap \mathbb{L}^\infty$ for every $\omega \in \Omega$. We denote

$$\underline{\Delta}_{\mathrm{L}}^{(d)}(\mu, \nu) = \sup_{f \in \underline{\Lambda^{(1)}}} (\mu(f) - \nu(f)).$$

We now state the parametrized Kantorovich-Rubinstein Theorem of Dedecker, Prieur and Raynaud De Fitte (2004) [47], which is the main tool to prove the coupling result of Lemma 5.3. The proof of that theorem mainly relies on ideas contained in Rüschendorf (1985) [171].

Theorem 5.1. (Parametrized Kantorovich–Rubinstein Theorem) *Let $\mu, \nu \in \mathcal{Y}^{d,1}(\Omega, \mathcal{A}, \mathbb{P}; \mathcal{X})$ and let $\omega \mapsto \mu_\omega$ and $\omega \mapsto \nu_\omega$ be disintegrations of μ and ν respectively.*

1. *Let $G : \omega \mapsto \Delta_{\mathrm{KR}}^{(d)}(\mu_\omega, \nu_\omega) = \Delta_{\mathrm{L}}^{(d)}(\mu_\omega, \nu_\omega)$ and let $\mathcal{A}^*$ be the universal completion of $\mathcal{A}$. There exists an $\mathcal{A}^*$–measurable mapping $\omega \mapsto \lambda_\omega$ from Ω to $\mathcal{P}(\mathcal{X} \times \mathcal{X})$ such that λ_ω belongs to $D(\mu_\omega, \nu_\omega)$ and*

$$G(\omega) = \int_{\mathcal{X} \times \mathcal{X}} d(x, y)\, d\lambda_\omega(x, y).$$

2. The following equalities hold:

$$\underline{\Delta}_{\mathrm{KR}}^{(d)}(\mu,\nu) = \int_{\Omega\times\mathcal{X}\times\mathcal{X}} d(x,y)\,d\lambda(\omega,x,y) = \underline{\Delta}_{\mathrm{L}}^{(d)}(\mu,\nu),$$

where λ is the element of $\mathcal{Y}(\Omega,\mathcal{A},\mathbb{P};\mathcal{X}\times\mathcal{X})$ defined by $\lambda(A\times B\times C) = \int_A \lambda_\omega(B\times C)\,d\mathbb{P}(\omega)$ for any A in $\mathcal{A}$, B and C in $\mathcal{B}_{\mathcal{X}}$. In particular, λ belongs to $\underline{D}(\mu,\nu)$, and the infimum in the definition of $\underline{\Delta}_{\mathrm{KR}}^{(d)}(\mu,\nu)$ is attained for this λ.

Remark 5.4. *This theorem is proved in a more general frame in Dedecker, Prieur and Raynaud De Fitte (2004) [47]. It also allows more general coupling results when working with more general cost functions than the metric d of $\mathcal{X}$.*

Proof of Lemma 5.3. First notice that the assumption that $\int d(x,x_0)\mathbb{P}_X(dx) < \infty$ for some $x_0 \in \mathcal{X}$ means that the unique measure of $\mathcal{Y}(\Omega,\mathcal{A},\mathbb{P};\mathcal{X}\times\mathcal{X})$ with disintegration $\mathbb{P}_{X|\mathcal{M}}(.,\omega)$ belongs to $\mathcal{Y}^{d,1}(\Omega,\mathcal{A},\mathbb{P};\mathcal{X})$. We now prove that if Q is any element of $\mathcal{Y}^{d,1}(\Omega,\mathcal{A},\mathbb{P};\mathcal{X})$, there exists a $\sigma(\delta)\vee\sigma(X)\vee\mathcal{X}$–measurable random variable Y such that $Q_\bullet$ is a regular conditional probability of Y given $\mathcal{M}$, and

$$\mathbb{E}\big(d(X,Y)\big|\,\mathcal{M}\big) = \sup_{f\in\Lambda^{(1)}\cap\mathbb{L}^\infty}\left|\int f(x)\mathbb{P}_{X|\mathcal{M}}(dx) - \int f(x)Q_\bullet(dx)\right| \quad \mathbb{P}\text{-a.s.} \tag{5.3.6}$$

Applying (5.3.6) with $Q = \mathbb{P}\otimes\mathbb{P}_X$, we get the result of Lemma 5.3. □

Proof of Equation (5.3.6). We apply Theorem 5.1 to the probability space $(\Omega,\mathcal{M},\mathbb{P})$ and to the disintegrated measures $\mu_\omega(\cdot) = \mathbb{P}_{X|\mathcal{M}}(\cdot,\omega)$ and $\nu_\omega = Q_\omega$. From point 1 of Theorem 5.1 we infer that there exists a mapping $\omega\mapsto\lambda_\omega$ from Ω to $\mathcal{P}(\mathcal{X}\times\mathcal{X})$, measurable for $\mathcal{M}^*$ and $\mathcal{B}_{\mathcal{P}(\mathcal{X}\times\mathcal{X})}$, such that λ_ω belongs to $D(\mathbb{P}_{X|\mathcal{M}}(\cdot,\omega),Q_\omega)$ and $G(\omega) = \int_{\mathcal{X}\times\mathcal{X}} d(x,y)\lambda_\omega(dx,dy)$. On the measurable space $(\mathbb{M},\mathcal{T}) = (\Omega\times\mathcal{X}\times\mathcal{X},\mathcal{M}^*\otimes\mathcal{B}_{\mathcal{X}}\otimes\mathcal{B}_{\mathcal{X}})$ we put the probability

$$\pi(A\times B\times C) = \int_A \lambda_\omega(B\times C)\mathbb{P}(d\omega)\,.$$

If $I = (I_1,I_2,I_3)$ is the identity on $\mathbb{M}$, we see that a regular conditional distribution of (I_2,I_3) given I_1 is $\mathbb{P}_{(I_2,I_3)|I_1=\omega} = \lambda_\omega$. Since $\mathbb{P}_{X|\mathcal{M}}(\cdot,\omega)$ is the first margin of λ_ω, a regular conditional probability of I_2 given I_1 is $\mathbb{P}_{I_2|I_1=\omega}(\cdot) = \mathbb{P}_{X|\mathcal{M}}(\cdot,\omega)$. Let $\lambda_{\omega,x} = \mathbb{P}_{I_3|I_1=\omega,I_2=x}$ be a regular conditional distribution of I_3 given (I_1,I_2), so that $(\omega,x)\mapsto\lambda_{\omega,x}$ is measurable for $\mathcal{M}^*\otimes\mathcal{B}_{\mathcal{X}}$ and $\mathcal{B}_{\mathcal{P}(\mathcal{X})}$. From the unicity (up to $\mathbb{P}$-a.s. equality) of regular conditional probabilities, it follows that

$$\lambda_\omega(B\times C) = \int_B \lambda_{\omega,x}(C)\mathbb{P}_{X|\mathcal{M}}(dx,\omega) \quad \mathbb{P}\text{-a.s.} \tag{5.3.7}$$

Assume that we can find a random variable $\tilde{Y}$ from Ω to $\mathcal{X}$, measurable for $\sigma(U) \vee \sigma(X) \vee \mathcal{M}^*$ and $\mathcal{B}_{\mathcal{X}}$, such that $\mathbb{P}_{\tilde{Y}|\sigma(X)\vee\mathcal{M}^*}(\cdot,\omega) = \lambda_{\omega,X(\omega)}(\cdot)$. Since $\omega \mapsto \mathbb{P}_{X|\mathcal{M}}(\cdot,\omega)$ is measurable for $\mathcal{M}^*$ and $\mathcal{B}_{\mathcal{P}(\mathcal{X})}$, one can check that $\mathbb{P}_{X|\mathcal{M}}$ is a regular conditional probability of X given $\mathcal{M}^*$. For A in $\mathcal{M}^*$, B and C in$\mathcal{B}_{\mathcal{X}}$, we thus have

$$\begin{aligned} E\left(\mathbf{1}_A \mathbf{1}_{X\in B} \mathbf{1}_{\tilde{Y}\in C}\right) &= E\left(\mathbf{1}_A E\left(\mathbf{1}_{X\in B} E\left(\mathbf{1}_{\tilde{Y}\in C}|\sigma(X)\vee\mathcal{M}^*\right)|\mathcal{M}^*\right)\right) \\ &= \int_A \Big(\int_B \lambda_{\omega,x}(C)\mathbb{P}_{X|\mathcal{M}}(dx,\omega)\Big)\mathbb{P}(d\omega) \\ &= \int_A \lambda_\omega(B\times C)\mathbb{P}(d\omega)\,. \end{aligned}$$

We infer that λ_ω is a regular conditional probability of $(X,\tilde{Y})$ given $\mathcal{M}^*$. By definition of λ_ω, we obtain that

$$E\left(d(X,\tilde{Y})|\mathcal{M}^*\right) = \sup_{f\in\Lambda^{(1)}\cap\mathbb{L}^\infty} \left|\int f(x)\mathbb{P}_{X|\mathcal{M}}(dx) - \int f(x)Q_\bullet(dx)\right| \quad \mathbb{P}\text{-a.s.} \tag{5.3.8}$$

As $\mathcal{X}$ is Polish, there exists a $\sigma(\delta)\vee\sigma(X)\vee\mathcal{M}$-measurable modification Y of $\tilde{Y}$, so that (5.3.8) still holds for $\mathbb{E}(d(X,Y)|\mathcal{M}^*)$. We obtain (5.3.6) by noting that $E\left(d(X,Y)|\mathcal{M}^*\right) = E\left(d(X,Y)|\mathcal{M}\right)$ $\mathbb{P}$-a.s. It remains to build $\tilde{Y}$. Since $\mathcal{X}$ is Polish, there exists a one to one map f from $\mathcal{X}$ to a Borel subset of $[0,1]$, such that f and f^{-1} are measurable for $\mathcal{B}([0,1])$ and $\mathcal{B}_{\mathcal{X}}$. Define $F(t,\omega) = \lambda_{\omega,X(\omega)}(f^{-1}([0,t]))$. The map $F(\cdot,\omega)$ is a distribution function with generalized inverse $F^{-1}(\cdot,\omega)$ and the map $(u,\omega) \mapsto F^{-1}(u,\omega)$ is $\mathcal{B}([0,1])\otimes\mathcal{M}^*\vee\sigma(X)$-measurable. Let $T(\omega) = F^{-1}(\delta(\omega),\omega)$ and $\tilde{Y} = f^{-1}(T)$. It remains to see that $\mathbb{P}_{\tilde{Y}|\sigma(X)\vee\mathcal{M}^*}(\cdot,\omega) = \lambda_{\omega,X(\omega)}(\cdot)$. For any A in $\mathcal{M}^*$, B in $\mathcal{B}_{\mathcal{X}}$ and t in $\mathbb{R}$, we have

$$E\left(\mathbf{1}_A\mathbf{1}_{X\in B}\mathbf{1}_{\tilde{Y}\in f^{-1}([0,t])}\right) = \int_A \mathbf{1}_{X(\omega)\in B}\mathbf{1}_{\delta(\omega)\le F(t,\omega)}\mathbb{P}(d\omega).$$

Since δ is independent of $\sigma(X)\vee\mathcal{M}$, it is also independent of $\sigma(X)\vee\mathcal{M}^*$. Hence

$$\begin{aligned} E\left(\mathbf{1}_A\mathbf{1}_{X\in B}\mathbf{1}_{\tilde{Y}\in f^{-1}([0,t])}\right) &= \int_A \mathbf{1}_{X(\omega)\in B}F(t,\omega)\mathbb{P}(d\omega) \\ &= \int_A \mathbf{1}_{X(\omega)\in B}\lambda_{\omega,X(\omega)}(f^{-1}([0,t]))\mathbb{P}(d\omega). \end{aligned}$$

Since $\{f^{-1}([0,t])/t\in[0,1]\}$ is a separating class, the result follows. □

5.4 Exponential and Moment inequalities

The first theorem of this section extends Bennett's inequality for independent sequences to the case of τ_1-dependent sequences. For any positive integer q, we obtain an upper bound involving two terms: the first one is the classical Bennett's bound at level λ for a sum $\sum_{i=1}^{n} \xi_i$ of independent variables ξ_i such that $\mathrm{Var}\left(\sum_{i=1}^{n} \xi_i\right) = v_q$ and $\|\xi_i\|_\infty \leq qM$, and the second one is equal to $n\lambda^{-1} \tau_q(q+1)$. Using Item 2. of Lemma 5.1, we obtain the same inequalities as those established by Rio (2000) [161] for strongly mixing sequences. This is not surprising, we follow the proof of Rio and we use Lemma 5.2 instead of Rio's coupling lemma. Note that the same approach has been previously used by Bosq (1993) [26], starting from Bradley's coupling lemma (1983) [29]. Theorem 5.2 and Theorem 5.3 below are due to Dedecker and Prieur (2004) [45].

5.4.1 Bennett-type inequality

Theorem 5.2. *Let $(X_i)_{i>0}$ be a sequence of real-valued random variables such that $\|X_i\|_\infty \leq M$, and $\mathcal{M}_i = \sigma(X_k, 1 \leq k \leq i)$. Let $S_k = \sum_{i=1}^{k}(X_i - \mathbb{E}(X_i))$ and $\overline{S}_n = \max_{1\leq k\leq n} |S_k|$. Let q be some positive integer, v_q some nonnegative number such that*

$$v_q \geq \|X_{q[n/q]+1} + \cdots + X_n\|_2^2 + \sum_{i=1}^{[n/q]} \|X_{(i-1)q+1} + \cdots + X_{iq}\|_2^2 .$$

and h the function defined by $h(x) = (1+x)\log(1+x) - x$.

1. *For $\lambda > 0$, $\mathbb{P}(|S_n| \geq 3\lambda) \leq 4\exp\Big(-\frac{v_q}{(qM)^2} h\Big(\frac{\lambda qM}{v_q}\Big)\Big) + \frac{n}{\lambda}\tau_{1,q}(q+1)$.*

2. *For $\lambda \geq Mq$,*

$$\mathbb{P}(\overline{S}_n \geq (\mathbf{1}_{q>1} + 3)\lambda) \leq 4\exp\Big(-\frac{v_q}{(qM)^2} h\Big(\frac{\lambda qM}{v_q}\Big)\Big) + \frac{n}{\lambda}\tau_{1,q}(q+1) .$$

Proof of Theorem 5.2. We proceed as in Rio (2000) [161] page 83. For $1 \leq i \leq [n/q]$, define the variables $U_i = S_{iq} - S_{iq-q}$ and $U_{[n/q]+1} = S_n - S_{q[n/q]}$. Let $(\delta_j)_{1\leq j\leq [n/q]+1}$ be independent random variables uniformly distributed over $[0,1]$ and independent of $(U_i)_{1\leq j\leq [n/q]+1}$. We apply Lemma 5.2: For any $1 \leq i \leq [n/q]+1$, there exists a measurable function F_i such that $U_i^* = F_i(U_1, \ldots, U_{i-2}, U_i, \delta_i)$ satisfies the conclusions of Lemma 5.2, with $\mathcal{M} = \sigma(U_l, l \leq i-2)$. The sequence $(U_i^*)_{1\leq j\leq [n/q]+1}$ has the following properties:

a. For any $1 \leq i \leq [n/q]+1$, the random variable U_i^* is distributed as U_i.

b. The variables $(U_{2i}^*)_{2\leq 2i\leq [n/q]+1}$ are independent and so are the variables $(U_{2i-1}^*)_{1\leq 2i-1\leq [n/q]+1}$.

c. Moreover $\|U_i - U_i^*\|_1 \leq \tau_1(\sigma(U_l, l \leq i-2), U_i)$.

Since for $1 \le i \le [n/q]$ we have $\tau_1(\sigma(U_l, l \le i-2), U_i) \le q\tau_{1,q}(q+1)$, we infer that

$$\begin{aligned} \text{for } 1 \le i \le [n/q], \quad \|U_i - U_i^*\|_1 &\le q\tau_{1,q}(q+1) \\ \text{and} \quad \|U_{[n/q]+1} - U_{[n/q]+1}^*\|_1 &\le (n - q[n/q])\tau_{1,n-q[n/q]}(q+1)\,. \end{aligned} \tag{5.4.1}$$

Proof of 1. Clearly

$$|S_n| \le \sum_{i=1}^{[n/q]+1} |U_i - U_i^*| + \Big|\sum_{i=1}^{([n/q]+1)/2} U_{2i}^*\Big| + \Big|\sum_{i=1}^{[n/q]/2+1} U_{2i-1}^*\Big|. \tag{5.4.2}$$

Combining (5.4.1) with the fact that $\tau_{1,n-q[n/q]}(q+1) \le \tau_{1,q}(q+1)$, we obtain

$$\mathbb{P}\Big(\sum_{i=1}^{[n/q]+1} |U_i - U_i^*| \ge \lambda\Big) \le \frac{n}{\lambda}\tau_{1,q}(q+1)\,. \tag{5.4.3}$$

The result follows by applying Bennett's inequality to the two other sums in (5.4.2). The proof of the second item is omitted. It is similar to the proof of Theorem 6.1 in Rio (2000) [161], page 83, for α-mixing sequences.

Proceeding as in Theorem 5.2, we establish Fuk-Nagaev type inequalities (see Fuk and Nagaev (1971) [89]) for sums of τ_1-dependent sequences. Applying Item 2. of Lemma 5.1, we obtain the same inequalities (up to some numerical constant) as those established by Rio (2000) [161] for strongly mixing sequences.

Notations 5.1. *For any non-increasing sequence $(\delta_i)_{i\ge 0}$ of nonnegative numbers, define $\delta^{-1}(u) = \sum_{i\ge 0} \mathbf{1}_{u<\delta_i} = \inf\{k \in \mathbb{N}/\delta_k \le u\}$. Note that δ^{-1} is the generalized inverse (see (2.2.14)) of the càdlàg function $x \mapsto \delta_{[x]}$, $[\cdot]$ denoting the integer part.*

Theorem 5.3. *Let $(X_i)_{i>0}$ be a sequence of centered and square integrable random variables, and define $(\mathcal{M}_i)_{i>0}$ and $\overline{S}_n$ as in Theorem 5.2. Let X be some positive random variable such that $Q_X \ge \sup_{k\ge 1} Q_{|X_k|}$ and*

$$s_n^2 = \sum_{i=1}^{n}\sum_{j=1}^{n} |\mathrm{Cov}(X_i, X_j)|\,.$$

Let $R = ((\tau/2)^{-1} \circ G_X^{-1})Q_X$ and $S = R^{-1}$. For any $\lambda > 0$ and $r \ge 1$,

$$\mathbb{P}(\overline{S}_n \ge 5\lambda) \le 4\Big(1 + \frac{\lambda^2}{rs_n^2}\Big)^{-r/2} + \frac{4n}{\lambda}\int_0^{S(\lambda/r)} Q_X(u)du. \tag{5.4.4}$$

Proof of Theorem 5.3. Let q be any positive integer, and $M > 0$. As for the proof of Theorem 5.2 we define $U_i = S_{iq} - S_{iq-q}$, for $1 \le i \le [n/q]$. We also define $\overline{U}_i = (U_i \wedge qM) \vee (-qM)$. Let $\varphi_M(x) = (|x| - M)_+$. Following the proof of Rio (2000) [161] for strongly mixing sequences, we first prove that

$$\overline{S}_n \le \max_{1\le j\le [n/q]} |\sum_{i=1}^{j} \overline{U}_i| + qM + \sum_{k=1}^{n} \varphi_M(X_k) . \tag{5.4.5}$$

To prove (5.4.5), we just have to notice that, if $\overline{S}_n = S_{k_0}$, then for $j_0 = [k_0/q]$,

$$\overline{S}_n \le |\sum_{i=1}^{j_0} \overline{U}_i| + \sum_{i=1}^{j_0} |U_i - \overline{U}_i| + \sum_{k=j_0+1}^{k_0} |X_k| , \tag{5.4.6}$$

and then, as φ_M is convex, that

$$\sum_{i=1}^{j_0} |U_i - \overline{U}_i| \le \sum_{k=1}^{qj_0} \varphi_M(X_k), \tag{5.4.7}$$

and, by definition of φ_M, that

$$\sum_{k=j_0+1}^{k_0} |X_k| \le (k_0 - qj_0)M + \sum_{k=j_0+1}^{k_0} \varphi_M(X_k). \tag{5.4.8}$$

Now, to be able to apply Theorem 5.2, we need to center the variables $\overline{U}_i$. Then as the random variables U_i are centered, we get

$$\begin{aligned}\max_{1\le j\le [n/q]} |\sum_{i=1}^{j} \overline{U}_i| &\le \max_{1\le j\le [n/q]} |\sum_{i=1}^{j} (\overline{U}_i - \mathbb{E}(\overline{U}_i))| + \sum_{i=1}^{[n/q]} \mathbb{E}(|U_i - \overline{U}_i|) \\ &\le \max_{1\le j\le [n/q]} |\sum_{i=1}^{j} (\overline{U}_i - \mathbb{E}(\overline{U}_i))| + \sum_{k=1}^{n} \mathbb{E}(\varphi_M(X_k)) ,\end{aligned}$$

using the convexity of φ_M. Hence we have proved that

$$\overline{S_n} \le \max_{1\le j\le [n/q]} |\sum_{i=1}^{j} (\overline{U}_i - \mathbb{E}(\overline{U}_i))| + qM + \sum_{k=1}^{n} (\mathbb{E}(\varphi_M(X_k)) + \varphi_M(X_k)). \tag{5.4.9}$$

Let us now choose the size q of the blocks and the constant of truncation M. Let $v = S(\lambda/r)$, $q = (\tau/2)^{-1} \circ G_X^{-1}(v)$ and $M = Q_X(v)$. Clearly, we have that $qM = R(v) = R(S(\lambda/r)) \le \lambda/r$. Since $M = Q_X(v)$,

$$\mathbb{P}\Big(\sum_{k=1}^{n} (\mathbb{E}(\varphi_M(X_k)) + \varphi_M(X_k)) \ge \lambda\Big) \le \frac{2n}{\lambda} \int_0^v Q_X(u)du . \tag{5.4.10}$$

We are now interested in the term $\mathbb{P}\left(\max_{1\le j\le [n/q]} |\sum_{i=1}^{j}(\overline{U}_i - \mathbb{E}(\overline{U}_i))| \ge 3\lambda\right)$. We apply Theorem 5.2 to the sequence $(\overline{U}_i - \mathbb{E}(\overline{U}_i))_{i\in\mathbb{Z}}$ with $n' = [n/q]$ and $q' = 1$. We have $\overline{U}_i = h(U_i)$ where $h : \mathbb{R} \to \mathbb{R}$ is a Lipschitz function such that $\mathrm{Lip}\,(h) \le 1$. Hence, we have $\tau_1(\sigma(\overline{U}_l, l \le i-2), \overline{U}_i) \le q\tau_{1,q}(q+1)$. Since $s_n^2 \ge \|\overline{U}_1\|_2^2 + \cdots + \|\overline{U}_{[n/q]}\|_2^2$ we obtain:

$$\mathbb{P}\Big(\max_{1\le j\le [n/q]}\Big|\sum_{i=1}^{j}(\overline{U}_i - \mathbb{E}(\overline{U}_i))\Big| \ge 3\lambda\Big) \le 4\Big(1+\frac{\lambda^2}{rs_n^2}\Big)^{-r/2} + \frac{n}{\lambda}\tau_{1,\infty}(q+1). \quad (5.4.11)$$

To conclude, let us notice that the choice of q implies that $\tau_{1,\infty}(q+1) \le 2\int_0^v Q_X(u)du$. Hence, since $qM \le \lambda$, combining (5.4.11), (5.4.10) and (5.4.9), we get the result. □

5.4.2 Burkholder's inequalities

The next result extends Theorem 2.5 of Rio (2000) [161] to non-stationary sequences.

Proposition 5.4. *Let $(X_i)_{i\in\mathbb{N}}$ be a sequence of centered and square integrable random variables, and $\mathcal{M}_i = \sigma(X_j, 0 \le j \le i)$. Define $S_n = X_1 + \cdots + X_n$ and*

$$b_{i,n} = \max_{i\le l\le n}\Big\|X_i\sum_{k=i}^{l}\mathbb{E}(X_k|\mathcal{M}_i)\Big\|_{p/2}.$$

For any $p \ge 2$, the following inequality holds

$$\|S_n\|_p \le \Big(2p\sum_{i=1}^{n} b_{i,n}\Big)^{1/2}. \quad (5.4.12)$$

Proof. We proceed as in Rio (2000) [161] pages 46-47. For any t in $[0,1]$ and $p \ge 2$, let $h_n(t) = \|S_{n-1} + tX_n\|_p^p$. Our induction hypothesis at step $n-1$ is the following: for any $k < n$

$$h_k(t) \le (2p)^{p/2}\Big(\sum_{i=1}^{k-1} b_{i,k} + tb_{k,k}\Big)^{p/2}.$$

This assumption is true at step 1. Assuming that it holds for $n-1$, we have to check it at step n. Setting $G(i,n,t) = X_i(t\mathbb{E}(X_n|\,\mathcal{M}_i) + \sum_{k=i}^{n-1}\mathbb{E}(X_k|\,\mathcal{M}_i))$ and applying Theorem (2.3) in Rio (2000) with $\psi(x) = |x|^p$, we get

$$\frac{h_n(t)}{p^2} \le \sum_{i=1}^{n-1}\int_0^1 \mathbb{E}\big(|S_{i-1}+sX_i|^{p-2}G(i,n,t)\big)ds + \int_0^t \mathbb{E}(|S_{n-1}+sX_n|^{p-2}X_n^2)ds\,. \quad (5.4.13)$$

Note that the function $t \to \mathbb{E}(|G(i,n,t)|^{p/2})$ is convex, so that for any t in $[0,1]$, $\mathbb{E}(|G(i,n,t)|^{p/2}) \le \mathbb{E}(|G(i,n,0)|^{p/2}) \vee \mathbb{E}(|G(i,n,1)|^{p/2}) \le b_{i,n}^{p/2}$. Applying Hölder's inequality, we obtain

$$\mathbb{E}\big(|S_{i-1}+sX_i|^{p-2}G(i,n,t)\big) \le (h_i(s))^{(p-2)/p}\,\|G(i,n,t)\|_{p/2} \le (h_i(s))^{(p-2)/p}\, b_{i,n}\,.$$

This bound together with (5.4.13) and the induction hypothesis yields

$$\begin{aligned} h_n(t) &\le p^2\Big(\sum_{i=1}^{n-1} b_{i,n}\int_0^1 (h_i(s))^{(p-2)/p}ds + b_{n,n}\int_0^t (h_n(s))^{(p-2)/p}ds\Big) \\ &\le p^2\Big(\sum_{i=1}^{n-1}(2p)^{\frac{p}{2}-1}b_{i,n}\int_0^1\Big(\sum_{j=1}^{i} b_{j,n}+sb_{i,n}\Big)^{\frac{p}{2}-1}ds + b_{n,n}\int_0^t (h_n(s))^{1-\frac{2}{p}}ds\Big)\,. \end{aligned}$$

Integrating with respect to s we find

$$b_{i,n}\int_0^1\Big(\sum_{j=1}^{i} b_{j,n}+sb_{i,n}\Big)^{\frac{p}{2}-1}ds = \frac{2}{p}\Big(\sum_{j=1}^{i} b_{j,n}\Big)^{\frac{p}{2}} - \frac{2}{p}\Big(\sum_{j=1}^{i-1} b_{j,n}\Big)^{\frac{p}{2}},$$

and summing in j we finally obtain

$$h_n(t) \le \Big(2p\sum_{j=1}^{n-1} b_{j,n}\Big)^{\frac{p}{2}} + p^2 b_{n,n}\int_0^t (h_n(s))^{1-\frac{2}{p}}ds\,. \tag{5.4.14}$$

Clearly the function $u(t) = (2p)^{p/2}(b_{1,n}+\cdots+tb_{n,n})^{p/2}$ solves the equation associated to Inequality (5.4.14). A classical argument ensures that $h_n(t) \le u(t)$ which concludes the proof.

Corollary 5.3. *Let $(X_i)_{i\in\mathbb{N}}$ and $(\mathcal{M}_i)_{i\in\mathbb{N}}$ be as in Proposition 5.4. Define $\gamma_{1,i} = \sup_{k\ge 0}\gamma_1(\mathcal{M}_k, X_{i+k})$, $\tilde\alpha_i = \sup_{k\ge 0}\tilde\alpha(\mathcal{M}_k, X_{i+k})$ and $\tilde\phi_i = \sup_{k\ge 0}\tilde\phi(\mathcal{M}_k, X_{i+k})$.*

1. *Let X be any random variable such that $Q_X \ge \sup_{k\ge 1} Q_{X_k}$, and let $\gamma_{1,n}^{-1}(u) = \sum_{k=0}^{n}\mathbf{1}_{u\le\lambda_{1,k}}$ and $\tilde\alpha_n^{-1}(u) = \sum_{k=0}^{n}\mathbf{1}_{u\le\tilde\alpha_k}$. For $p\ge 2$ we have the inequalities*

$$\begin{aligned} \|S_n\|_p &\le \sqrt{2pn}\Big(\int_0^{\|X\|_1}(\gamma_{1,n}^{-1}(u))^{p/2}Q_X^{p-1}\circ G_X(u)du\Big)^{1/p} \\ &\le \sqrt{2pn}\Big(\int_0^1(\tilde\alpha_n^{-1}(u))^{p/2}Q_X^{p}du\Big)^{1/p}. \end{aligned}$$

2. *Let $M_q = \sup_{k\ge 1}\|X_i\|_q$. For $q\ge p\ge 2$ we have the inequality*

$$\|S_n\|_p \le 2\Big(pM_qM_{qp/(2q-p)}\sum_{k=0}^{n-1}(n-k)\tilde\phi_k^{(q-1)/q}\Big)^{1/2}.$$

Proof of 1. Let $r = p/(p-2)$. By duality there exists an $\mathcal{M}_i$-mesurable Y such that $\|Y\|_r = 1$, and

$$b_{i,n} \leq \sum_{k=i}^{n} \mathbb{E}(|YX_i \mathbb{E}(X_k|\mathcal{M}_i)|)\,.$$

Let $\lambda_i = G_X(\gamma_{1,i})$. Applying (5.2.1) and Fréchet's inequality (1957) [88], we obtain

$$b_{i,n} \leq \sum_{k=i}^{n} \int_0^{\gamma_{1,k-i}} Q_{YX_i} \circ G_X(u)du \leq \sum_{k=i}^{n} \int_0^{\lambda_{k-i}} Q_Y(u)Q_X^2(u)du\,.$$

Using the duality once more, we get

$$b_{i,n}^{p/2} \leq \int_0^1 \Big(\sum_{k=0}^{n} \mathbf{1}_{u\leq\lambda_k}\Big)^{p/2} Q_X^p(u)du = \int_0^{\|X\|_1} (\gamma_{1,n}^{-1}(u))^{p/2} Q_X^{p-1} \circ G_X(u)du\,.$$

The first inequality follows. To prove the second one, note that $\lambda_k \leq \tilde{\alpha}_k$.

Proof of 2. First, note that $b_{i,n} \leq \sum_{k=i}^{n} \|X_i\mathbb{E}(X_k|\ \mathcal{M}_i)\|_{p/2}$. Let $r = p/(p-2)$. By duality, there exist a $\mathcal{M}_i$-measurable variable Y such that $\|Y\|_r = 1$ and

$$\|X_i\mathbb{E}(X_k|\ \mathcal{M}_i)\|_{p/2} = |\mathrm{Cov}(YX_i, X_k)|\,.$$

Applying inequality (5.2.7), and next Hölder's inequality, we obtain that

$$\|X_i\mathbb{E}(X_k|\ \mathcal{M}_i)\|_{p/2} \leq 2\tilde{\phi}_{k-i}^{(q-1)/q}\|YX_i\|_{q/(q-1)}\|X_k\|_q \leq 2M_q M_{qp/(2q-p)}\tilde{\phi}_{k-i}^{(q-1)/q}\,.$$

The result follows. □

5.4.3 Rosenthal inequalities using Rio techniques

We suppose that the sequence $(X_n)_{n\in\mathbb{N}}$ fulfills the following covariance inequality,

$$|\mathrm{Cov}(f(X_1,\ldots,X_n), g(X_1,\ldots,X_n))| \leq \sum_{i\in I}\sum_{j\in J} \|\frac{\partial f}{\partial x_i}\|_\infty \|\frac{\partial g}{\partial x_j}\|_\infty\, |\mathrm{Cov}(X_i,X_j)|\,, \tag{5.4.15}$$

for all real valued functions f and g defined on $\mathbb{R}^n$ having bounded first differentials and depending respectively on $(x_i)_{i\in I}$ and on $(x_i)_{i\in J}$ where I and J are disjoints subsets of $\mathbb{N}$. Sequences fulfilling (5.4.15) with $\sup_{i\in I}\|\frac{\partial f}{\partial x_i}\|_\infty < \infty$ and $\sup_{j\in J}\|\frac{\partial g}{\partial x_j}\|_\infty$ are κ-dependent with $\kappa(r) = \sup_{|i-j|\geq r} |\mathrm{Cov}(X_i,X_j)|$, they are also

ζ-dependent with $\zeta(r) = \sup_{i \in \mathbb{N}} \sum_{\{j/\, |i-j| \geq r\}} |\mathrm{Cov}(X_i, X_j)|$.

Our task in this section, is to extend Doukhan and Portal (1983) [73] method in order to provide moment bounds of order r, where r is any positive real number not less than two. The main result of this paragraph is the following theorem.

Theorem 5.4. *Let $r > 2$ be a fixed real number. Let (X_n) be a strictly stationary sequence of centered r.v's fulfilling (5.4.15). Suppose moreover that this sequence is bounded by M. Then there exists a positive constant C_r depending only on r, such that*

$$\mathbb{E}|S_n|^r \leq C_r \left(s_n^r + \sum_{k=1}^{n} \sum_{i=0}^{k-1} M^{r-2}(i+1)^{r-2} |\mathrm{Cov}(X_1, X_{1+i})| \right), \tag{5.4.16}$$

where $s_n^2 := n \sum_{i=0}^{n} |\mathrm{Cov}(X_1, X_{1+i})|$.

Theorem 5.4 gives, in particular, a unifying Rosenthal-type inequality for at least two models: associated or negatively associated processes.
An immediate consequence of Theorem 5.4 is the following Marcinkiewicz-Zygmund bound.

Corollary 5.4. *Let $r > 2$ be a fixed real number. Let (X_n) be a strictly stationary sequence of centered r.v's bounded by M and fulfilling (5.4.15). Suppose that*

$$|\mathrm{Cov}(X_1, X_{1+i})| = \mathcal{O}(i^{-r/2}), \quad \text{as} \quad i \to +\infty. \tag{5.4.17}$$

Then

$$\mathbb{E}|S_n|^r = \mathcal{O}(n^{r/2}). \tag{5.4.18}$$

For bounded associated sequences, condition (5.4.17) is shown to be optimal for the Marcinkiewicz-Zygmund bound (5.4.18) (cf. Birkel (1988) [22]).

Proof of Theorem 5.4. The method is a generalization of the Lindeberg decomposition to an order $r > 2$. This method was first developed by Rio (1995) [158] for mixing sequences and for $r \in]2, 3]$. The restriction to sequences fulfilling the bound (5.4.15) is only for the sake of clarity and the method can be adapted successfully to other dependent sequences (cf. Proposition 5.5 below). We give here the great lines of the proof and we refer to Louhichi (2003) [125] for more details.
Let $p \geq 2$ be a fixed integer. Let Φ_p be the class of functions $\phi : \mathbb{R}^+ \to \mathbb{R}^+$ such that $\phi(0) = \phi'(0) = \cdots = \phi^{(p-1)}(0) = 0$ and $\phi^{(p)}$ is non decreasing and concave. Let ϕ be a function of the set Φ_p. Theorem 5.4 is proved if we suitably control $\mathbb{E}\phi(|S_n|)$ (since the function $x \mapsto x^r$, for $r \in]p, p+1]$ is one of those functions ϕ). Such a control will be done into the following steps.

Step 1. The purpose of Step 1 is to reduce the control of $\mathbb{E}\phi(|S_n|)$ to that of a suitable *polynomial* function of $|S_n|$. For this, define $g_p : \mathbb{R}^+ \times \mathbb{R} \to \mathbb{R}^+$ by,

$$g_p(t,x) := \frac{1}{(p+1)!}\left[x^{p+1}\mathbf{1}_{0\le x\le t} + (x^{p+1} - (x-t)^{p+1})\mathbf{1}_{t<x}\right], \tag{5.4.19}$$

for any $x \ge 0$ and $g_p(t,x) = g_p(t,-x)$. The following lemma is a generalization of Equality (4.3) of Rio (1995) [158], which was written for $p = 2$.

Lemma 5.4. *Let $p \ge 2$ be a fixed integer. Let $\phi \in \Phi_p$. Suppose that $\phi^{(p+1)}$ exists and that $\lim_{x\to\infty}\phi^{(p+1)}(x) = 0$. Then*

$$\phi(x) = \int_0^{+\infty} g_p(t,x)\nu_p(dt),$$

where ν_p is the Stieltjes measure of $-\phi_p^{(p+1)}$ defined by $\nu_p(dt) = -d\phi^{(p+1)}(t)$.

Lemma 5.4 reduces then the estimation of $\mathbb{E}\phi(|S_n|)$ to that of $\mathbb{E}g_p(t,S_n)$.

Step 2. The purpose of Step 2 is then to give bounds of $\mathbb{E}f(S_n)$, for real-valued functions f belonging to a suitable set containing the functions $x \to g_p(t,x)$. For this, we denote by $\mathcal{C}_p$ the class of real-valued, p times continuously differentiable functions f such that $f(0) = \cdots = f^{(p)}(0) = 0$. Let $\mathcal{F}_p(b_1,b_2)$ be the subclass of $\mathcal{C}_{p+1}$ such that $\|f^{(p)}\|_\infty \le b_1$ and that $\|f^{(p+1)}\|_\infty \le b_2$, where $\|f^{(i)}\|_\infty = \sup_{x\in\mathbb{R}}|f^{(i)}(x)|$ and $f^{(i)}$ is the differential of order i of f.
In this step, we give an estimation of $\mathbb{E}f(S_n)$, for $f \in \mathcal{F}_p(b_1,b_2)$. Let us note that the function g_p as defined by (5.4.19) belongs to the set $\mathcal{F}_p(t,1)$
We first exhibit the mains terms of our calculations.
Notations. We denote by $\sum_{(p-2)}$ the sum over $i_1,\ldots,i_{p-2}$ such that $0 := i_0 \le i_1 \le \cdots \le i_{p-2} \le k-1$, that is $\sum_{0\le i_1\le\cdots\le i_{p-2}\le k-1}$. We define

$$\mathbb{E}_{p-2,k}(\Delta f) \quad = \quad \sup_{0\le u\le 1}\sum_{(p-2)}\left|\mathbb{E}X_kX_{k-i_1}\cdots X_{k-i_{p-2}}\Delta_{p-2,k}(f,u)\right|,$$

where

$$\begin{aligned}\Delta_{p-2,k}(f) := \Delta_{p-2,k}(f,u) &= \left[f(S_{k-i_{p-2}-1} + uX_{k-i_{p-2}}) - f(S_{k-i_{p-2}-1})\right]\\ &= uX_{k-i_{p-2}}\int_0^1 f'\left(S_{k-i_{p-2}-1} + uvX_{k-i_{p-2}}\right)dv.\end{aligned}$$

We set for $p \ge 2$,

$$\mathbb{E}_{p-2,k}(f) \quad = \quad \sum_{(p-2)}|\mathbb{E}X_kX_{k-i_1}\cdots X_{k-i_{p-2}}f(S_{k-i_{p-2}-1})|.$$

Finally, recall that $i_0 = 0$, $\mathbb{E}_{0,k}(f) = |\mathbb{E}X_k f(S_{k-1})|$ and that

$$\mathbb{E}_{0,k}(\Delta f) = \sup_{0\le u\le 1} |\mathbb{E}X_k \Delta_{0,k}(f,u)|.$$

For any real-valued function f of the set $\mathcal{F}_p(b_1, b_2)$, the quantity $|\mathbb{E}(f(S_n))|$ is evaluated by means of the main terms $\mathbb{E}_{p-2,k}(f^{(p-1)})$ and $\mathbb{E}_{p-2,k}(\Delta f^{(p-1)})$ as shows the following lemma.

Lemma 5.5. *Let $p \ge 2$ be a fixed integer. Let (X_n) be a sequence of r.v's fulfilling (5.4.15), centered and bounded by M. There exists a positive constant C_p depending only on p, such that for any $f \in \mathcal{F}_p(b_1, b_2)$,*

$$\begin{aligned} |\mathbb{E}(f(S_n))| \;&\le\; C_p \Bigg\{ s_n^p(b_1 \wedge b_2 s_n) + (b_1 \wedge b_2 M)M^{p-2} \sum_{k=1}^{n}\sum_{i=0}^{k-1} |\mathrm{Cov}(X_k, X_{k-i})| \\ &+ \sum_{k=1}^{n} \mathbb{E}_{p-2,k}(f^{(p-1)}) + \sum_{k=1}^{n} \mathbb{E}_{p-2,k}(\Delta f^{(p-1)}) \Bigg\}. \end{aligned}$$

From now C_p denotes a positive constant depending only on p and that will be different from line to line.

Evaluation of the main terms $\mathbb{E}_{p-2,k}(f)$ and $\mathbb{E}_{p-2,k}(\Delta f)$. The object of this step is to evaluate the main terms $\mathbb{E}_{p-2,k}(f)$ and $\mathbb{E}_{p-2,k}(\Delta f)$ of Lemma 5.5. This evaluation involves the following covariance quantities:

$$M_{m,n} := M^{m-2} \sum_{k=1}^{n}\sum_{i=0}^{k-1} (i+1)^{m-2} |\mathrm{Cov}(X_1, X_{i+1})|, \quad \text{for } 2 \le m \le p, \qquad (5.4.20)$$

$$M_{m,n}(b_1, b_2) := \sum_{k=1}^{n}\sum_{r=0}^{k-1} (b_1 \wedge b_2(r+1)M)(r+1)^{m-2} M^{m-2} |\mathrm{Cov}(X_1, X_{r+1})|. \qquad (5.4.21)$$

Let us note that $M_{2,n}$ is close to Var S_n and that $M_{2,n} = n\mathrm{Var}\, X_1$ in the i.i.d. case. Those covariance quantities satisfy the following analogous of Hölder's inequality:

$$M_{m,n} M_{r-m,n} \le s_n^{2r/(r-2)} M_{r,n}^{(r-4)/(r-2)} \le s_n^r + M_{r,n}, \qquad (5.4.22)$$

for any $r > 4$, $2 < m < r$. We now state the basic technical lemma of the proof of Theorem 5.4.

Lemma 5.6. *Let f be a real valued function of the set $\mathcal{F}_1(b_1, b_2)$. Let (X_n) be a centered sequence of random variables fulfilling (5.4.15). Suppose that (X_n)*

is uniformly bounded by M. Then, for any integer $p \geq 2$, there exists a positive constant C_p depending only on p, for which

$$\sum_{k=1}^{n} \mathbb{E}_{p-2,k}(\Delta f) + \sum_{k=1}^{n} \mathbb{E}_{p-2,k}(f) \tag{5.4.23}$$
$$\leq C_p \left\{ s_n^p(b_1 \wedge b_2 s_n) + \sum_{m=2}^{p-2} M_{m,n} M_{p-m,n}(b_1, b_2) + M_{p,n}(b_1, b_2) \right\},$$

where the sum $\sum_{m=2}^{p-2}$ equals to 0, whenever $p \in \{2, 3\}$.

End of the proof of Theorem 5.4. Finally, we combine the three previous steps in order to finish the proof of Theorem 5.4. Let us explain. We first make use of Lemma 5.4, together with Fubini's theorem, to obtain,

$$\mathbb{E}\phi(|S_n|) = \int_0^{+\infty} \mathbb{E} g_p(t, S_n)\, \nu_p(dt). \tag{5.4.24}$$

We recall that the functions $x \mapsto g_p(t,x)$ and $x \mapsto g_p^{(p-1)}(t,x)$ belong respectively to $\mathcal{F}_p(t,1)$ and to $\mathcal{F}_1(t,1)$ (in fact if $f \in \mathcal{F}_p(t,1)$, then $f^{(p-1)} \in \mathcal{F}_1(t,1)$). Hence we deduce, applying Lemma 5.5 to the function $x \mapsto g_p(t,x)$ and Lemma 5.6 to the function $x \mapsto g_p^{(p-1)}(t,x)$,

$$\mathbb{E} g_p(t, S_n) \leq C_p \left\{ \sum_{m=2}^{p-2} M_{m,n} M_{p-m,n}(t,1) + M_{p,n}(t,1) + s_n^p(t \wedge s_n) \right\}. \tag{5.4.25}$$

Taking into account Lemma 5.4 and the fact $g^{(p)}(x) = x \wedge t$, we deduce that

$$\phi^{(p)}(x) = \int_0^{+\infty} (t \wedge x) \nu_p(dt). \tag{5.4.26}$$

Inequalities (5.4.24), (5.4.25) and (5.4.26) yield:

$$\mathbb{E}\phi(|S_n|) \leq C_p \Big\{ s_n^p \phi^{(p)}(s_n)$$
$$+ \sum_{k=1}^{n} \sum_{i=0}^{k-1} (M(i+1))^{p-2} \phi^{(p)}(M(i+1)) |\mathrm{Cov}(X_1, X_{1+i})| \tag{5.4.27}$$
$$+ \sum_{m=2}^{p-2} M_{m,n} \Big(\sum_{k=1}^{n} \sum_{i=0}^{k-1} (M(i+1))^{p-m-2} \phi^{(p)}(M(i+1)) |\mathrm{Cov}(X_1, X_{1+i})| \Big) \Big\}.$$

Now, we use the concavity property of the function $\phi^{(p)}$,

$$\phi(x) = \frac{x^p}{(p-1)!} \int_0^1 (1-t)^{p-1} \phi^{(p)}(tx) dt \geq x^p \phi^{(p)}(x) \int_0^1 \frac{t(1-t)^{p-1}}{(p-1)!} dt,$$

to deduce that

$$x^p \phi^{(p)}(x) \le C_p \phi(x). \tag{5.4.28}$$

We conclude, combining inequalities (5.4.27) and (5.4.28),

$$\begin{aligned} \mathbb{E}\phi(|S_n|) \quad &\le \quad C_p \left\{ \sum_{k=1}^{n} \sum_{i=0}^{k-1} (M(i+1))^{-2} \phi(M(i+1)) |\mathrm{Cov}(X_1, X_{1+i})| + \phi(s_n) \right. \\ &+ \left. \sum_{m=2}^{p-2} M_{m,n} \left(\sum_{k=1}^{n} \sum_{i=0}^{k-1} (M(i+1))^{-m-2} \phi(M(i+1)) |\mathrm{Cov}(X_1, X_{1+i})| \right) \right\} \end{aligned}$$

The last inequality applied to $\phi(x) = x^r$, for $r \in]p, p+1]$, leads to

$$\mathbb{E}|S_n|^r \le C_r \left\{ \sum_{k=1}^{n} \sum_{i=0}^{k-1} (M(i+1))^{r-2} |\mathrm{Cov}(X_1, X_{1+i})| + s_n^r + \sum_{m=2}^{p-2} M_{m,n} M_{r-m,n} \right\}.$$

The proof of Theorem 5.4 is now complete, using the last inequality together with (5.4.22) (recall that $M_{2,n} \le s_n^2$). □

In the case where the sequence $(X_n)_{n \in \mathbb{N}^*}$ is θ−dependent, the proof of the inequalities of Theorem 5.4 can be adapted. We then get a variation of the technical proof of Theorem 5.4 written just above. Hence it will be omitted here. Let us state the inequalities we obtained in the case where $(X_n)_{n \in \mathbb{N}^*}$ is θ−dependent.

Proposition 5.5. *Let r be a fixed real number > 2. Let (X_n) be a stationary sequence of $\theta_{1,\infty}$−dependent centered random variables. Suppose moreover that this sequence is bounded by 1. Let $S_n := X_1 + X_2 + \cdots + X_n$, for $n \ge 1$ and $S_0 = X_0 = 0$. Then there exists a positive constant C_r depending only on r, such that*

$$\mathbb{E}|S_n|^r \le C_r \left(\tilde{s}_n^r + M_{r,n} \right), \tag{5.4.29}$$

where $M_{r,n} := n \sum_{i=0}^{n-1} (i+1)^{r-2} \theta(i)$, and $\tilde{s}_n^2 := M_{2,n} = n \sum_{i=0}^{n-1} \theta(i)$.

We refer to Prieur (2002) [155] for a detailed proof of Proposition 5.5.

5.4.4 Rosenthal inequalities for τ_1-dependent sequences

We give here a corollary of Theorem 5.3.

Corollary 5.5. *Let $(X_i)_{i>0}$ be a sequence of centered random variables belonging to $\mathbb{L}^p$ for some $p \ge 2$. Define $(\mathcal{M}_i)_{i>0}$, $\overline{S}_n$, Q_X and s_n as in Theorem 5.3. Recall that τ^{-1} has been defined in Notations 5.1. We have*

$$\|\overline{S}_n\|_p^p \le a_p s_n^p + n b_p \int_0^{\|X\|_1} ((\tau/2)^{-1}(u))^{p-1} Q_X^{p-1} \circ G_X(u) du \,,$$

where $a_p = 4p5^p(p+1)^{p/2}$ *and* $(p-1)b_p = 4p5^p(p+1)^{p-1}$*. Moreover we have that*

$$s_n^2 \leq 4n \int_0^{\|X\|_1} (\tau/2)^{-1}(u) Q_X \circ G_X(u) du\,.$$

Proof of Corollary 5.5. It suffices to integrate (5.4.4) (as done in Rio (2000) [161] page 88) and to note that

$$\int_0^1 Q(u)(R(u))^{p-1}(u)du = \int_0^{\|X\|_1} ((\tau/2)^{-1}(u))^{p-1} Q_X^{p-1} \circ G_X(u) du\,.$$

The bound for s_n^2 holds with $\theta_{1,1}$ instead of $\tau_{1,\infty}$ (see Dedecker and Doukhan (2002) [43]).

5.4.5 Rosenthal inequalities under projective conditions

We recall two moment inequalities given in Dedecker (2001) [42]. We use these inequalities in Chapter 10 to prove the tightness of the empirical process for $\tilde{\alpha}$, $\tilde{\beta}$ and $\tilde{\phi}$ dependent sequences.

Proposition 5.6. *Let* $(X_i)_{i\in\mathbb{Z}}$ *be a stationary sequence of centered and square integrable random variables and let* $S_n = X_1 + \cdots + X_n$*. Let* $\mathcal{M}_i = \sigma(X_j, j \leq i)$*. The following upper bound holds*

$$\|S_n\|_p \leq (pnV_\infty)^{1/2} + \Big(3p^2 n\Big(\|X_0^3\|_{p/3} + M_1(p) + M_2(p) + M_3(p)\Big)\Big)^{1/3},$$

where $V_N = \mathbb{E}(X_0^2) + 2\sum_{k=1}^{N} |\mathbb{E}(X_0 X_k)|$ *and*

$$\begin{aligned}
M_1(p) &= \sum_{l=1}^{+\infty}\sum_{m=0}^{l-1} \|X_0 X_m \mathbb{E}(X_{l+m}|\, \mathcal{M}_m)\|_{p/3}\\
M_2(p) &= \sum_{l=1}^{+\infty}\sum_{m=l}^{+\infty} \|X_0 \mathbb{E}(X_m X_{l+m} - \mathbb{E}(X_m X_{l+m})|\, \mathcal{M}_0)\|_{p/3}\\
M_3(p) &= \frac{1}{2}\sum_{k=1}^{+\infty} \|X_0 \mathbb{E}(X_k^2 - \mathbb{E}(X_k^2)|\, \mathcal{M}_0)\|_{p/3}\,.
\end{aligned}$$

Proposition 5.7. *We keep the same notations as in Proposition 5.6. For any positive integer* N*, the following upper bound holds*

$$\|S_n\|_p \leq (pn\,(V_{N-1} + 2M_0(p)))^{1/2} + \Big(3p^2 n\Big(\|X_0^3\|_{p/3} + \tilde{M}_1(p) + \tilde{M}_2(p) + M_3(p)\Big)\Big)^{1/3},$$

where

$$\begin{aligned} M_0(p) &= \sum_{l=N}^{+\infty} \|X_0 \mathbb{E}(X_l | \mathcal{M}_0)\|_{p/2} \\ \tilde{M}_1(p) &= \sum_{l=1}^{N-1} \sum_{m=0}^{l-1} \|X_0 X_m \mathbb{E}(X_{l+m} | \mathcal{M}_m)\|_{p/3} \\ \tilde{M}_2(p) &= \sum_{l=1}^{N-1} \sum_{m=l}^{+\infty} \|X_0 \mathbb{E}(X_m X_{l+m} - \mathbb{E}(X_m X_{l+m}) | \mathcal{M}_0)\|_{p/3} \,. \end{aligned}$$

5.5 Maximal inequalities

Our first result is an extension of Doob's inequality for martingales. This maximal inequality is stated in the nonstationary case.

Proposition 5.8. *Let $(X_i)_{i\in\mathbb{Z}}$ be a sequence of square-integrable and centered random variables, adapted to a nondecreasing filtration $(\mathcal{F}_i)_{i\in\mathbb{Z}}$. Let λ be any nonnegative real number and $G_k = (S_k^* > \lambda)$. We have*

$$\mathbb{E}((S_n^* - \lambda)_+^2) \le 4 \sum_{k=1}^{n} \mathbb{E}(X_k^2 \mathbf{1}_{G_k}) + 8 \sum_{k=1}^{n-1} \|X_k \mathbf{1}_{G_k} \mathbb{E}(S_n - S_k | \mathcal{F}_k)\|_1 \,.$$

Proof of Proposition 5.8. We proceed as in Garsia (1965) [90]:

$$(S_n^* - \lambda)_+^2 = \sum_{k=1}^{n} ((S_k^* - \lambda)_+^2 - (S_{k-1}^* - \lambda)_+^2). \tag{5.5.1}$$

Since the sequence $(S_k^*)_{k\ge 0}$ is nondecreasing, the summands in (5.5.1) are nonnegative. Now

$$((S_k^* - \lambda)_+ - (S_{k-1}^* - \lambda)_+)((S_k^* - \lambda)_+ + (S_{k-1}^* - \lambda)_+) > 0$$

if and only if $S_k > \lambda$ and $S_k > S_{k-1}^*$. In that case $S_k = S_k^*$, whence

$$(S_k^* - \lambda)_+^2 - (S_{k-1}^* - \lambda)_+^2 \le 2(S_k - \lambda)((S_k^* - \lambda)_+ - (S_{k-1}^* - \lambda)_+).$$

Consequently

$$\begin{aligned} (S_n^* - \lambda)_+^2 &\le 2 \sum_{k=1}^{n} (S_k - \lambda)(S_k^* - \lambda)_+ - 2 \sum_{k=1}^{n} ((S_k - \lambda)(S_{k-1}^* - \lambda)_+) \\ &\le 2(S_n - \lambda)_+ (S_n^* - \lambda)_+ - 2 \sum_{k=1}^{n} (S_{k-1}^* - \lambda)_+ X_k. \end{aligned}$$

Noting that $2(S_n-\lambda)_+(S_n^*-\lambda)_+ \le \frac{1}{2}(S_n^*-\lambda)_+^2 + 2(S_n-\lambda)_+^2$, we infer

$$(S_n^*-\lambda)_+^2 \le 4(S_n-\lambda)_+^2 - 4\sum_{k=1}^n (S_{k-1}^*-\lambda)_+ X_k. \tag{5.5.2}$$

In order to bound $(S_n-\lambda)_+^2$, we adapt the decomposition (5.5.1) and next we apply Taylor's formula:

$$\begin{aligned}(S_n-\lambda)_+^2 &= \sum_{k=1}^n \big((S_k-\lambda)_+^2 - (S_{k-1}-\lambda)_+^2\big)\\ &= 2\sum_{k=1}^n (S_{k-1}-\lambda)_+ X_k + 2\sum_{k=1}^n X_k^2 \int_0^1 (1-t)\mathbf{1}_{S_{k-1}+tX_k>\lambda}\,dt.\end{aligned}$$

Since $\mathbf{1}_{S_{k-1}+tX_k>\lambda} \le \mathbf{1}_{S_k^*>\lambda}$, it follows that

$$(S_n-\lambda)_+^2 \le 2\sum_{k=1}^n (S_{k-1}-\lambda)_+ X_k + \sum_{k=1}^n X_k^2 \mathbf{1}_{S_k^*>\lambda}. \tag{5.5.3}$$

Hence, by (5.5.2) and (5.5.3)

$$(S_n^*-\lambda)_+^2 \le 4\sum_{k=1}^n \big(2(S_{k-1}-\lambda)_+ - (S_{k-1}^*-\lambda)_+\big)X_k + 4\sum_{k=1}^n X_k^2 \mathbf{1}_{S_k^*>\lambda}.$$

Let $D_0 = 0$ and $D_k = 2(S_k-\lambda)_+ - (S_k^*-\lambda)_+$ for $k>0$. Clearly

$$D_{k-1}X_k = \sum_{i=1}^{k-1} (D_i - D_{i-1})X_k.$$

Hence

$$(S_n^*-\lambda)_+^2 \le 4\sum_{i=1}^{n-1} (D_i - D_{i-1})(S_n - S_i) + 4\sum_{k=1}^n X_k^2 \mathbf{1}_{S_k^*>\lambda}. \tag{5.5.4}$$

Since the random variables $D_i - D_{i-1}$ are $\mathcal{F}_i$-measurable, we have:

$$\begin{aligned}\mathbb{E}((D_i - D_{i-1})(S_n - S_i)) &= \mathbb{E}((D_i - D_{i-1})\mathbb{E}(S_n - S_i \mid \mathcal{F}_i))\\ &\le \mathbb{E}|(D_i - D_{i-1})\mathbb{E}(S_n - S_i \mid \mathcal{F}_i)|.\end{aligned} \tag{5.5.5}$$

It remains to bound $|D_i - D_{i-1}|$. If $(S_i^*-\lambda)_+ = (S_{i-1}^*-\lambda)_+$, then

$$|D_i - D_{i-1}| = 2|(S_i-\lambda)_+ - (S_{i-1}-\lambda)_+| \le 2|X_i|\mathbf{1}_{S_i^*>\lambda},$$

because $D_i - D_{i-1} = 0$ whenever $S_i \le \lambda$ and $S_{i-1} \le \lambda$. Otherwise $S_i = S_i^* > \lambda$ and $S_{i-1} \le S_{i-1}^* < S_i$, which implies that

$$D_i - D_{i-1} = (S_i - \lambda)_+ + (S_{i-1}^* - \lambda)_+ - 2(S_{i-1} - \lambda)_+.$$

Hence $D_i - D_{i-1}$ belongs to $[0, 2((S_i - \lambda)_+ - (S_{i-1} - \lambda)_+)]$. In any case

$$|D_i - D_{i-1}| \le 2|X_i|\mathbf{1}_{S_i^* > \lambda},$$

which together with (5.5.4) and (5.5.5) implies Proposition 5.8. □

Consider the projection operators P_i: for any f in $\mathbb{L}^2$, $P_i(f) = \mathbb{E}(f|\ \mathcal{F}_i) - \mathbb{E}(f|\ \mathcal{F}_{i-1})$. Combining Proposition 5.8 and a decomposition due to McLeish, we obtain the following maximal inequality.

Proposition 5.9. *Let $(X_i)_{i\in\mathbb{Z}}$ be a sequence of square-integrable and centered random variables, and $(\mathcal{F}_i)_{i\in\mathbb{Z}}$ be any nondecreasing filtration. Define the σ-algebras $\mathcal{F}_{-\infty} = \bigcap_{i\in\mathbb{Z}} \mathcal{F}_i$ and $\mathcal{F}_\infty = \sigma(\bigcap_{i\in\mathbb{Z}} \mathcal{F}_i)$. Define the random variables $S_n = X_1 + \cdots + X_n$ and $S_n^* = \max\{0, S_1, \ldots, S_n\}$. For any i in $\mathbb{Z}$, let $(Y_{i,j})_{j\ge 1}$ be the martingale $Y_{i,j} = \sum_{k=1}^{j} P_{k-i}(X_k)$ and $Y_{i,n}^* = \max\{0, Y_{i,1}, \ldots, Y_{i,n}\}$. Let λ be any nonnegative real number and $G(i,k,\lambda) = \{Y_{i,k}^* > \lambda\}$. Assume that the sequence is regular: for any integer k, $\mathbb{E}(X_k|\mathcal{F}_{-\infty}) = 0$ and $\mathbb{E}(X_k|\mathcal{F}_\infty) = X_k$. For any two sequences of nonnegative numbers $(a_i)_{i\ge 0}$ and $(b_i)_{i\ge 0}$ such that $K = \sum a_i^{-1}$ is finite and $\sum b_i = 1$ we have*

$$\mathbb{E}\left((S_n^* - \lambda)_+^2\right) \le 4K \sum_{i=0}^{\infty} a_i \Big(\sum_{k=1}^{n} \mathbb{E}(P_{k-i}^2(X_k)\mathbf{1}_{G(i,k,b_i\lambda)})\Big).$$

Proof of Proposition 5.9. Since the sequence is regular, we decompose

$$X_k = \sum_{i=-\infty}^{+\infty} P_{k-i}(X_k).$$

Consequently $S_j = \sum_{i\in\mathbb{Z}} Y_{i,j}$ and therefore: $(S_j - \lambda)_+ \le \sum_{i\in\mathbb{Z}} (Y_{i,j} - b_i\lambda)_+$. Applying Hölder inequality and taking the maximum on both sides, we get

$$(S_n^* - \lambda)_+^2 \le K \sum_{i\in\mathbb{Z}} a_i (Y_{i,n}^* - b_i\lambda)_+^2.$$

Taking the expectation and applying Proposition 5.8 to the martingale $(Y_{i,n})_{n\ge 1}$, we obtain Proposition 5.9. □

Chapter 6

Applications of strong laws of large numbers

We consider in this chapter a stochastic algorithm with weakly dependent input noise (according to Definition 2.2). In particular, the case of γ_1-dependence is considered. The ODE (ordinary differential equation) method is generalized to such situation. For this, we use tools for causal and non causal sequences developed in the previous chapters. Illustrations to the linear regression frame and to the law of large numbers for triangular arrays of weighted dependent random variables are also given.

6.1 Stochastic algorithms with non causal dependent input

We consider the $\mathbb{R}^d$-valued stochastic algorithm, defined on a probability space $(\Omega, \mathcal{A}, \mathbb{P})$ and driven by the recurrence equation

$$Z_{n+1} = Z_n + \gamma_n h(Z_n) + \zeta_{n+1}, \tag{6.1.1}$$

where
- h is a continuous function from an open set $G \subseteq \mathbb{R}^d$ to $\mathbb{R}^d$,
- (γ_n) a decreasing to zero deterministic real sequence satisfying

$$\sum_{n\geq 0} \gamma_n = \infty. \tag{6.1.2}$$

- (ζ_n) is a "small" stochastic disturbance.

The ordinary differential equation (ODE) method associates (we refer for instance to Benveniste *et al.* (1987) [15], Duflo (1996) [82], Kushner and Clark

(1978) [114]) the possible limit sets of (6.1.1) with the properties of the associated ODE

$$\frac{dz}{dt} = h(z). \tag{6.1.3}$$

These sets are compact connected invariant and "chain-recurrent" in the Benaïm sense for the ODE (*cf.* Benaïm (1996) [14]). These sets are more or less complicated. Various situations may then happen. The most simple case is an equilibrium : z is a solution of $h(z) = 0$, but equilibria cycle, or a finite set of equilibria is linked to the ODE's trajectories, connected sets of equilibria or, periodic cycles for the ODE may also happen...
In order to use the ODE method, we suppose that (Z_n) is *a.s.* bounded and

$$\zeta_{n+1} \quad = \quad c_n(\xi_{n+1} + r_{n+1}), \tag{6.1.4}$$

where (c_n) denotes a nonnegative deterministic sequence such that

$$\gamma_n = \mathcal{O}(c_n), \quad \sum c_n^2 < \infty, \tag{6.1.5}$$

(ξ_n) and (r_n) are $\mathbb{R}^d$-valued sequences, defined on $(\Omega, \mathcal{A}, \mathbb{P})$, and adapted with respect to an increasing sequence of σ-fields $(\mathcal{F}_n)_{n\geq 0}$ and satisfying almost surely (*a.s.*) on $A \subset \Omega$,

$$\sum_{n=0}^{\infty} c_n \xi_{n+1} < \infty \quad a.s. \qquad \text{and} \tag{6.1.6}$$

$$\lim_{n\to\infty} r_n = 0 \quad a.s. \tag{6.1.7}$$

The classical theory of algorithms is related to a noise (ξ_n) which is a martingale difference sequence. Our aim is to replace this condition about the noise by weakly dependence conditions as being introduced in Chapter 2.
In Section 6.1.1, we suppose that the sequence (ξ_n) is $(\Lambda^{(1)} \cap \mathbb{L}^\infty, \Psi)$-dependent according to Definition 2.2, where

$$\Psi(f, g) = C(d_f, d_g)(\mathrm{Lip}(f) + \mathrm{Lip}(g)),$$

for some function $C : \mathbb{N}^* \times \mathbb{N}^* \to \mathbb{R}$.
Various examples of this situation may be found in Chapter 3, they include general Bernoulli shifts, stable Markov chains such as, $\xi_t = G(\xi_{t-1}, \ldots, \xi_{t-p}) + \zeta_t$, $\xi_t = \left(a_0 + \sum_{j\geq 1} a_j \xi_{t-j}\right) \zeta_t$ generated by some i.i.d. sequence (ζ_t), or ARCH(∞) models.
In Section 6.1.2 below, we consider a weakly dependent noise in the sense of the γ_1-weak coefficients of Dedecker and Doukhan (2003) [43] defined by (2.2.17).

Note that (cf. Remark 2.5) a causal version of $(\mathcal{F}, \mathcal{G}, \Psi)$-dependence implies γ_1-dependence, where the left hand side in definition of weak dependence writes $\leq C(v)\mathrm{Lip}(g)\epsilon(r)$. Counter-examples of γ_1-dependent sequences which are not θ-dependent may also be found there.

The notion of γ_1-dependence is generalized to $\mathbb{R}^d$ valued sequences.

Proposition 6.1. *The two following assertions are equivalent:*
(i) A $\mathbb{R}^d$-valued sequence (X_n) is γ_1-dependent,
(ii) Each component (X_n^ℓ) $(\ell = 1, \ldots, d)$ of (X_n) is γ_1- dependent.

Proof. Clearly, $\|\mathbb{E}(X^\ell_{n+r} - \mathbb{E}(X^\ell_{n+r})|\mathcal{F}_n)\|_1 \leq \|\mathbb{E}(X_{n+r} - \mathbb{E}(X_{n+r})|\mathcal{F}_n)\|_1$, hence (i) implies (ii). The second implication follows from,

$$\|\mathbb{E}(X_{n+r} - \mathbb{E}(X_{n+r})|\mathcal{F}_n)\|_1 = \mathbb{E}\sqrt{\sum_{\ell=1}^{d}\left(\mathbb{E}(X^\ell_{n+r} - \mathbb{E}(X^\ell_{n+r})|\mathcal{F}_n)\right)^2}. \qquad \square$$

The two forthcoming sections are devoted to provide moment inequalities of the Marcinkiewicz-Zygmund type adapted to deduce the relation (6.1.6) in those two frames. The following sections are devoted to study the examples of Robbins-Monro and Kiefer-Wolfowitz algorithms and to obtain sufficient conditions for the complete convergence of triangular arrays, extending on Chow (1966) [37]. Finally, the last section is devoted to the specific of the linear regression algorithm with dependent entries. In [36], Chen (1985) has also studies this topic. He works in a more general matrix valued framework. Assuming only the stationarity and the ergodicity of entries, he derives the a.s. convergence of the algorithm. We get the same result with a γ_1-dependence assumption, but this assumption, more restrictive, allow us to reach, thanks to a moment technic, a precise $n^{-1/2}$-convergence rate.

6.1.1 Weakly dependent noise

Let (ξ_n) be a sequence of centered random variables. Let S_n be the sum $\sum_{i=1}^n \xi_i$ and $C_q = \max_{u+v\leq q} C(u,v)$. Suppose that an analogous of the bounds (4.3.2) and (4.3.3) are satisfied by the process ξ:

$$\sup|\mathrm{Cov}(\xi_{t_1}\cdots\xi_{t_m}, \xi_{t_{m+1}}\cdots\xi_{t_q})| \leq C_q q^\gamma M^{q-2}\epsilon(r), \tag{6.1.8}$$

where the supremum is taken over all $\{t_1, \ldots, t_q\}$ such that $1 \leq t_1 \leq \cdots \leq t_q$, and $1 \leq m < q$ such that $t_{m+1} - t_m = r$, or

$$|\mathrm{Cov}(\xi_{t_1}\cdots\xi_{t_m}, \xi_{t_{m+1}}\cdots\xi_{t_q})| \leq (C_q \vee 2)\int_0^{\epsilon(r)\wedge 1} Q_{\xi_{t_1}}(x)\cdots Q_{\xi_{t_q}}(x)dx. \tag{6.1.9}$$

Those bounds respectively imply moment inequalities in Theorems 4.2 and 4.3. Denoting $\Sigma_n = \sum_{i=1}^n c_{i-1}\xi_i$, and using similar techniques as in § 4.3 (see Doukhan and Louhichi (1999) [67]) one has,

Proposition 6.2. *Let $p \geq 2$ be some fixed integer and let (ξ_n) be a centered sequence of real random variables such that (6.1.8) holds for all $q \leq p$. Then for $n \geq 2$,*

$$|\mathbb{E}\Sigma_n^p| \leq \frac{(2p-2)!}{(p-1)!}\left\{\left(C_p p^\gamma M^{p-2}\sum_{i=1}^n c_{i-1}^p \sum_{r=0}^{n-1}(r+1)^{p-2}\epsilon(r)\right)\right. \\ \left. \vee \left(C_2 2^\gamma \sum_{i=1}^n c_{i-1}^2 \sum_{r=0}^{n-1}\epsilon(r)\right)^{p/2}\right\}.$$

This result is mainly adapted to bounded sequences.

Proof of proposition 6.2. The proof is done in Brandière and Doukhan (2004) [32]. We have, using arguments from Doukhan and Louhichi's (1999) [67] as done for the proof of Theorem 4.2,

$$\mathbb{E}\left(\sum_{i=1}^n c_i\xi_i\right)^p \leq p! \sum_{1\leq t_1\leq\cdots\leq t_p\leq n} c_{t_1}\cdots c_{t_p}|\mathbb{E}(\xi_{t_1}\cdots\xi_{t_p})|. \tag{6.1.10}$$

Denote $A_p(n) = \sum_{1\leq t_1\leq\cdots\leq t_p\leq n} c_{t_1}\cdots c_{t_p}|\mathbb{E}(\xi_{t_1}\cdots\xi_{t_p})|$, so for any $t_2 \leq t_m \leq t_{p-1}$,

$$\begin{aligned} A_p(n) \;&\leq \sum_{1\leq t_1\leq\cdots\leq t_p\leq n} c_{t_1}\cdots c_{t_p}|\mathbb{E}(\xi_{t_1}\cdots\xi_{t_m})\mathbb{E}(\xi_{t_{m+1}}\cdots\xi_{t_p})| \\ &+ \sum_{1\leq t_1\leq\cdots\leq t_p\leq n} c_{t_1}\cdots c_{t_p}|\mathrm{Cov}(\xi_{t_1}\cdots\xi_{t_m}, \xi_{t_{m+1}}\cdots\xi_{t_p})|. \end{aligned}$$

Denote

$$\begin{aligned} A_p^1(n) &= \sum_{1\leq t_1\leq\cdots\leq t_p\leq n} c_{t_1}\cdots c_{t_p}|\mathbb{E}(\xi_{t_1}\cdots\xi_{t_m})\mathbb{E}(\xi_{t_{m+1}}\ldots\xi_{t_p})|, \\ A_p^2(n) &= \sum_{1\leq t_1\leq\cdots\leq t_p\leq n} c_{t_1}\cdots c_{t_p}|\mathrm{cov}(\xi_{t_1}\cdots\xi_{t_m}, \xi_{t_{m+1}}\cdots\xi_{t_p})|. \end{aligned}$$

Since the sequence (c_n) is decreasing to 0, we deduce, as in Doukhan and Louhichi (1999) [67],

$$A_p^1(n) \leq A_m(n)A_{p-m}(n). \tag{6.1.11}$$

By (6.1.8) we obtain $A_p^2(n) \leq \sum_{t_1=1}^n c_{t_1}^p \sum_{r=0}^{n-1} C_p p^\gamma M^{p-2}(r+1)^{p-2}\epsilon(r)$, and the expression $\sum_{i=1}^n c_i^p \sum_{r=0}^{n-1} C_p p^\gamma M^{p-2}(r+1)^{p-2}\epsilon(r) = V_p(n)$, verifies, for any integer $2 \leq q \leq p-1$: $V_q(n) \leq V_p^{\frac{q-2}{p-2}}(n) V_2^{\frac{p-q}{p-2}}(n)$. Now, Lemma 4.7 (see also Lemma 12 of Doukhan and Louhichi (1999) [67]) leads to $A_p(n) \leq \frac{1}{p}\begin{pmatrix} 2p-2 \\ p-1 \end{pmatrix}(V_2^{\frac{p}{2}}(n) \vee V_p(n))$, hence

$$\mathbb{E}\left(\sum_{i=1}^n c_i \xi_i\right)^p \leq \frac{(2p-2)!}{(p-1)!}(V_2^{\frac{p}{2}}(n) \vee V_p(n)).$$

This ensures the result. □

The following result is appropriate to more general real-valued random variables but require a moment assumption and a tail condition.

Proposition 6.3. *Let $p > 2$ be a fixed integer and (ξ_n) be a centered sequence of random variables. Assume that for all $2 < q \leq p$, Inequality (6.1.9) holds with*

$$M_q \leq M_p^{\frac{q-2}{p-2}} M_2^{\frac{p-q}{p-2}} \tag{6.1.12}$$

and there exists a constant $c > 0$ such that

$$\exists k > p, \qquad \forall i \geq 0: \qquad \mathbb{P}(|\xi_i| > t) \leq \frac{c}{t^k}. \tag{6.1.13}$$

Then for $n \geq 2$,

$$\begin{aligned} |\mathbb{E}\Sigma_n^p| \leq \frac{(2p-2)!}{(p-1)!} c^{1/k} \Bigg\{ & \left(M_p \sum_{i=1}^n c_{i-1}^p \sum_{r=0}^{n-1} (r+1)^{p-2} \epsilon(r)^{\frac{k-p}{k}} \right) \\ & \vee \left(M_2 \sum_{i=1}^n c_{i-1}^2 \sum_{r=0}^{n-1} \epsilon(r)^{\frac{k-2}{k}} \right)^{p/2} \Bigg\} \end{aligned} \tag{6.1.14}$$

Note that (6.1.13) holds as soon as the ξ_n's have a k-th order moment such that $\sup_{i\geq 0} \mathbb{E}|\xi_i|^k \leq c$.

Now we argue as in Billingsley (1968) [20]: if (6.1.8) holds for some p such that

$$\begin{aligned} \Bigg\{ & \left(C_p p^\gamma M^{p-2} \sum_{i=1}^n c_{i-1}^p \sum_{r=0}^{n-1} (r+1)^{p-2} \epsilon(r) \right) \\ & \qquad \vee \left(C_2 2^\gamma \sum_{i=1}^n c_{i-1}^2 \sum_{r=0}^{n-1} \epsilon(r) \right)^{p/2} \Bigg\} < \infty \end{aligned}$$

then for any $t > 0$, $\lim_{n\to\infty} \mathbb{P}(\sup_{k\geq 1} |\Sigma_{n+k} - \Sigma_n| > t) = 0$. Thus (Σ_n) is *a.s.* a Cauchy sequence, hence it converges. In the same way, if (6.1.9) holds for some p such that

$$\left(\sum_{i=1}^{n} c_{i-1}^{p} \sum_{r=0}^{\infty} (r+1)^{p-2} \epsilon(r)^{\frac{k-p}{k}}\right) \vee \left(\sum_{i=1}^{n} c_{i-1}^{2} \sum_{r=0}^{n-1} \epsilon(r)^{\frac{k-2}{k}}\right)^{p/2} < \infty, \quad (6.1.15)$$

then (Σ_n) converges with probability 1.

Proof of proposition 6.3. Using the same notations as in the previous proof, by (6.1.9)

$$V_p(n) \leq M_p \sum_{i=1}^{n} c_i^p \int_0^1 (\epsilon^{-1}(u) \wedge n)^{p-1} Q_i^p(u) du,$$

where $\epsilon(u) = \epsilon_{[u]}$ ($[u]$ denotes the integer part of u). Denote

$$W_p(n) = M_p \sum_{i=1}^{n} c_i^p \int_0^1 (\epsilon^{-1}(u) \wedge n)^{p-1} Q_i^p(u) du.$$

If (6.1.12) is verified, then

$$W_q(n) \leq W_p^{\frac{q-2}{p-2}}(n) W_2^{\frac{p-q}{p-2}}(n),$$

which completes the proof. □

6.1.2 γ_1−dependent noise

Let $(\xi_n)_{n\geq 0}$ be a sequence of integrable real-valued random variables, and $(\gamma_1(r))_{r\geq 0}$ be the associated mixingale-coefficients defined in (2.2.17). Then the following moment inequality holds.

Proposition 6.4. *Let $p > 2$ and $(\xi_n)_{n\in\mathbb{N}}$ be a sequence of centered random variables such that (6.1.13) holds. Then for any $n \geq 2$,*

$$|\mathbb{E}\Sigma_n^p| \leq \left(2pK_1 \sum_{i=1}^{n} c_i^{2-\frac{2(k-p)}{p(k-1)}} \sum_{j=0}^{n-1} \gamma_1(j)^{\frac{2(k-p)}{p(k-1)}}\right)^{p/2}, \quad (6.1.16)$$

where K_1 depends on r, p and c.

Notice that here $p \in \mathbb{R}$, and is not necessarily an integer. If now (6.1.16) holds for some p such that

$$\sum_{i=1}^{\infty} c_i^{2-m} \sum_{j=0}^{\infty} \gamma_1(j)^m < \infty, \quad (6.1.17)$$

where $m = \frac{2(k-p)}{p(k-1)} < 1$, then (Σ_n) converges with probability 1. The result extends to $\mathbb{R}^d$. Indeed, if we consider a centered $\mathbb{R}^d$-valued and γ_1-dependent sequence $(\xi_n)_{n\geq 0}$, one has as previously

$$\mathbb{E}\|\Sigma_n\|^p = \mathbb{E}\sum_{\ell=1}^{d}\left(\sum_{i=1}^{n} c_i\xi_i^\ell\right)^p,$$

and if each component $(\xi_n^\ell)_{n\geq 0}$ $(\ell = 1,\dots,d)$ is γ_1-dependent and verifies (6.1.13) and (6.1.17), $\mathbb{E}\|\Sigma_n\|^p < \infty$ and we conclude as before that $(\Sigma_n)_{n\geq 0}$ converges a.s.

Proof of proposition 6.4. Proceeding as in Dedecker and Doukhan (2003) [43], we deduce

$$|\mathbb{E}(\Sigma_n^p)| \leq \left(2p\sum_{i=1}^{n} b_{i,n}\right)^{\frac{p}{2}},$$

where

$$b_{i,n} = \max_{i\leq \ell\leq n}\left\| c_i\xi_i\sum_{k=0}^{\ell-i}\mathbb{E}(c_{i+k}\xi_{i+k}|\mathcal{F}_i)\right\|_{\frac{p}{2}}.$$

Let $q = p/(p-2)$, then there exists Y such that $\|Y\|_q = 1$.
Applying Proposition 1 of Dedecker and Doukhan (2003) [43], we obtain

$$b_{i,n} \leq \sum_{k=0}^{n-i}\int_0^{\gamma_1(k)} Q_{\{Yc_i\xi_i\}}\circ G_{\{c_{i+k}\xi_{i+k}\}}(u)du,$$

where G_X is the inverse of $x \mapsto \int_0^x Q_X(u)du$. Since $G_{\{c_i\xi_i\}}(u) = G_{\xi_i}(\frac{u}{c_i}) = G(\frac{u}{c_i})$, we get

$$b_{i,n} \leq \sum_{k=0}^{n-i}\int_0^{\gamma_1(k)} Q_{\{Yc_i\xi_i\}}\circ G\left(\frac{u}{c_{i+k}}\right)du \leq \sum_{k=0}^{n-i} c_{i+k}\int_0^{\frac{\gamma_1(k)}{c_{i+k}}} Q_{\{Yc_i\xi_i\}}\circ G(u)du,$$

and the Fréchet inequality (1957) [88] yields

$$\begin{aligned} b_{i,n} &\leq \sum_{k=0}^{n-i} c_{i+k}\int_0^{G(\frac{\gamma_1(k)}{c_{i+k}})} Q_Y(u)Q_{\{c_i\xi_i\}}(u)Q(u)du \\ &\leq \sum_{k=0}^{n-i} c_ic_{i+k}\int_0^1 \mathbf{1}_{\left\{u\leq G(\frac{\gamma_1(k)}{c_{i+k}})\right\}}Q^2(u)Q_Y(u)du \end{aligned}$$

where $Q = Q_{\xi_i}$. Using Hölder's inequality, we also obtain

$$b_{i,n} \le c_i \sum_{k=0}^{n-i} c_{i+k} \left(\int_0^1 \mathbf{1}_{\{u \le G(\frac{\gamma_1(k)}{c_{i+k}})\}} Q^p(u) du \right)^{\frac{2}{p}}.$$

By (6.1.13), $Q(u) \le c^{\frac{1}{r}} u^{-\frac{1}{r}}$ and setting $K = \frac{r-1}{rc^{\frac{1}{r}}}$ yields

$$\begin{aligned} b_{i,n} &\le c_i \sum_{k=0}^{n-i} c_{i+k} \left(\int_0^1 \mathbf{1}_{\left\{u \le G(\frac{\gamma_1(k)}{c_{i+k}})\right\}} c^{\frac{p}{r}} u^{-\frac{p}{r}} du \right)^{\frac{2}{p}} \\ &\le c_i \sum_{k=0}^{n-i} c_{i+k} \left(K \frac{\gamma_1(k)}{c_{i+k}} \right)^{\frac{r}{r-1}(1-\frac{p}{r})\frac{2}{p}}. \end{aligned}$$

Noting that $(c_n)_{n \ge 0}$ is decreasing, the result follows with $K_1 = K^{\frac{2(r-p)}{p(r-1)}}$. □

Equip $\mathbb{R}^d$ with its p-norm $\|(x_1, \ldots, x_d)\|_p^p = x_1^p + \cdots + x_d^p$. Let $(\xi_n)_{n \ge 0}$ be an $\mathbb{R}^d$-valued and $(\mathcal{F}, \Psi)$- dependent sequence. Set $\xi_n = (\xi_n^1, \ldots, \xi_n^d)$ then $\|\sum_{i=1}^n c_i \xi_i\|_p^p = \sum_{\ell=1}^d (\sum_{i=1}^n c_i \xi_i^\ell)^p$. If each component $(\xi_n^\ell)_{n \ge 0}$ is $(\mathcal{F}, \Psi)$-dependent and such that a relation like (6.1.15) holds, then $\mathbb{E}\|\Sigma_n\|_p^p < \infty$. Arguing as before, we deduce that the sequence $(\Sigma_n)_{n \ge 0}$ converges with probability 1. □

6.2 Examples of application

6.2.1 Robbins-Monro algorithm

The Robbins-Monro algorithm is used for dosage, to obtain level a of a function f which is usually unknown. It is also used in mechanics, for adjustments, as well as in statistics to fit a median (Duflo (1996) [82], page 50). It writes

$$Z_{n+1} = Z_n - c_n(f(Z_n) - a) + c_n \xi_{n+1}, \tag{6.2.1}$$

with $\sum c_n = \infty$ and $\sum c_n^2 < \infty$. It is usually assumed that the prediction error (ξ_n) is an identically distributed and independent random variables, but this does not look natural. Weak dependence seems more reasonable. Hence the previous results, ensure the convergence a.s. of this algorithm, under the usual assumptions and the conditions yielding the a.s. convergence of $\sum_0^n c_n \xi_{n+1}$. Under the assumptions of Proposition 6.2, if for some integer $p > 2$

$$\sum_{r=0}^{\infty} (r+1)^{p-2} \epsilon(r) < \infty, \tag{6.2.2}$$

then the algorithm (6.2.1) converges *a.s.*
If the assumptions of Proposition 6.3 hold, then the convergence a.s. of the algorithm (6.2.1) is ensured as soon as, for some $p > 2$,

$$\left(\sum_{r=0}^{\infty}(r+1)^{p-2}\epsilon(r)^{\frac{k-p}{k}}\right) \vee \left(\sum_{r=0}^{\infty}\epsilon(r)^{\frac{k-2}{k}}\right) < \infty.$$

Under the assumptions of Proposition 6.4, as soon as (6.1.17) is satisfied, the algorithm (6.2.1) converges with probability 1.

6.2.2 Kiefer-Wolfowitz algorithm

It is also a dosage algorithm. Here we want to reach the minimum z^* of a real function V which is $\mathcal{C}^2$ on an open set G of $\mathbb{R}^d$. The Kiefer-Wolfowiftz algorithm (Duflo (1996) [82], page 53) is stated as:

$$Z_{n+1} = Z_n - 2c_n \nabla V(Z_n) - \zeta_{n+1} \tag{6.2.3}$$

where $\zeta_{n+1} = \frac{c_n}{b_n}\xi_{n+1} + c_n b_n^2 q(n, Z_n)$, $\|q(n, Z_n)\| \leq K$ (for some $K > 0$), $\sum c_n = \infty$, $\sum_n c_n b_n^2 < \infty$ and $\sum_n (c_n/b_n)^2 < \infty$ (for instance, $c_n = \frac{1}{n}, b_n = n^{-b}$ with $0 < b < \frac{1}{2}$).
Usually, the prediction error (ξ_n) is assumed to be i.i.d, centered, square integrable and independent of Z_0. The previous results improve on this assumption until weakly dependent innovations. It is now enough to ensure the *a.s.* convergence of $\sum \frac{c_n}{b_n}\xi_{n+1}$. The weak dependence assumptions are the same as for the Robbins-Monro algorithm. Concerning the γ_1-weak dependence, the condition (6.1.17) is replaced by

$$\sum_{i=1}^{\infty}\left(\frac{c_i}{b_i}\right)^{2-m}\sum_{i=1}^{\infty}\gamma_1(i)^m < \infty.$$

6.3 Weighted dependent triangular arrays

In this section, we consider a sequence $(\xi_i)_{i\geq 1}$ of random variables and a triangular array of non-negative constant weights $\{(c_{ni})_{1\leq i\leq n}; n \geq 1\}$. Let

$$U_n = \sum_{i=1}^{n} c_{ni}\xi_i.$$

If the ξ_i's are i.i.d., Chow (1966) [37] has established the following complete convergence result.

Theorem (Chow (1966) [37]) *Let $(\xi_i)_i$ be independent and identically distributed random variables with $\mathbb{E}\xi_i = 0$ and $\mathbb{E}|\xi_i|^q < \infty$ for some $q \geq 2$. If for some constant K (non depending on n), $n^{1/q}\max_{1\leq i\leq n}|c_{ni}| \leq K$, and $\sum_{i=1}^n c_{ni}^2 \leq K$ then,*

$$\forall t > 0, \qquad \sum_{n=1}^{\infty} \mathbb{P}(n^{-1/q}|U_n| \geq t) < \infty.$$

The last inequality is a result of complete convergence of $n^{-1/q}|U_n|$ to 0. This notion was introduced by Hsu and Robbins (1947) [109]. Complete convergence implies the almost sure convergence from the Borel-Cantelli Lemma.
Li *et al.*(1995) [120] extend this result to arrays $(c_{ni})_{\{n\geq 1\ i\in Z\}}$ for $q = 2$. Quote also Yu (1990) [196], who obtains a result analogue to Chow's for martingale differences. Ghosal and Chandra (1998) [91] extend the previous results and prove some similar results to these of Li *et al.*(1995) [120] for martingales differences. As in [120], their main tool is Hoffmann-Jorgensen Inequality (Hoffmann-Jorgensen (1974) [107]). Peligrad and Utev (1997) [143] propose a central limit theorem for partial sums of a sequence $U_n = \sum_{i=1}^n c_{ni}\xi_i$ where $\sup_n c_{ni}^2 < \infty$, $\max_{1\leq i\leq n}|c_{ni}| \to 0$ as $n \to \infty$ and ξ_i's are in turn, pairwise mixing martingale difference, mixing sequences or associated sequences. Mcleish (1975) [128], De Jong (1996) [56], and, more recently Shinxin (1999) [177], extend the previous results in the case of L_q-mixingale arrays. Those results have various applications. They are used for the proof of strong convergence of kernel estimators. Li *et al.*(1995) [120] results are extended to our weak dependent frame. A straightforward consequence of Proposition 6.3 is the following.

Corollary 6.1. *Under the assumptions of Proposition 6.3, if q is an even integer such that $k > q > p$, and if for some constant K, non depending on n we assume that $\sum_{i=1}^n c_{n,i-1}^2 < K$, and if $\epsilon(r) = \mathcal{O}(r^{-\alpha})$, with $\alpha > (\frac{q-1}{k-q})k$, or $\epsilon(r) = \mathcal{O}(e^{-r})$, then for all positive real number t,*

$$\sum_n \mathbb{P}(n^{-1/p}|U_n| \geq t) < \infty.$$

Proof. Proposition 6.3 implies

$$\begin{aligned}\mathbb{E}|U_n|^q \;&\leq\; \frac{(2q-2)!}{(q-1)!}c^{1/k}\left(\left(M_q\sum_{i=1}^n c_{n,n-i}^q\sum_{r=0}^{n-1}(r+1)^{q-2}\epsilon(r)^{(k-q)/k}\right)\right.\\ &\vee \left.\left(M_2\left(\sum_{i=1}^n c_{n,n-i}^2\sum_{r=0}^{n-1}\epsilon(r)^{(k-2)/k}\right)^{p/2}\right)\right).\end{aligned}$$

If $\sum_{i=1}^n c_{n,i-1}^2 < K$ and $\epsilon(r) = \mathcal{O}(r^{-\alpha})$, with $\alpha > (\frac{q-1}{k-q})k$, then there is some $K_1 > 0$ with $\mathbb{E}|U_n|^q < K_1$, ¿the result follows from $\mathbb{P}(n^{-1/p}|U_n| > t) \leq \dfrac{\mathbb{E}|U_n|^q}{t^q n^{q/p}}$.

If $\sum_{i=1}^{n} c_{n,i-1}^2 < K$ and $\epsilon(r) = \mathcal{O}(e^{-\theta r})$, then $\mathbb{E}|U_n|^q < K_2$ for a real constant K_2 and $\sum_n \mathbb{P}\left(n^{-1/p}|U_n| > t\right) < \infty$. □

The following corollary is a direct consequence of proposition 6.4.

Corollary 6.2. *Suppose that all the assumptions of Proposition 6.4 are satisfied. If $q > p$, $k > q > 1$, and $\sum_{i=1}^{\infty} c_{ni}^{2-n} \sum_{j=0}^{\infty} \gamma_1(j)^m < \infty$ where $m = \frac{2}{q}(\frac{k-q}{k-1})$, then for any positive real number t,*

$$\sum_n \mathbb{P}(n^{-1/p}|U_n| \geq t) < \infty.$$

Proof. We have from Proposition 6.4, $\mathbb{E}|U_n|^q \leq \left(2qK_1 \sum_{i=1}^{n} c_{ni}^{2-m} \sum_{j=0}^{n-1} \gamma_1(j)^m\right)^{q/2}$.

Now the relation $\sum_{i=1}^{\infty} c_{ni}^{2-n} \sum_{j=0}^{\infty} \gamma_1(j)^m < \infty$ implies $\sum_n \mathbb{P}(n^{-1/p}|U_n| > t) < \infty$. This concludes the proof. □

6.4 Linear regression

We observe a stationary bounded sequence, $(y_n, x_n) \in \mathbb{R} \times \mathbb{R}^d$, defined on a probability space $(\Omega, \mathcal{A}, \mathcal{P})$.

We look for the vector Z^* which minimizes the linear prediction error of y_n with x_n. We identify the $\mathbb{R}^d$-vector x_n and its column matrix in the canonical basis. So

$$Z^* = \arg\min_{Z \in \mathbb{R}^d} \mathbb{E}[(y_n - x_n^T Z)^2].$$

This problem leads to study the gradient algorithm

$$Z_{n+1} = Z_n + c_n(y_{n+1} - x_{n+1}^T Z_n)x_{n+1},$$

where $c_n = \frac{g}{n}$ with $g > 0$ (so (c_n) verifies (6.1.2) and (6.1.5)). Let $C_{n+1} = x_{n+1}x_{n+1}^T$, we obtain:

$$Z_{n+1} = Z_n + c_n(y_{n+1}x_{n+1} - C_{n+1}Z_n). \tag{6.4.1}$$

Denote $U = \mathbb{E}(y_{n+1}x_{n+1})$, $C = \mathbb{E}(C_{n+1})$, $Y_n = Z_n - C^{-1}U$ and $h(Y) = -CY$, then (6.4.1) becomes :

$$Y_{n+1} = Y_n + c_n h(Y_n) + c_n \zeta_{n+1}, \quad \text{with} \tag{6.4.2}$$

$$\zeta_{n+1} = (y_{n+1}x_{n+1} - C_{n+1}C^{-1}U) + (C - C_{n+1})Y_n. \tag{6.4.3}$$

Note that the solutions of (6.1.3) are the trajectories

$$z(t) = z_0 e^{-Ct},$$

so every trajectory converges to 0, the unique equilibrium point of the differentiable function h ($Dh(0) = -C$ and 0 is an attractive zero of h).
Denoting $\mathcal{F}_n = (\sigma(Y_i) \,/\, i \le n)$, we also define the following assumption **A-lr**: *C is not singular, (C_n) and (y_n, x_n) are γ_1-dependent sequences with $\gamma_1(r) = \mathcal{O}(a^r)$ for $a < 1$.*

Note that if $(y_n, x_n)_{n\in\mathbb{N}}$ is $\theta_{1,1}$-dependent (see Definition 2.3) then **A-lr** is satisfied. First, note that if a $\mathbb{R}^d$-valued sequence (X_n) is $\theta_{1,1}$-dependent, any $\mathbb{R}^j$-valued sequence ($j = 1, \dots, d-1$) $(Y_n) = (X_n^{t_1}, \dots, X_n^{t_j})$ is $\theta_{1,1}$-dependent. So, if (y_n, x_n) is $\theta_{1,1}$-dependent, then so are (y_n) and (x_n^j) ($j = 1, \dots, d$). Let f a bounded 1-Lipschitz function, defined on $\mathbb{R}$ and g the function defined on $\mathbb{R}^2$ by $g(x, y) = f(xy)$. It is enough to prove that g is a Lipschitz function defined on $\mathbb{R}^2$.

$$\begin{aligned}\frac{|g(x,y) - g(x',y')|}{|x - x'| + |y - y'|} &\le \frac{|xy - x'y'|}{|x - x'| + |y - y'|}\\ &\le \frac{|x||y - y'| + |y'||x - x'|}{|x - x'| + |y - y'|}\\ &\le \max(|x|, |y'|),\end{aligned}$$

and g is Lipschitz as soon as x and y are bounded. Thus, since (x_n) and (y_n) are bounded, the result follows. □

Denoting $M = \sup_n \|x_n\|^2$, one has,

Proposition 6.5. *Under Assumption* **A-lr** *(Y_n) is a.s. bounded and the perturbation (ζ_n) of algorithm (6.4.2) splits into three terms of which two are γ_1-dependent and one is a rest leading to zero. So the ODE method assures the a.s convergence of Y_n to zero (hence $Z^* = C^{-1}U$). Moreover if $g < \frac{1}{2M}$ then*

$$\sqrt{n}Y_n = \mathcal{O}(1), \qquad a.s. \tag{6.4.4}$$

Proof of Proposition 6.5. To start with, we prove that $Y_n \to 0$ a.s by assuming that (Y_n) is a.s bounded. Then we justify this assumption and finally we prove (6.4.4).
The perturbation ζ_{n+1} splits into two terms : $(y_{n+1}x_{n+1} - C_{n+1}C^{-1}U)$ and $(C - C_{n+1})Y_n$. The first term is centered and obvious γ_1-dependent with a dependence coefficient $\gamma_1(r)$. Now $\gamma_1(r) = \mathcal{O}(a^r)$ thanks to Assumption **A-lr**. It remains to study $(C - C_{n+1})Y_n$.
Study of $(C - C_{n+1})Y_n$: write $(C - C_{n+1})Y_n = \xi_{n+1} + r_{n+1}$ with
$\xi_{n+1} = (C - C_{n+1})Y_n - \mathbb{E}[(C - C_{n+1})Y_n]$ and $r_{n+1} = \mathbb{E}[(C - C_{n+1})Y_n]$. We will prove that the sequence (ξ_n) is γ_1-dependent with an appropriate dependent coefficient and that $\lim_{n\to\infty} r_n = 0$. Notice that

$$r_{n+1} = \mathbb{E}[(C - C_{n+1}) \sum_{j=\frac{n}{2}}^{n-1} (Y_{j+1} - Y_j)] + \mathbb{E}[(C - C_{n+1})Y_{\frac{n}{2}}],$$

and since $Y_{j+1} - Y_j = -c_j C_{j+1} Y_j + c_j(y_{j+1}x_{j+1} - C_{j+1}C^{-1}U)$, we obtain

$$r_{n+1} = \mathbb{E}(C - C_{n+1})Y_{\frac{n}{2}} - \sum_{j=\frac{n}{2}}^{n-1} \mathbb{E}(C - C_{n+1})c_j C_{j+1} Y_j$$
$$+ \sum_{j=\frac{n}{2}}^{n-1} \mathbb{E}(C - C_{n+1})c_j(y_{j+1}x_{j+1} - C_{j+1}C^{-1}U).$$

If $\frac{n}{2}$ is not an integer, we replace it by $\frac{n-1}{2}$. In the same way, in the first sum we replace Y_j by $\sum_{i=j/2}^{j-1}(Y_{i+1} - Y_i) + Y_{j/2}$ with the same remark as above if $j/2$ is not an integer. So

$$r_{n+1} = \sum_{j=\frac{n}{2}}^{n-1} \mathbb{E}(C - C_{n+1})c_j(y_{j+1}x_{j+1} - C_{n+1}C^{-1}U) + \mathbb{E}(C - C_{n+1})Y_{\frac{n}{2}}$$
$$-\sum_{j=\frac{n}{2}}^{n-1} \mathbb{E}(C - C_{n+1})c_j C_{j+1} \left[\sum_{i=j/2}^{j-1} -c_i C_{i+1} Y_i + c_i(y_{i+1}x_{i+1} - C_{i+1}C^{-1}U) + Y_{j/2} \right]$$

Expectations conditionally with respect to $\mathcal{F}_{j+1}$ of each term of the second sum and with respect $\mathcal{F}_{\frac{n}{2}}$ of the last term give, by assuming that (Y_n) is bounded:

$$\|r_{n+1}\| \leq \|A\| + K_1 \sum_{j=\frac{n}{2}}^{n-1} c_j \gamma_1(n + 1 - j)\gamma_1(j + 1) + K_2\gamma_1(n/2 + 1), \quad (6.4.5)$$

where A denote the last sum of the previous representation of r_n, and K_i (for $i = 1, 2, \ldots$) non-negative constants. Moreover,

$$A = - \sum_{j=\frac{n}{2}}^{n-1} \mathbb{E}(C - C_{n+1})c_j(C_{j+1} - C)[$$
$$\times \sum_{i=j/2}^{j-1} -c_i C_{i+1} Y_i + c_i(y_{i+1}x_{i+1} - C_{i+1}C^{-1}U)]$$
$$- \sum_{j=\frac{n}{2}}^{n-1} \mathbb{E}(C - C_{n+1})c_j(C_{j+1} - C)Y_{j/2}$$

$$+\ \sum_{j=\frac{n}{2}}^{n-1}\mathbb{E}(C-C_{n+1})c_jC[\sum_{i=j/2}^{j-1}-c_iC_{i+1}Y_i+c_i(y_{i+1}x_{i+1}-C_{i+1}C^{-1}U)]$$
$$+\ \sum_{j=\frac{n}{2}}^{n-1}\mathbb{E}(C-C_{n+1})c_jCY_{j/2}.$$

Expectations conditionally successively with respect to $\mathcal{F}_{j+1}$ and $\mathcal{F}_{i+1}$ of each term of the first and the third sum and with respect $\mathcal{F}_{j+1}$ then $\mathcal{F}_{\frac{j}{2}}$ of the second and the fourth sum give, by assuming that (Y_n) is bounded,

$$\|A\|\leq K_3\sum_{j=\frac{n}{2}}^{n-1}c_j\gamma_1(n-j)\sum_{i=j/2}^{j-1}c_i\gamma_1(j-i)+K_4\sum_{j=\frac{n}{2}}^{n-1}c_j\gamma_1(n-j)\gamma_1(j/2).\quad(6.4.6)$$

Since $c_j=\frac{g}{j}$, (6.4.5) and (6.4.6) involve that r_n is $\mathcal{O}(n^{-2})$, so r_n converges to zero. On the other hand, for $r\geq 6$:

$$\begin{aligned}\mathbb{E}(\xi_{n+r}|\mathcal{F}_n) &= \mathbb{E}[(C-C_{n+r})Y_{n+r-1}|\mathcal{F}_n]-\mathbb{E}(\xi_{n+r}),\\ &= \sum_{j=n+\frac{r}{2}}^{n+r-2}\mathbb{E}[(C-C_{n+r})(Y_{j+1}-Y_j)|\mathcal{F}_n]\\ &+ \mathbb{E}[(C-C_{n+r})Y_{n+\frac{r}{2}}|\mathcal{F}_n]-r_{n+r}.\end{aligned}$$

Note also that if $\frac{r}{2}$ is not an integer, we replace it by $\frac{r+1}{2}$. Using the same technics as above, we obtain

$$\begin{aligned}&\mathbb{E}\|\mathbb{E}(\xi_{n+r}|\mathcal{F}_n)\|\\ &\leq K_5\left(\sum_{j=n+\frac{r}{2}}^{n+r-2}c_j\gamma_1(n+r-j-1)\sum_{i=j/2}^{j-1}c_i\gamma_1(j-i)+\gamma_1(r/2)\right)+\mathcal{O}((n+r)^{-2})\end{aligned}$$

hence,

$$\mathbb{E}\|\mathbb{E}(\xi_{n+r}|\mathcal{F}_n)\|\leq\mathcal{O}((n+\frac{r}{2})^{-2})+K_5\gamma_1(\frac{r}{2})+\mathcal{O}((n+r)^{-2}),$$

and $\|\mathbb{E}(\xi_{n+r}|\mathcal{F}_n)\|_1=\gamma_1(r)$, with $\gamma_1(r)=\mathcal{O}(r^{-2})$. So (ξ_n) is γ_1-dependent and since (6.1.17) is satisfied, the ODE method may be used and Y_n converges to 0 $a.s.$

Now we prove that Y_n is $a.s$ bounded. Let $V(Y)=Y^TCY=\|\sqrt{C}Y\|^2$. Since C is not singular, V is a Lyapounov function and $\nabla V(Y)=2CY$ is a Lipschitz function, so we have

$$V(Y_{n+1})\leq V(Y_n)+(Y_{n+1}-Y_n)^T\nabla V(Y_n)+K_6\|Y_{n+1}-Y_n\|^2.$$

Furthermore $\|Y_{n+1}-Y_n\|^2 \le 2c_n^2(\|y_{n+1}x_{n+1}-C_{n+1}C^{-1}U\|^2+2c_n^2\|C_{n+1}Y_n\|^2)$. Since (y_n,x_n) is bounded, (C_n) and $\|y_{n+1}x_{n+1}-C_{n+1}C^{-1}U\|^2$ are also bounded. Moreover

$$\|C_{n+1}Y_n\|^2 \le K_7\|Y_n\|^2 \le \frac{K_7}{\lambda_{\min}(C)}V(Y)$$

where $\lambda_{\min}(C)$ is the smallest eigenvalue of C. So

$$V(Y_{n+1}) \le V(Y_n)(1+K_8c_n^2)+K_9c_n^2+2(Y_{n+1}-Y_n)^TCY_n.$$

The last term becomes

$$\begin{aligned}2(Y_{n+1}-Y_n)^TCY_n &= -2c_n\|CY_n\|^2+2c_n(y_{n+1}x_{n+1}-C_{n+1}C^{-1}U)^TCY_n\\ &\quad +2c_nY_n^T(C-C_{n+1})CY_n,\\ &\le -2c_n\|CY_n\|^2+c_nK_{10}\|CY_n\|+2c_nY_n^T(C-C_{n+1})CY_n\\ &\le -2c_n\|CY_n\|^2+c_nK_{10}\|CY_n\|+2c_nu_{n+1}V(Y_n),\end{aligned}$$

where $u_n=\max\{X_i^T(C-C_n)X_i\ 1\le i\le d\}$ and $\{X_1,\dots,X_d\}$ is an orthogonal basis of unit eigenvectors of C.

We now obtain

$$\begin{aligned}V(Y_{n+1}) &\le V(Y_n)(1+K_8c_n^2+2c_nu_{n+1})\\ &+ K_9c_n^2-c_n(2\|CY_n\|^2-K_{10}\|CY_n\|).\end{aligned}\tag{6.4.7}$$

Note that under the assumption **A-lr** (u_n) is a γ_1-dependent sequence with a weakly dependent coefficient $\gamma_1(r)=\mathcal{O}(a^r)$ and $\sum_{n=0}^\infty c_nu_{n+1}<\infty$.
If $V(Y_n)\ge \frac{K_{10}^2}{4\lambda_{\min}(C)}$, then $\|CY_n\|\ge K_{10}/2$ and $-(2\|CY_n\|^2-K_{10}\|CY_n\|)\le 0$.

Denote $T=\inf\{n/\ V(Y_n)\le \frac{K_{10}^2}{4\lambda_{\min}(C)}\}$. By the Robbins-Sigmund theorem, $V(Y_n)$ converges a.s. to a finite limit on $\{T=+\infty\}$, so (Y_n) is bounded since V is a Lyapounov function.
On $\{\liminf_n V(Y_n)\le \frac{K_{10}^2}{4\lambda_{\min}(C)}\}$, $V(Y_n)$ does not converge to ∞ and using Theorem 2 of Delyon (1996) [58], we deduce that $V(Y_n)$ converges to a finite limit, as soon as :

$$\forall k>0,\quad \sum c_n^2\ \|h(Y_n)+\zeta_{n+1}\|^2\ \mathbf{1}_{\{V(Y_n)<k\}}<\infty \tag{6.4.8}$$

$$\forall k>0,\quad \sum c_n\langle\zeta_{n+1},\nabla V(Z_n)\rangle\ \mathbf{1}_{\{V(Y_n)<k\}}<\infty. \tag{6.4.9}$$

Using relation $\sum c_n^2<\infty$ and the fact that on $\{V(Y_n)<k\}$, $\|h(Y_n)+\zeta_{n+1}\|^2$ is bounded, we deduce (6.4.8). To prove (6.4.9), it is enough, by Proposition 6.4, to prove that $\langle\zeta_{n+1},\nabla V(Y_n)\rangle\ \mathbf{1}_{\{V(Y_n)<k\}}=e_{n+1}$ is a γ_1-dependent sequence with a dependent coefficient satisfying (6.1.17). But to use the result of Proposition 6.4, it is necessary to center e_{n+1}. So we are going to prove that $\sum c_n\mathbb{E}e_{n+1}<\infty$

and that $(e_{n+1} - \mathbb{E}e_{n+1})$ is a γ_1-dependent sequence with a dependent coefficient $\gamma_1(r)$ equals to $\mathcal{O}(r^{-2})$.
Study of $\mathbb{E}(e_{n+1})$. First of all, we must note a few elements. Denoting I the unit matrix of $\mathbb{R}^d$, $Y_n = (I - c_{n-1}C_n)Y_{n-1} + c_{n-1}(x_n y_n - C_n C^{-1}U)$. Note that $\lambda_{\max}(C_n) = \|x_n\|^2 \leq M$ ($\lambda_{\max}(C_n) =:$ the largest eigenvalue of C_n). For n large enough $c_{n-1}M < 1$ and $(I - c_{n-1}C_n)$ is not singular. So, if $M_1 = \sup_n\{x_n y_n - C_n C^{-1}U\}$, then we obtain

$$\|Y_{n-1}\| \leq \frac{1}{1 - c_{n-1}\lambda}(\|Y_n\| + c_{n-1}M_1) \leq (1 + bc_{n-1})(\|Y_n\| \wedge M_1)$$

where $b \geq 0$ does not depend on n. Moreover

$$V(Y_n) < k \Longrightarrow \|Y_n\|^2 < \frac{k}{\lambda_{\min}(C)},$$

and

$$\|Y_n\| < k' \Longrightarrow V(Y_n) < \lambda_{\max}(C)k'^2.$$

So that,

$$\mathbf{1}_{\{V(Y_n)<k\}} = \mathbf{1}_{\{\|Y_n\|<k_n\}} = \mathbf{1}_{\{\|Y_{n-j}\|<k_{n-j}\}},$$

where

$$k_{n-j} \leq (1 + c_{n-1})^j \left(\sqrt{\frac{k}{\lambda_{\min}(C)}} \wedge M_1\right).$$

And since $c_n = \frac{g}{n}$, for any $0 \leq j \leq n$, $(1 + ac_{n-1})^j$ is bounded independently of n, so is k_{n-j}. And $\mathbb{E}(e_{n+1}) = \mathbb{E}(x_{n+1}y_{n+1} - C_{n+1}C^{-1}U)^T CY_n \mathbf{1}_{\{V(Y_n)<k\}} + \mathbb{E}Y_n^T(C - C_{n+1})CY_n \mathbf{1}_{\{V(Y_n)<k\}}$. We have,

$$\begin{aligned}
\mathbb{E}(e_{n+1}) &= \sum_{j=\frac{n}{2}}^{n-1} \mathbb{E}(y_{n+1}x_{n+1} - C_{n+1}C^{-1}U)^T C(Y_{j+1} - Y_j)\ \mathbf{1}_{\{\|Y_j\|<k_j\}} \\
&+ \mathbb{E}(y_{n+1}x_{n+1} - C_{n+1}C^{-1}U)^T CZ_{\frac{n}{2}} \\
&+ \sum_{j=\frac{n}{2}}^{n-1} \mathbb{E}(Y_{j+1} - Y_j)^T(C - C_{n+1})C(Y_{j+1} - Y_j)\ \mathbf{1}_{\{\|Y_j\|<k_j\}} \\
&+ 2\sum_{j=\frac{n}{2}}^{n-1}\sum_{i=j+1}^{n-2} \mathbb{E}(Y_{j+1} - Y_j)^T(C - C_{n+1}) \\
&\quad \times C(Y_{i+1} - Y_i)\ \mathbf{1}_{\{\|Y_j\|<k_j\}\cap\{\|Y_i\|\leq k_i\}} \\
&+ 2\sum_{j=\frac{n}{2}}^{n-1} \mathbb{E}Y_{n/2}^T(C - C_{n+1})C(Y_{j+1} - Y_j)\ \mathbf{1}_{\{\|Y_j\|<k_j\}\cap\{\|Y_{n/2}\|<k_{n/2}\}} \\
&+ \mathbb{E}Y_{n/2}^T(C - C_{n+1})CY_{n/2}\ \mathbf{1}_{\{\|Y_{n/2}\|<k_{n/2}\}}.
\end{aligned}$$

Note that if $\frac{n}{2}$ is not an integer, we replace it by $\frac{n-1}{2}$. Using always the same technique, we obtain that

$$\mathbb{E}e_{n+1} = \mathcal{O}(n^{-2}) + \mathcal{O}(a^{\frac{n}{2}}) + \mathcal{O}(n^{-2}) + \mathcal{O}(n^{-2}) + \mathcal{O}(a^{\frac{n}{2}}),$$

hence $\sum c_n \mathbb{E}e_n < \infty$.

Study of $(e_n - \mathbb{E}e_n)$. We now prove that this sequence is γ_1-dependent with a relevant dependent coefficient. Write

$$\mathbb{E}(e_{n+r} - \mathbb{E}e_{n+r}|\mathcal{F}_n) = D_{n+r} + G_{n+r} - \mathbb{E}e_{n+r},$$

with

$$\begin{aligned}
D_{n+R} &= \mathbb{E}[(y_{n+r}x_{n+r} - C_{n+1}C^{-1}U)^T CY_{n+r-1}\, \mathbf{1}_{\{V(Y_{n+r-1})<k\}}\, |\mathcal{F}_n],\\
G_{n+r} &= \mathbb{E}(Y^T_{n+r-1}(C - C_{n+r})CY_{n+r-1}\, \mathbf{1}_{\{V(Y_{n+r-1})<k\}}\, |\mathcal{F}_n]\\
D_{n+r} &= \sum_{j=n+\frac{r}{2}}^{n+r-2} \mathbb{E}[((y_{n+r}x_{n+r} - C_{n+1}C^{-1}U)^T C(Y_{j+1} - Y_j)\, \mathbf{1}_{\{\|Y_j\|<k_j\}}\, |\mathcal{F}_n]\\
&+ \mathbb{E}[((y_{n+r}x_{n+r} - C_{n+1}C^{-1}U)^T CY_{n+\frac{r}{2}}\, \mathbf{1}_{\{\|Y_{n+\frac{r}{2}}\|<k_{n+\frac{r}{2}}\}}\, |\mathcal{F}_n],
\end{aligned}$$

Here again, if $\frac{r}{2}$ is not a integer, we replace it by $\frac{r-1}{2}$. Again, the same techniques as for r_n give

$$\mathbb{E}\|D_{n+r}\| = \mathcal{O}((n+r)^{-2}) + \mathcal{O}(a^{n+\frac{r}{2}})$$

We study G_{n+r} in the same way and $\mathbb{E}\|G_{n+r}\| = \mathcal{O}((n+r)^{-2})$, and since $\mathbb{E}e_{n+r} = \mathcal{O}((n+r)^{-2})$, (6.1.17) is satisfied and the result is proved.

Proof of (6.4.4) For $n > N$, denote by $\Pi_n^N = (I - c_nC_{n+1})\cdots(I - c_NC_{N+1})$. Since $g < \frac{1}{2M}$, for $N \geq 1$, Π_n^N is no singular and

$$Y_{n+1} = \Pi_n^N Y_N + \sum_{j=N}^{n} c_j \Pi_n^N (\Pi_j^N)^{-1}\xi_{j+1}^1,$$

where $\xi_{j+1}^1 = y_{j+1}x_{j+1} - C_{j+1}C^{-1}U$. We obtain, since $Y_n \to 0$,

$$-Y_N = \sum_{j=N}^{\infty} c_j (\Pi_j^N)^{-1}\xi_{j+1}^1.$$

$(\Pi_j^N)^{-1} = (I - c_NC_{N+1})^{-1}\cdots(I - c_jC_{j+1})^{-1}$ and

$$\|(\Pi_j^N)^{-1}\| \leq \frac{1}{\prod_{i=N}^{j}(1 - c_iM)}.$$

Hence

$$\|(\Pi_j^N)^{-1}\| = \mathcal{O}\left(\exp\left(M\sum_{i=N}^{j} c_i\right)\right) = \mathcal{O}\left(\left(\frac{j}{N}\right)^{gM}\right), \tag{6.4.10}$$

$$\|\sqrt{N}Y_N\| = \left\|\sum_{j=N}^{\infty} \frac{g}{\sqrt{j}}\sqrt{\frac{N}{j}}(\Pi_j^N)^{-1}\xi_{j+1}^1\right\|.$$

Since $g < \frac{1}{2M}$, (6.4.10) involves that the sum converges. Indeed (6.1.17) is verified with $k = 5$ and $p = 3$ (so $m = \frac{1}{3}$) and since ξ_{j+1}^1 is γ_1-dependent with a mixingale coefficient $\gamma_1(r) = \mathcal{O}(a^r)$. Hence the result is proved. □

Chapter 7

Central Limit theorem

In this chapter, we give sufficient conditions for the central limit theorem in the non causal and causal contexts. In Sections 7.1 and 7.2, we give sufficient conditions for κ and λ dependent sequences, for random variables having moments of order $2+\zeta$. The proof is based on a decomposition which combines Bernstein blocks with the Lindeberg method. In Section 7.3, we prove a central limit theorem for random fields, under an exponential decay of the covariance of Lipshitz functions of the variables. The proof is based on Stein's method, as described in Bolthausen (1982) [24]. In Section 7.4, we focus on the causal case: in Theorem 7.5, we give necessary and sufficient conditions for the conditional central limit theorem. This notion is more precise than convergence in distribution and implies the stable convergence in the sense of Rényi (1963) [156]. In the last Section 7.5, we give some applications of Theorem 7.5: in particular, we give sufficient conditions for the central limit theorem for γ, $\tilde{\alpha}$ and $\tilde{\phi}$-dependent sequences.

7.1 Non causal case: stationary sequences

In all the section, we shall consider a centered and stationary real-valued sequence $(X_n)_{n\in\mathbb{Z}}$ such that

$$\mu = \mathbb{E}|X_0|^m < \infty, \qquad \text{for a real number } m = 2+\zeta > 2. \tag{7.1.1}$$

We also define

$$\sigma^2 = \sum_{k\in\mathbb{Z}} \mathrm{Cov}(X_0, X_k),$$

whenever it exists. Let $S_n = X_1 + \cdots + X_n$. The following results come from Doukhan and Wintenberger (2005) [77].

Theorem 7.1 (κ-dependence). *Assume that the κ-weakly dependent stationary process satisfies (7.1.1) and that $\kappa(r) = \mathcal{O}(r^{-\kappa})$ for $\kappa > 2 + \frac{1}{\zeta}$. Then σ^2 is well defined and $n^{-1/2}S_n$ converges in distribution to the normal distribution with mean 0 and variance σ^2.*

Remark 7.1. *Under the more restrictive ζ-dependence condition, Bulinski and Shashkin (2005) [33] obtain the central limit theorem with the sharper assumption $\zeta(r) = \mathcal{O}(r^{-\kappa})$ for $\kappa > 1 + 1/(m-2)$ (beware notations in this remark). The difference between the two conditions is natural, since it may be proved for ζ-weakly dependent sequences that $\zeta(r) \geq \sum_{s \geq r} \kappa(s)$. This simple bound, checked from the definitions, explains the loss in the rate of convergence of $\kappa(r)$ to 0.*

The following result relaxes the previous dependence assumptions to the cost of a faster decay for the dependence coefficients.

Theorem 7.2 (λ-dependence). *Assume that the λ-weakly dependent stationary process satisfies (7.1.1) and that $\lambda(r) = \mathcal{O}(r^{-\lambda})$ for $\lambda > 4 + \frac{2}{\zeta}$. Then the conclusion of Theorem 7.1 holds.*

As stressed in section 3.1.3 the coefficient λ is very useful to work out the case of Bernoulli shifts with weakly dependent innovations $(\xi_i)_i$.

Theorem 7.3. *Denote by $\lambda_\xi(r)$ the λ-dependence coefficients of the sequence $(\xi_i)_i$. Assume that $H : \mathbb{R}^{\mathbb{Z}} \to \mathbb{R}$ satisfies the condition (3.1.11) for some $m > 2$ such that $lm \leq m' - 1$ with $\mathbb{E}|\xi_0|^{m'} < \infty$, and some sequence $b_i \geq 0$ such that $\sum_i |i| b_i < \infty$. Then $X_n = H(\xi_{n-i}, i \in \mathbb{Z})$ satisfy the central limit theorem in the following cases:*

- ***Geometric case:*** *$b_r = \mathcal{O}\left(e^{-rb}\right)$ and $\lambda_\xi(r) = \mathcal{O}\left(e^{-rc}\right)$.*
- ***Mixed case:*** *$b_r = \mathcal{O}\left(e^{-rb}\right)$ and $\lambda_\xi(r) = \mathcal{O}\left(r^{-c}\right)$ with $c > 4 + 2/(m-2)$.*
- ***Riemanian case:*** *If $b_r = \mathcal{O}\left(r^{-b}\right)$ for some $b > 2$ and $\lambda_\xi(r) = \mathcal{O}\left(r^{-c}\right)$ with*
$$c > \frac{(10-4m)b(m'-1)}{(2-m)(b-2)(m'-1-l)}.$$

The constants $b > 0$ and $c > 0$ obtained are different for each case.

This theorem is useful to derive the weak invariance principle in many cases (see again Section 3.1.3). We now look with more detail the following example.

Example. Consider the two sided sequence $X_t = \sum_{-\infty}^{\infty} a_i \, \xi_{t-i}$ with ARCH(∞) innovations:

$$\xi_t = \tilde{\xi}_t \left(a' + \sum_{j=1}^{\infty} a'_j \xi_{t-j} \right),$$

where the process $(\tilde{\xi}_i)_i$ is i.i.d.. Under Riemanian decays ($a_r = \mathcal{O}(r^{-a})$ and $a'_r = \mathcal{O}(r^{-a'})$), we infer from Theorem 7.3 that the central limit theorem holds as soon as:

$$a' > \frac{(10-4m)a(m'-1)}{(2-m)(a-2)(m'-1-l)} + 1.$$

Remark 7.2. *The technique of the proofs is based on Lindeberg method and we prove in fact that* $|\mathbb{E}\left(f(S_n/\sqrt{n}) - f(\sigma N)\right)| \le C n^{-c^*}$ *for* $f(x) = e^{itx}$*, where the constants* $c^*, C > 0$ *depend only on the parameters* ζ *and* κ *or* λ *respectively, and where* $c^* < \frac{1}{2}$ *(see Proposition 2 section 7.2.2 for more details). When* κ *or* λ *tends to infinity, we have* $c^* = \zeta/(4+\zeta)$*. For* $\zeta \ge 2$ *and* κ *or* λ *tends to infinity, we notice that* $c^* \to \frac{1}{3}$*.*

Using a smoothing lemma, this also yields an analogue bound for the uniform distance in the real case ($d = 1$)*:*

$$\sup_{t \in \mathbb{R}} \left| \mathbb{P}\left(\frac{1}{\sqrt{n}} S_n \le t \right) - \mathbb{P}\left(\sigma N \le t\right) \right| \le C n^{-c'}.$$

A first and easy way to control c' *is to set* $c' = c^*/4$ *but the corresponding rate is really a bad one. Petrov (1995) [144] obtains the exponent* $\frac{1}{2}$ *in the i.i.d. case and Rio (2000) [161] reaches the exponent* $\frac{1}{3}$ *for strongly mixing sequences. In proposition 3 section 7.2.2, we achieve* $c' = c^*/3$*. Analogous bad convergence rates have been settled in the case of weakly dependent random fields in [63].*

The following subsections are devoted to the proofs. We first describe in detail the Lindeberg method with Bernstein blocks in Section 7.2 (another version of the Lindeberg method will be presented is the causal framework in Section 7.4.3). The main tools are the controls of the variance of S_n and of $\|S_n\|_{2+\delta}$ obtained in Lemma 4.2 and 4.3 of section 4.2. Rates of convergence for the central limit theorem are obtained in 7.2.2.

7.2 Lindeberg method

Let $x_1, \dots, x_k$ be random variables with values in $\mathbb{R}^d$ (equipped with the Euclidean norm $\|(x_1, \dots, x_d)\| = \sqrt{x_1^2 + \cdots + x_d^2}$), centered at expectation and such that for some $0 < \delta \le 1$:

$$\sum_{i=1}^{k} \mathbb{E}\|x_i\|^{2+\delta} \le A < \infty \tag{7.2.1}$$

We consider independent random variables $y_1, \dots, y_k$, independent of the variables $x_1, \dots, x_k$ and such that $y_i \sim \mathcal{N}_d(0, \operatorname{Var} x_i)$.
Denote by $\mathcal{C}_b^3$ the set of bounded functions from $\mathbb{R}^d$ to $\mathbb{R}$ with bounded and continuous partial derivatives up to order 3. The Lindeberg method relies on the following lemma

Lemma 7.1 (Bardet *et al.*, 2006 [10]). *For any $f \in \mathcal{C}_b^3$, let:*

$$\begin{aligned} \Delta &= |\mathbb{E}\,(f(x_1 + \cdots + x_k) - f(y_1 + \cdots + y_k))| && (7.2.2)\\ \text{and } T_j &= \sum_{i=1}^k \operatorname{Cov}\left(f_i{}^{(j)}{}_i(x_1 + \cdots + x_{i-1}), x_i^j\right), && j = 1, 2. \ (7.2.3)\\ f_i(t) &= |\mathbb{E}\, f(t + y_{i+1} + \cdots + y_k) \end{aligned}$$

Let

$$\begin{aligned} \operatorname{Cov}(f'(x), y) &= \sum_{\ell=1}^d \operatorname{Cov}\left(\frac{\partial f}{\partial x_\ell}(x), y_\ell\right), \text{ and}\\ \operatorname{Cov}(f''(x), y^2) &= \sum_{k=1}^d \sum_{\ell=1}^d \operatorname{Cov}\left(\frac{\partial^2 f}{\partial x_k \partial x_\ell}(x), y_k y_\ell\right), \end{aligned}$$

where by convention the empty sums are equal to 0. The following upper bound holds:

$$\Delta \le |T_1| + \frac{1}{2}|T_2| + 4\|f''\|_\infty^{1-\delta}\|f'''\|_\infty^\delta A.$$

Remark 7.3. *If k tends to ∞, we denote the variables by $(x_{i,k})_{1\le i\le k}$ and we set $A = A(k), T_j = T_j(k)$. Assume that $A(k)$ and $T_j(k)$ tend to 0 as k tends to ∞, and assume moreover that $\sigma_k^2 = \sum_{i=1}^k \mathbb{E} x_{i,k}^2$ converges to σ^2 as k tends to ∞. Then*

$$S_k = \sum_{i=1}^k x_{i,k} \to_{k\to\infty} \mathcal{N}(0, \sigma^2), \qquad \text{in distribution.}$$

Condition $A(k) \to 0$ implies the usual Lindeberg condition, condition $\sigma_k^2 \to \sigma^2$ is only the convergence of variances, while the conditions $T_j(k) \to 0$ are the only one related to dependence.

Examples. Assume that $(X_t)_{t\in\mathbb{Z}}$ is a stationary times series the following examples are widely developed in [10].

- The first example of application of such a situation is the Bernstein block method used below for proving the central limit theorem.
- Functional estimation also enters this frame. In that case we write $x_{i,k} = f_k(X_i)$ for functions f_k which approximate the Dirac distribution, see chapter 11.

- A final example is provided by subsampling, in which $x_{i,k} = X_{im_n}$ for $1 \le im_n \le n$ and $1 \ll m_n \ll n$; this example fits chapter 13 but we defer the reader to [10] for shortness.

Proof of lemma 7.1. We first notice that:

$$\begin{aligned} \Delta &\le \Delta_1 + \cdots + \Delta_k, \quad \text{where} && (7.2.4) \\ \Delta_i &= |\mathbb{E}\,(f_i(w_i + x_i) - f_i(w_i + y_i)))|, \qquad i = 1, \ldots, k \\ w_i &= x_1 + \cdots + x_{i-1} \\ & && (7.2.5) \end{aligned}$$

Let $x, w \in \mathbb{R}^d$. The Taylor formula writes in the two distinct following ways (for suitable values w_1, w_2):

$$\begin{aligned} f(w+x) &= f(w) + xf'(w) + \frac{1}{2}f''(w_1)(x,x) \\ &= f(w) + xf'(w) + \frac{1}{2}f''(w)(x,x) + \frac{1}{6}f'''(w_2)(x,x,x) \end{aligned}$$

here $f^{(j)}(x)(y_1, \ldots, y_j)$ stands for the value of the symmetric j-linear form $f^{(j)}$ at $(y_1, \ldots, y_j)$. Let $\|f^{(j)}\|_\infty = \sup_x \|f^{(j)}(x)\|$ with

$$\|f^{(j)}(x)\| = \sup_{\|y_1\|, \ldots, \|y_j\| \le 1} |f^{(j)}(x)(y_1, \ldots, y_j)|.$$

Thus for $w, x, y \in \mathbb{R}^d$ we may write:

$$\begin{aligned} f(w+x) - f(w+y) &= f'(w)(x-y) + \frac{1}{2}f''(w)(x^2 - y^2) \\ &\quad + \frac{(f''(w_1) - f''(w))(x,x) - (f''(w_1') - f''(w))(y,y)}{2} \\ &= (x-y)f'(w) + \frac{1}{2}f''(w)(x^2 - y^2) \\ &\quad + \frac{f'''(w_2)(x,x,x) - f'''(w_2')(y,y,y)}{6} \end{aligned}$$

Thus

$$\begin{aligned} T &= f(w+x) - f(w+y) - f'(w)(x-y) - \frac{1}{2}f''(w)(x^2 - y^2) \quad \text{satisfies} \\ |T| &\le 2(\|x\|^2 + \|y\|^2)\|f''\|_\infty \wedge \frac{2}{3}(\|x\|^3 + \|y\|^3)\|f'''\|_\infty \\ &\le 2\|f''\|_\infty \left\{ \|x\|^2 \left(1 \wedge \frac{\|f'''\|_\infty}{3\|f''\|_\infty}\|x\|\right) + y^2 \left(1 \wedge \frac{\|f'''\|_\infty}{3\|f''\|_\infty}|y|\right)\right\} \\ &\le \frac{2}{3^\delta}\|f''\|_\infty^{1-\delta}\|f'''\|_\infty^\delta \left\{\|x\|^{2+\delta} + \|y\|^{2+\delta}\right\}, \end{aligned}$$

the last inequality following from the inequality $1 \wedge a \leq a^\delta$, valid for $a \geq 0$ and $\delta \in [0,1[$. Here we set $f''(w)(x^2 - y^2) = f''(w)(x,x)) - f''(w)(y,y)$ for notational convenience. This relation together with the decomposition (7.2.4) and the upper bound $\|f_i^{(j)}\|_\infty \leq \|f^{(j)}\|_\infty$ (valid for $1 \leq i \leq k$, and $0 \leq j \leq 3$) entails

$$\begin{aligned}
\Delta &\leq |T_1| + \frac{1}{2}|T_2| + \frac{2}{3^\delta}\|f''\|_\infty^{1-\delta}\|f'''\|_\infty^\delta \sum_{i=1}^k \{\mathbb{E}\|x_i\|^{2+\delta} + \mathbb{E}\|y_i\|^{2+\delta}\} \\
&\leq |T_1| + \frac{1}{2}|T_2| + 2\frac{1+3^{\frac{\delta}{4}-\frac{1}{2}}}{3^\delta}\|f''\|_\infty^{1-\delta}\|f'''\|_\infty^\delta A \\
&\leq |T_1| + \frac{1}{2}|T_2| + 4\|f''\|_\infty^{1-\delta}\|f'''\|_\infty^\delta A
\end{aligned}$$

where we have used the bound $\mathbb{E}\|y_i\|^{2+\delta} \leq (\mathbb{E}\|y_i\|^4)^{(2+\delta)/4} \leq (3(\mathbb{E}\|x_i\|^2)^2)^{\frac{1}{2}+\frac{\delta}{4}}$.□

7.2.1 Proof of the main results

In this section we first prove theorem 7.1 and 7.2, and then we give rates for this central limit results. Some useful moment inequalities are proved in section 4.1. They are essential in the following proof.

Proof of Theorems 7.1 and 7.2. Let $S = \frac{1}{\sqrt{n}}S_n$ and consider $p = p(n)$ and $q = q(n)$ in such a way that

$$\lim_{n\to\infty} \frac{1}{q(n)} = \lim_{n\to\infty} \frac{q(n)}{p(n)} = \lim_{n\to\infty} \frac{p(n)}{n} = 0\,.$$

Let $k = k(n) = \left[\frac{n}{p(n)+q(n)}\right]$ and

$$Z = \frac{1}{\sqrt{n}}(U_1 + \cdots + U_k)\,, \qquad \text{with} \quad U_j = \sum_{i \in B_j} X_i,$$

where $B_j =](p+q)(j-1), (p+q)(j-1)+p] \cap \mathbb{N}$ is a subset of p successive integers from $\{1,\ldots,n\}$ such that, for $j \neq j'$, B_j and $B_{j'}$ are at least distant of $q = q(n)$. We note B'_j the block between B_j and B_{j+1} and $V_j = \sum_{i\in B'_j} X_i$. V_k is the last block of X_i between the end of B_k and n. Let $\sigma_p^2 = \mathrm{Var}\,(U_1)/p$, and

$$Y = \frac{V'_1 + \cdots + V'_k}{\sqrt{n}}, \qquad V'_j \sim \mathcal{N}(0, p\cdot\sigma_p^2),$$

where the Gaussian variables V'_j are independent and independent of the sequence $(X_n)_{n\in\mathbb{Z}}$. We fix $t \in \mathbb{R}^d$ and we define $f : \mathbb{R}^d \to \mathbb{C}$ with $f(x) = e^{it\cdot x}$. Then:

$$\mathbb{E}(f(S) - f(\sigma N)) = \mathbb{E}(f(S) - f(Z)) + \mathbb{E}(f(Z) - f(Y)) + \mathbb{E}(f(Y) - f(\sigma N)).$$

The Lindeberg method consists in proving that this expression converges to 0 as $n \to \infty$. The first and the last term in this inequality are referred to as auxiliary terms in this Bernstein-Lindeberg method. They come from the replacement of the individual initial - non-Gaussian and Gaussian respectively - random variables. The second term is analogue to that obtained with decoupling and turns the proof of the central limit theorem to the independent case. The third term is referred to as the main term and following the proof under independence it will be bounded above by using a Taylor expansion. Because of the dependence structure, in the corresponding bounds, some additional covariance terms will appear.

Auxiliary terms. Using Taylor expansions up to the second order, we bound:

$$|\mathbb{E}(f(S) - f(Z))| \leq \frac{\|f''\|_\infty^2}{2}\mathbb{E}|S - Z|^2 \qquad \text{and,}$$
$$|\mathbb{E}(f(Y) - f(\sigma N))| \leq \frac{\|f''\|_\infty^2}{2}\mathbb{E}|Y - \sigma N|^2$$

Firstly:

$$\mathbb{E}|Z - S|^2 \preceq \frac{1}{n}\mathbb{E}\left(V_1 + \cdots + V_k\right)^2$$

Note that the set under the sums of X_i in $V_1 + \cdots + V_k$ has cardinality smaller than $(k+1)q + p$. Using the bounds (4.2.6) and (4.2.5), we infer that under conditions (4.2.3) and (4.2.2) respectively,

$$\mathbb{E}|Z - S|^2 \preceq \frac{(k+1)q + p}{n}.$$

We notice that Y follows the distribution of $\sqrt{\frac{kp}{n}}\sigma_p N$ and then working with Gaussian random variables:

$$\mathbb{E}|Y - \sigma N|^2 \leq \left|\frac{kp}{n} - 1\right|\sigma_p^2 + \left|\sigma_p^2 - \sigma^2\right|.$$

Since $|kp/n - 1| \preceq q/p$, we need to bound

$$|\sigma_p^2 - \sigma^2| \leq \sum_{|i|<p}\frac{|i|}{p}|\mathbb{E}(X_0X_i)| + \sum_{|i|>p}|\mathbb{E}(X_0X_i)|.$$

Let $a_i = |\mathbb{E}(X_0X_i)|$. Under conditions (4.2.3) and (4.2.2) respectively, the series $\sum_{i=0}^\infty a_i$ converge and $s_j = \sum_{i=j}^\infty a_i$ converges to 0 as j converges to infinity. Consequently

$$|\sigma_p^2 - \sigma^2| \leq 2\sum_{i=0}^{p-1}\frac{i}{p}\cdot a_i + 2s_p \leq \frac{2}{p}\sum_{i=0}^{p-1}s_i + 2s_p.$$

Cesaro lemma implies that this term converges to 0. Hence $|\mathbb{E}f(S) - f(Z)| + |\mathbb{E}f(Y) - f(\sigma N)|$ tends to 0 as $n \uparrow \infty$.

To precise the rate of convergence, we now assume that $a_i = \mathcal{O}(i^{-\alpha})$ with $\alpha > 1$ we see that

$$|\sigma_p^2 - \sigma^2| \preceq p^{1-\alpha}.$$

The convergence rate is then given by $\frac{q}{p} + \frac{p}{n} + p^{1-\alpha}$ if $\mathbb{E}(X_0 X_i) = \mathcal{O}(i^{-\alpha})$. Since $\mathbb{E}(X_0 X_i) = \mathrm{Cov}(X_0, X_i)$, we then use equations (4.2.5) and (4.2.6) and we find $\alpha = \kappa$ or $\alpha = \lambda(m-2)/(m-1)$ depending of the weak-dependence setting. With $p = n^a$, $q = n^b$, those bounds become:

$n^{b-a} + n^{a-1} + n^{a(1-\kappa)}$, in the κ-weak dependence setting,

$n^{b-a} + n^{a-1} + n^{a(1-\lambda(m-2)/(m-1))}$, under λ-weak dependence.

Main terms. It remains to control the expression $|T_1| + \frac{1}{2}|T_2|$ and A in lemma 7.2.2 with now, for $1 \le j \le k$:

$$x_j = \frac{1}{\sqrt{n}} U_j, \qquad y_j = \frac{1}{\sqrt{n}} V_j$$

- The terms T_j, are bounded by using the weak-dependence properties. The expressions of this bound are obtained by rewriting

$$|\mathrm{Cov}(F(X_m, m \in B_i, i < j), G(X_m, m \in B_j))|.$$

Note that $\|F\|_\infty \le 1$ and that we can compute a bound for $\mathrm{Lip}\, F$ with $F(x_1, \ldots, x_{kp}) = f\left(\frac{1}{\sqrt{n}} \sum_{i<j} x_i\right)$ (with possible repetitions in the sequence $(x_1, \ldots, x_{kp})$):

$$\begin{aligned}\left| f\left(\frac{1}{\sqrt{n}} \sum_{i<j} x_i\right) - f\left(\frac{1}{\sqrt{n}} \sum_{i<j} y_i\right)\right| &\le \left|1 - \exp it \cdot \left(\frac{1}{\sqrt{n}} \sum_{i<j} (y_i - x_i)\right)\right| \\ &\le \frac{\|t\|_2}{\sqrt{n}} \sum_{i=1}^{kp} \|y_i - x_i\|_2.\end{aligned}$$

For $G(x_1, \ldots, x_p) = f\left(\sum_{i=1}^p x_i/\sqrt{n}\right)$, we have $\|G\|_\infty \le 1$ and $\mathrm{Lip}\, G \preceq 1/\sqrt{n}$. We then distinguish the two cases, noting that the gap between blocks is at least q. Since the bounds of the different covariances do not depend of j, we then obtain the controls:

  - In the κ dependence setting:

$$|T_1| + \frac{1}{2}|T_2| \preceq kp \cdot \kappa(q)$$

– In the λ dependence setting:

$$|T_1| + \frac{1}{2}|T_2| \preceq kp(1+\sqrt{k/p}) \cdot \lambda(q).$$

Reminding that $p = n^a$, $q = n^b$, $\kappa(r) = \mathcal{O}(r^{-\kappa})$ or $\lambda(r) = \mathcal{O}(r^{-\lambda})$, the bounds become $n^{1-\kappa b}$ or $n^{1+(1/2-a)_+-\lambda b}$ in respectively κ or λ context.

- Using the stationarity of the sequence X_n we obtain:

$$|A| \quad \preceq \quad n^{-1-\frac{\delta}{2}}\left(\mathbb{E}|S_p|^{2+\delta} \vee p^{1+\frac{\delta}{2}}\right).$$

We then use the result of lemma 4.3 to bound the moment $\mathbb{E}|S_p|^{2+\delta}$. If $\kappa > 2 + \frac{1}{\zeta}$, or $\lambda > 4 + \frac{2}{\zeta}$, where $\kappa(r) = \mathcal{O}(r^{-\kappa})$ or $\lambda(r) = \mathcal{O}(r^{-\lambda})$ then there exists $\delta \in]0, \zeta \wedge 1[$ and $C > 0$ such that:

$$\mathbb{E}|S_p|^{2+\delta} \leq Cp^{1+\delta/2}.$$

We then obtain:

$$A \preceq k(p/n)^{1+\delta/2}.$$

Reminding that $p = n^a$, the bound is of order $n^{(a-1)\delta/2}$ in both κ or λ-weak dependence setting.

7.2.2 Rates of convergence

We present two propositions that give rates of convergence in the central limit theorem.

Proposition 7.1. *Assume that the weakly dependent stationary process $(X_n)_n$ satisfies (7.1.1) then the difference between the characteristic functions is bounded by:*

$$\left|\mathbb{E}\left(f(S_n/\sqrt{n}) - f(\sigma N)\right)\right| \leq Cn^{-c^*},$$

where C is some positive constant and c^* depends of the weakly dependent coefficients as follows:

- *in the λ-dependence case with $\lambda(r) = \mathcal{O}(r^{-\lambda})$ for $\lambda > 4 + \frac{2}{\zeta}$, then*

$$c^* = \frac{A}{2}\frac{2\lambda - 1}{(2+A)(\lambda+1)},$$

where

$$A = \frac{\sqrt{(2\lambda - 6 - \zeta)^2 + 4(\lambda\zeta - 4\zeta - 2)} + \zeta + 6 - 2\lambda}{2} \wedge 1,$$

- *in the κ-dependence case with $\kappa(r) = \mathcal{O}(r^{-\kappa})$ for $\kappa > 2 + \frac{1}{\zeta}$, then*

$$c^* = \frac{(\kappa - 1)B}{\kappa(2 + B)}$$

where

$$B = \frac{\sqrt{(2\kappa - 3 - \zeta)^2 + 4(\kappa\zeta - 2\zeta - 1)} + \zeta + 3 - 2\kappa}{2} \wedge 1.$$

In the case $d = 1$, we use Theorem 5.1 of Petrov (1995) [144] to obtain:

Proposition 7.2 (A rate in the Berry Essen bound)**.** *Assume that the real weakly dependent stationary process $(X_n)_n$ satisfies the same assumptions than in Proposition 7.1. We obtain:*

$$\sup_x |F_n(x) - \Phi(x)| = \mathcal{O}\left(n^{-c^*/4}\right).$$

where c^* is defined in Proposition 7.1.

Proof of proposition 7.1. In the previous section, we have expressed the rates of the different terms. Let us recall these rates:

- In the λ-dependence case, we finally only have to consider the three largest rates: $(a-1)\delta/2$, $1+(1/2-a)_+-\lambda b$ and $b-a$. The previous optimal choice of a^* is smaller than $1/2$, then we have to consider the rate $3/2 - a - \lambda b$ and not $1 - \lambda b$. We find:

$$\begin{aligned} a^* &= \frac{(1+\lambda)\delta + 3}{(2+\delta)(\lambda+1)} \in \left]0, \frac{1}{2}\right[\\ b^* &= a^* \frac{3}{2(\lambda+1)} \in]0, a^*[\end{aligned}$$

  Finally, we obtain the rate n^{-c^*}.

- In the κ-dependence case:

  - Auxiliary terms: $b - a$, $a - 1$ and $a(1 - \kappa)$,
  - Main terms: $1 - \kappa b$ and $(a - 1)\delta/2$.

  The idea is to choose carefully a^* and $b^* \in]0, 1[$ such that the main rates are equal. Because $\delta < 1$, $a > b$, we directly see that $(a - 1)\delta/2 > a - 1$

and $1-\kappa b > a(1-\kappa)$, so that the only rate of the auxiliary term that it remains to consider is $b-a$. Finally, we obtain

$$\begin{aligned} a^* &= 1-\frac{2\kappa-2}{(2+\delta)\kappa+\delta} \in]0,1[\\ b^* &= a^*\frac{2+2\delta}{2+\delta+\delta\kappa} \in]0,a^*[\end{aligned}$$

Finally, we obtain the proposed rate. □

Proof of proposition 7.2. We have seen that for t fix, we control the distance between the characteristic functions of S and σN by a term proportional to $t^2 n^{-c^*}$. Here, t^2 appear because $|t|$ was included in the constants (not depending of n) of the bound of the Lipschitz coefficients. Let Φ be the distribution function of σN and F_n be the distribution function S. Theorem 5.1 p. 142 in Petrov (1995) [144] gives, for every $T>0$:

$$\sup_x |F_n(x)-\Phi(x)| \preceq n^{-c^*}T^3+\frac{1}{T}.$$

We optimize T to obtain a rate of convergence in the central limit theorem. □

7.3 Non causal random fields

Let $(B_n)_{n\in\mathbb{N}}$ be an increasing sequence of finite subsets of $\mathbb{Z}^d$ fulfilling

$$\mathbb{Z}^d=\bigcup_{n=1}^{\infty} B_n, \qquad \lim_{n\to+\infty}\frac{\#\partial B_n}{\#B_n}=0, \tag{7.3.1}$$

$$\partial B_n=\{i\in B_n \,/\, \exists j\notin B_n,\, d(i,j)=1\}$$

and for $i=(i(l))_{1\le l\le d}$ and $j=(j(l))_{1\le l\le d}$ in $\mathbb{Z}^d$, $d(i,j)=\max_{1\le l\le d}|i(l)-j(l)|$.

Let $(Y_i)_{i\in\mathbb{Z}^d}$ be a real valued random fields. Suppose that $(Y_i)_{i\in\mathbb{Z}^d}$ satisfies the following two assumptions:

A) A covariance inequality. Recall that for a real valued function h defined on $\mathbb{R}^n$

$$\mathrm{Lip}\,(h)=\sup_{x\ne y}\frac{|h(x)-h(y)|}{\sum_{i=1}^n |x_i-y_i|}\,.$$

Let R_1 and R_2 be two disjoints and finite subsets of $\mathbb{Z}^d$, and let f and g be two real valued functions defined respectively on $\mathbb{R}^{\#R_1}$ and $\mathbb{R}^{\#R_2}$ such that

$\text{Lip}(f) < \infty$ and $\text{Lip}(g) < \infty$. We suppose that for any positive real number δ, there exists a constant C_δ (not depending on f g, R_1 and R_2) such that

$$\begin{aligned}&|\text{Cov}(f(Y_i,\, i \in R_1), g(Y_i,\, i \in R_2)| \\ &\quad \le C_\delta \text{Lip}(f)\text{Lip}(g)(\#R_1 + \#R_2)\exp(-\delta d(R_1, R_2)),\end{aligned} \tag{7.3.2}$$

where $d(R_1, R_2) = \min_{i \in R_1, j \in R_2} d(i,j)$. We refer the reader to the book of Liggett (1985) [122] for some interacting particle models fulfilling such a covariance inequality.

B) Weak stationarity. Suppose that for any i, $j \in \mathbb{Z}^d$

$$\text{Cov}(Y_i, Y_j) = \text{Cov}(Y_0, Y_{j-i}). \tag{7.3.3}$$

Theorem 7.4 gives a central limit theorem for the random fields $(Y_i)_{i\in\mathbb{Z}^d}$.

Theorem 7.4. *Let $(B_n)_{n\in\mathbb{N}}$ be an increasing sequence of finite subsets of $\mathbb{Z}^d$ fulfilling (7.3.1). Let $(Y_i)_{i\in\mathbb{Z}^d}$ be a real valued random field, satisfying (7.3.2) and (7.3.3). Suppose that, for any $i \in \mathbb{Z}^d$, $\mathbb{E}Y_i = 0$ and $\sup_{i\in\mathbb{Z}^d} \|Y_i\|_\infty < M$. Let $S_n = \sum_{i\in B_n} Y_i$. Then $\sum_{k\in\mathbb{Z}^d} |\text{Cov}(Y_0, Y_k)| < \infty$ and $(\#B_n)^{-1/2} S_n$ converges in distribution to a centered normal law with variance $\sigma^2 = \sum_{k\in\mathbb{Z}^d} \text{Cov}(Y_0, Y_k)$.*

Proof of Theorem 7.4. The proof of Theorem 7.4 follows from Proposition 7.3 and Proposition 7.4 below.

Proposition 7.3. *Let $(Y_i)_{i\in\mathbb{Z}^d}$ be a real valued random field such that $\mathbb{E}Y_i = 0$ and $\mathbb{E}Y_i^2 < \infty$ for any $i \in \mathbb{Z}^d$. Suppose that, for any i, $j \in \mathbb{Z}^d$, (7.3.3) holds and that, for any positive real number δ, there exists a positive constant C_δ such that*

$$|\text{Cov}(Y_i, Y_j)| \le C_\delta e^{-\delta d(i,j)}. \tag{7.3.4}$$

Let $(B_n)_n$ be a sequence of finite and increasing sets of $\mathbb{Z}^d$ fulfilling (7.3.1). Let $S_n = \sum_{i\in B_n} Y_i$. Then

$$\sum_{k\in\mathbb{Z}^d} |\text{Cov}(Y_0, Y_k)| < \infty \quad \textit{and} \quad \lim_{n\to+\infty} \frac{1}{\#B_n} \text{Var}\, S_n = \sum_{k\in\mathbb{Z}^d} \text{Cov}(Y_0, Y_k).$$

Proof of Proposition 7.3. The first conclusion of Proposition 7.3 follows from the bound (7.3.4), together with the following elementary calculations

$$\begin{aligned}
\sum_{k\in\mathbb{Z}^d} |\mathrm{Cov}(Y_0, Y_k)| &\le C\sum_{k\in\mathbb{Z}^d} \exp(-\delta d(0,k)) \\
&\le C\sum_{k\in\mathbb{Z}^d}\sum_{r=0}^{\infty} \exp(-\delta d(0,k))\mathbf{1}_{r\le d(0,k)<r+1} \\
&\le C\sum_{r=0}^{\infty} \exp(-\delta r)\sum_{k\in\mathbb{Z}^d} \mathbf{1}_{d(0,k)<r+1} \\
&\le C\sum_{r=0}^{\infty} \exp(-\delta r)\#\{k\in\mathbb{Z}^d \,/\, d(0,k)<r+1\} \\
&\le C\sum_{r=0}^{\infty} \exp(-\delta r) r^d \qquad (7.3.5)
\end{aligned}$$

where C is a positive constant depending only on δ and d.
We now prove the second part of Proposition 7.3. Thanks to (7.3.1), we can find a sequence $u=(u_n)$ of positive real numbers such that

$$\lim_{n\to\infty} u_n = +\infty, \qquad \lim_{n\to+\infty} \frac{\#\partial B_n}{\#B_n}\exp(u_n) = 0. \qquad (7.3.6)$$

Let $(\partial_u B_n)_n$ be the sequence of subsets of $\mathbb{Z}^d$ defined by

$$\partial_u B_n = \{s\in B_n/\ d(s,\partial B_n) < u_n\}.$$

The following bound

$$\#\partial_u B_n \le C_d(\#\partial B_n)u_n^d,$$

together with the suitable choice of the sequence (u_n) and the limit (7.3.1) ensures

$$\lim_{n\to\infty}\frac{\#\partial_u B_n}{\#B_n} = 0, \qquad (7.3.7)$$

we shall use this fact below without further comments. Let $B_n^u = B_n \setminus \partial_u B_n$. We decompose the quantity Var S_n as in Newman (1980) [135]:

$$\frac{1}{\#B_n}\mathrm{Var}\ S_n = \frac{1}{\#B_n}\sum_{i\in B_n}\sum_{j\in B_n}\mathrm{Cov}\,(Y_i,Y_j) = T_{1,n}+T_{2,n}+T_{3,n},$$

where

$$\begin{aligned} T_{1,n} &= \frac{1}{\#B_n} \sum_{i \in B_n^u} \sum_{j \in B_n \,/ d(i,j) \geq u_n} \operatorname{Cov}(Y_i, Y_j), \\ T_{2,n} &= \frac{1}{\#B_n} \sum_{i \in B_n^u} \sum_{j \in B_n \,: d(i,j) < u_n} \operatorname{Cov}(Y_i, Y_j), \\ T_{3,n} &= \frac{1}{\#B_n} \sum_{i \in \partial_u B_n} \sum_{j \in B_n} \operatorname{Cov}(Y_i, Y_j). \end{aligned}$$

Control of $T_{1,n}$. We have, since $|B_n^u| \leq |B_n|$ and applying (7.3.4)

$$\begin{aligned} |T_{1,n}| &\leq \sup_{i \in \mathbb{Z}^d} \sum_{j \in \mathbb{Z}^d \,: d(i,j) \geq u_n} |\operatorname{Cov}(Y_i, Y_j)| \\ &\leq C_\delta \sup_{i \in \mathbb{Z}^d} \sum_{\{j \in \mathbb{Z}^d / d(i,j) \geq u_n\}} e^{-\delta d(i,j)}. \end{aligned} \tag{7.3.8}$$

For any fixed $i \in \mathbb{Z}^d$, we argue as for (7.3.5) and we obtain

$$\begin{aligned} \sum_{\{j \in \mathbb{Z}^d / d(i,j) \geq u_n\}} e^{-\delta d(i,j)} &\leq C \sum_{r=[u_n]}^{\infty} e^{-\delta r} r^d \\ &\leq C e^{-\delta [u_n]} u_n^d. \end{aligned} \tag{7.3.9}$$

We obtain, collecting (7.3.8), (7.3.9) together with the first limit in (7.3.6):

$$\lim_{n \to +\infty} T_{1,n} = 0. \tag{7.3.10}$$

Control of $T_{3,n}$. We obtain using (7.3.3) and (7.3.4) :

$$|T_{3,n}| \leq \frac{\#\partial_u B_n}{\#B_n} \sum_{k \in \mathbb{Z}^d} |\operatorname{Cov}(Y_0, Y_k)|. \tag{7.3.11}$$

The last bound, together with the limit (7.3.7) and the first conclusion of Proposition 7.3, gives

$$\lim_{n \to \infty} T_{3,n} = 0. \tag{7.3.12}$$

Control of $T_{2,n}$. We deduce using the following implication, if $i \in B_n^u$ and j is not belonging to B_n then $d(i,j) \geq u_n$, that

$$T_{2,n} = \frac{1}{\#B_n} \sum_{i \in B_n^u} \sum_{\{j \in \mathbb{Z}^d / d(i,j) < u_n\}} \operatorname{Cov}(Y_i, Y_j)$$

Equality (7.3.3) ensures

$$\sum_{\{j \in \mathbb{Z}^d / d(i,j) < u_n\}} \mathrm{Cov}(Y_i, Y_j) = \sum_{\{k \in \mathbb{Z}^d / d(0,k) < u_n\}} \mathrm{Cov}\,(Y_0, Y_k)\,.$$

Hence

$$T_{2,n} = \frac{\#B_n^u}{\#B_n} \sum_{\{k \in \mathbb{Z}^d / d(0,k) < u_n\}} \mathrm{Cov}(Y_0, Y_k).$$

The last equality together with the first limit in (7.3.6) and (7.3.7), implies that

$$\lim_{n \to \infty} T_{2,n} = \sum_{k \in \mathbb{Z}^d} \mathrm{Cov}(Y_0, Y_k). \tag{7.3.13}$$

The second conclusion of Proposition 7.3 is proved by collecting the limits (7.3.10), (7.3.12) and (7.3.13). □

Proposition 7.4. *Let $(B_n)_{n\in\mathbb{N}}$ be an increasing sequence of finite subsets of $\mathbb{Z}^d$ such that $\#B_n$ tends to infinity as n goes to infinity. Let $(Y_i)_{i\in\mathbb{Z}^d}$ be a real valued random field satisfying (7.3.2). Suppose that $\mathbb{E}Y_i = 0$ for any $i \in \mathbb{Z}^d$, and that $\sup_{i\in\mathbb{Z}^d} \|Y_i\|_\infty < M$. Let $S_n = \sum_{i\in B_n} Y_i$. If there exists a finite real number σ^2 such that*

$$\lim_{n\to\infty} \frac{\mathrm{Var}\, S_n}{\#B_n} = \sigma^2, \tag{7.3.14}$$

then $(\#B_n)^{-1/2} S_n$ converges in distribution to a centered normal law with variance σ^2.

Proof of Proposition 7.4. We need the following notation. Let (m_n) be a sequence of positive integers to be fixed later. We suppose for the moment that $\lim_{n\to\infty} m_n = \infty$. For any $i \in \mathbb{Z}^d$, we define a neighborhood V_i of i in B_n as

$$V_i = B(i, m_n) \cap B_n, \tag{7.3.15}$$

where $B(i, m_n) = \{j \in \mathbb{Z}^d \,/\, d(i,j) < m_n\}$. Let V_i^c denote the complementary of V_i in B_n *i.e.* $V_i^c = B_n \setminus V_i$. For I a subset of B_n, we denote by

$$S(I) = \sum_{i\in I} Y_i, \quad S(I^c) = \sum_{i \in B_n \setminus I} Y_i.$$

Hence $S(B_n) = \sum_{i\in B_n} Y_i =: S_n$ for the sake of brevity.

Finally we denote by $\mathcal{F}(b_2, b_3)$ the set of the real valued function h defined on $\mathbb{R}$, three times differentiable, such that $h(0) = 0$, $b_2 := \|h''\|_\infty < +\infty$ and that $b_3 := \|h^{(3)}\|_\infty < +\infty$.

We also need the following proposition.

Proposition 7.5. *Let h be a fixed function of the set $\mathcal{F}(b_2, b_3)$. Let $(B_n)_{n\in\mathbb{N}}$ be a sequence of finite subsets of $\mathbb{Z}^d$. For any $i \in B_n$, let V_i be the set as defined by (7.3.15). Let $(Y_i)_{i\in\mathbb{Z}^d}$ be a real valued random field. Suppose that $\mathbb{E}Y_i = 0$ and $\mathbb{E}Y_i^2 < +\infty$ for any $i \in \mathbb{Z}^d$. Let $S_n = \sum_{i\in B_n} Y_i$. Then*

$$\left|\mathbb{E}(h(S_n)) - \operatorname{Var} S_n \int_0^1 t\mathbb{E}(h''(tS_n))dt\right|$$
$$\leq \int_0^1 \sum_{i\in B_n} |\operatorname{Cov}(Y_i, h'(tS_n(V_i^c)))|\, dt + 2\sum_{i\in B_n} \mathbb{E}|Y_i||S(V_i)|\,(b_2 \wedge b_3|S(V_i)|)$$
$$+b_2\mathbb{E}\left|\sum_{i\in B_n}(Y_iS(V_i) - \mathbb{E}(Y_iS(V_i)))\right| + b_2\sum_{i\in B_n}|\operatorname{Cov}(Y_i, S(V_i^c))|\,. \quad (7.3.16)$$

Remark 7.4. *For a random field $(Y_i)_{i\in\mathbb{Z}^d}$ of independent random variables such that $\sup_{i\in\mathbb{Z}^d}\mathbb{E}Y_i^4 < \infty$, Proposition 7.5 applied with $V_i = \{i\}$ implies that*

$$\left|\mathbb{E}(h(S_n)) - \operatorname{Var} S_n \int_0^1 t\mathbb{E}(h''(tS_n))dt\right| \leq$$
$$2\sum_{i\in B_n}\mathbb{E}|Y_i|^2\,(b_2 \wedge b_3|Y_i|) + b_2\sqrt{|B_n|}\sup_{i\in\mathbb{Z}^d}\|Y_i^2\|_2.$$

Proof of Proposition 7.5. We have,

$$\begin{aligned}
h(S_n) &= S_n\int_0^1 h'(tS_n)dt = \int_0^1\left(\sum_{i\in B_n} Y_ih'(tS_n)\right)dt\\
&= \int_0^1\left(\sum_{i\in B_n} Y_ih'(tS(V_i^c))\right)dt\\
&\quad + \int_0^1\left(\sum_{i\in B_n} Y_i\left(h'(tS_n) - h'(tS(V_i^c)) - tS(V_i)h''(tS_n)\right)\right)dt\\
&\quad + \sum_{i\in B_n} Y_iS(V_i)\int_0^1 th''(tS_n)dt - \sum_{i\in B_n}\mathbb{E}\left(Y_iS(V_i)\right)\int_0^1 th''(tS_n)dt\\
&\quad + \sum_{i\in B_n}\mathbb{E}\left(Y_iS(V_i)\right)\int_0^1 th''(tS_n)dt - \sum_{i\in B_n}\mathbb{E}\left(Y_iS_n\right)\int_0^1 th''(tS_n)dt\\
&\quad + \sum_{i\in B_n}\mathbb{E}\left(Y_iS_n\right)\int_0^1 th''(tS_n)dt. \qquad (7.3.17)
\end{aligned}$$

We take the expectation in the equality (7.3.17). The obtained formula, together with the following estimations, proves Proposition 7.5. □

$$
\begin{aligned}
&|h'(tS_n) - h'(tS(V_i^c)) - tS(V_i)h''(tS_n)| \\
&\quad \le |h'(tS_n) - h'(tS(V_i^c)) - tS(V_i)h''(tS(V_i^c))| + |S(V_i)||h''(tS_n) - h''(tS(V_i^c))| \\
&\quad \le 2|S(V_i)|\,(b_2 \wedge b_3|S(V_i)|)\,. \; \square
\end{aligned}
$$

The purpose now is to control the right hand side of the bound (7.3.16) for a random field $(Y_i)_{i\in\mathbb{Z}^d}$ fulfilling the covariance inequality (7.3.2) and the requirements of Proposition 7.4.

Corollary 7.1. *Let h be a fixed function of the set $\mathcal{F}(b_2, b_3)$. Let $(B_n)_{n\in\mathbb{N}}$ be a sequence of finite subsets of $\mathbb{Z}^d$. For any $i \in B_n$, let V_i be the set as defined by (7.3.15). Let $(Y_i)_{i\in\mathbb{Z}^d}$ be a real valued random field, fulfilling the covariance inequality (7.3.2). Suppose that, for any $i \in \mathbb{Z}^d$, $\mathbb{E}Y_i = 0$ and $\sup_{i\in\mathbb{Z}^d} \|Y_i\|_\infty < M$. Let $S_n = \sum_{i\in B_n} Y_i$. Then, for any positive real number δ, there exists a positive constant $C(\delta, M, d)$ independent of n, such that*

$$
\begin{aligned}
&\sup_{h\in\mathcal{F}(b_2,b_3)} \left|\mathbb{E}(h(S_n)) - \operatorname{Var} S_n \int_0^1 t\mathbb{E}(h''(tS_n))dt\right| \\
&\qquad \le C(\delta, M, d)\Big\{b_2(\#B_n)^2 e^{-\delta m_n} + b_3 m_n^d \#B_n \\
&+ b_2 m_n^d \sqrt{\#B_n}\, m\Big(\sum_{k=3m_n}^{\infty} k^d e^{-\delta(k-2m_n)}\Big)^{1/2} + b_2\sqrt{\#B_n}\, m_n^d\Big(\sum_{k=1}^{3m_n} e^{-\delta k} k^d\Big)^{1/2}\Big\}.
\end{aligned}
$$

Proof of Corollary 7.1. From now, we denote by C a positive constant that may be different from line to line, independent of n and depending, eventually, on M, δ and d. We have

$$V_i^c = \{j \in \mathbb{Z}^d \;/\; d(i,j) \ge m_n\} \cap B_n.$$

Hence

$$d(\{i\}, V_i^c) \ge m_n.$$

The last bound together with (7.3.2), proves that

$$
\begin{aligned}
\sum_{i\in B_n} |\operatorname{Cov}(Y_i, h'(tS_n(V_i^c)))| &\le Cb_2 \sum_{i\in B_n} (\#V_i^c + 1)e^{-\delta d(\{i\},V_i^c)} \\
&\le Cb_2(\#B_n)^2 e^{-\delta m_n}. \qquad (7.3.18)
\end{aligned}
$$

In the same way, we prove that

$$b_2 \sum_{i\in B_n} |\operatorname{Cov}(Y_i, S(V_i^c))| \le Cb_2(\#B_n)^2 e^{-\delta m_n}. \qquad (7.3.19)$$

Since $\#V_i \le \#B(0, m_n) \le Cm_n^d$ and $\sup_{j\in\mathbb{Z}^d}\sum_{k\in\mathbb{Z}^d}|\mathrm{Cov}(Y_j, Y_k)| < \infty$, we deduce that

$$\sum_{i\in B_n}\mathbb{E}|Y_i||S(V_i)|\,(b_2\wedge b_3|S(V_i)|) \le b_3 M\#B_n \sup_{i\in\mathbb{Z}^d}\mathbb{E}|S(V_i)|^2 \tag{7.3.20}$$
$$\le Cb_3\#B_n\, m_n^d \sum_{k\in\mathbb{Z}^d}|\mathrm{Cov}(Y_0, Y_k)|.$$

It remains to control

$$\mathbb{E}\left|\sum_{i\in B_n}(Y_iS(V_i) - \mathbb{E}(Y_iS(V_i)))\right|.$$

For this, we argue as in Bolthausen (1982) [24]. We have

$$\mathbb{E}\left|\sum_{i\in B_n}(Y_iS(V_i) - \mathbb{E}(Y_iS(V_i)))\right|^2 = \mathrm{Var}\,(\sum_{i\in B_n}Y_iS(V_i))$$
$$= \sum_{i\in B_n}\sum_{j\in B_n}\mathrm{Cov}(Y_iS(V_i), Y_jS(V_j)).$$

Hence

$$\mathbb{E}\left|\sum_{i\in B_n}(Y_iS(V_i) - \mathbb{E}(Y_iS(V_i)))\right|^2$$
$$\le \sum_{i\in B_n}\sum_{i'\in B(i,m_n)}\sum_{j\in B_n}\sum_{j'\in B(j,m_n)}|\mathrm{Cov}(Y_iY_{i'}, Y_jY_{j'})|. \tag{7.3.21}$$

Next, we have

$$|\mathrm{Cov}(Y_iY_{i'}, Y_jY_{j'})| \tag{7.3.22}$$
$$\le |\mathrm{Cov}(Y_iY_{i'}, Y_jY_{j'})|\,\mathbf{1}_{d(i,j)\ge 3m_n} + |\mathrm{Cov}(Y_iY_{i'}, Y_jY_{j'})|\,\mathbf{1}_{d(i,j)\le 3m_n}.$$

We begin with the first term. The covariance inequality (7.3.2) together with some elementary estimations, imply that

$$|\mathrm{Cov}(Y_iY_{i'}, Y_jY_{j'})|\,\mathbf{1}_{d(i,j)\ge 3m_n} \le \sum_{k=3m_n}^{\infty}|\mathrm{Cov}(Y_iY_{i'}, Y_jY_{j'})|\,\mathbf{1}_{k\le d(i,j)\le k+1}$$
$$\le C\sum_{k=3m_n}^{\infty}e^{-\delta d(\{i,i'\},\{j,j'\})}\mathbf{1}_{k\le d(i,j)\le k+1}$$
$$\le C\sum_{k=3m_n}^{\infty}e^{-\delta(k-2m_n)}\mathbf{1}_{d(i,j)\le k+1}.$$

To obtain the last bound, note that for any $i' \in B(i, m_n)$ and $j' \in B(j, m_n)$, we have

$$d(\{i,i'\},\{j,j'\}) + 2m_n \geq d(\{i,i'\},\{j,j'\}) + d(i,i') + d(j,j') \geq d(i,j).$$

Hence,

$$\begin{aligned}\sum_{i\in B_n}\sum_{i'\in B(i,m_n)}\sum_{j\in B_n}\sum_{j'\in B(j,m_n)} &|\mathrm{Cov}(Y_iY_{i'},Y_jY_{j'})|\,\mathbf{1}_{d(i,j)\geq 3m_n}\\ &\leq Cm_n^{2d}\sum_{k=3m_n}^{\infty}\sum_{i\in B_n}\sum_{j\in B_n} e^{-\delta(k-2m_n)}\mathbf{1}_{j\in B(i,k+1)}\\ &\leq C\#B_n\, m_n^{2d}\sum_{k=3m_n}^{\infty} k^d e^{-\delta(k-2m_n)}. \qquad (7.3.23)\end{aligned}$$

We control the second term in (7.3.22). Inequality (7.3.2) and the fact that $d(\{i\},\{i',j,j'\}) \leq d(\{i\},\{i'\})$, imply that

$$\begin{aligned}&|\mathrm{Cov}(Y_iY_{i'},Y_jY_{j'})|\,\mathbf{1}_{d(i,j)\leq 3m_n}\\ &\quad\leq |\mathrm{Cov}(Y_i,Y_{i'}Y_jY_{j'})|\,\mathbf{1}_{d(i,j)\leq 3m_n} + |\mathrm{Cov}(Y_i,Y_{i'})|\,|\mathrm{Cov}(Y_j,Y_{j'})|\,\mathbf{1}_{d(i,j)\leq 3m_n}\\ &\quad\leq Ce^{-\delta d(\{i\},\{i',j,j'\})}\mathbf{1}_{d(i,j)\leq 3m_n}.\end{aligned}$$

Using the last bound, we infer that

$$\begin{aligned}&|\mathrm{Cov}(Y_iY_{i'},Y_jY_{j'})|\,\mathbf{1}_{d(i,j)\leq 3m_n}\\ &\quad< \sum_{k=1}^{3m_n}|\mathrm{Cov}(Y_iY_{i'},Y_jY_{j'})|\,\mathbf{1}_{d(i,j)\leq 3m_n}\mathbf{1}_{k-1\leq d(\{i\},\{i',j,j'\})\leq k}\\ &\quad\leq C\sum_{k=1}^{3m_n} e^{-\delta(k-1)}\mathbf{1}_{d(i,j)\leq 3m_n}\mathbf{1}_{d(\{i\},\{i',j,j'\})\leq k}. \qquad (7.3.24)\end{aligned}$$

Next, we have

$$\mathbf{1}_{d(\{i\},\{i',j,j'\})\leq k} \leq \mathbf{1}_{d(\{i\},\{i'\})\leq k} + \mathbf{1}_{d(\{i\},\{j\})\leq k} + \mathbf{1}_{d(\{i\},\{j'\})\leq k},$$

and consequently

$$\sum_{i\in B_n}\sum_{i'\in B(i,m_n)}\sum_{j\in B_n}\sum_{j'\in B(j,m_n)} \mathbf{1}_{d(i,j)\leq 3m_n}\mathbf{1}_{d(\{i\},\{i',j,j'\})\leq k} \leq C\#B_n\, m_n^{2d}k^d. \qquad (7.3.25)$$

Combining (7.3.24) and (7.3.25), we obtain that

$$\sum_{i\in B_n}\sum_{i'\in B(i,m_n)}\sum_{j\in B_n}\sum_{j'\in B(j,m_n)} |\mathrm{Cov}(Y_iY_{i'},Y_jY_{j'})|\,\mathbf{1}_{d(i,j)\le 3m_n}$$

$$\le C\#B_n\, m_n^{2d}\sum_{k=1}^{3m_n} e^{-\delta k}k^d. \quad (7.3.26)$$

We collect the bounds (7.3.21), (7.3.23) and (7.3.26) and we obtain,

$$\begin{aligned}\mathbb{E}\left|\sum_{i\in B_n}(Y_iS(V_i)-\mathbb{E}(Y_iS(V_i)))\right| &\le C\sqrt{\#B_n}\left\{m_n^d\left(\sum_{k=3m_n}^{\infty}k^de^{-\delta(k-2m_n)}\right)^{1/2}\right.\\ &\quad\left.+m_n^d\left(\sum_{k=1}^{3m_n}e^{-\delta k}k^d\right)^{1/2}\right\}. \quad (7.3.27)\end{aligned}$$

Finally, the bounds (7.3.18), (7.3.19), (7.3.20), (7.3.27), together with Proposition 7.5 prove Corollary 7.1. □

End of the proof of Proposition 7.4. We apply Corollary 7.1 to the real and imaginary parts of the function $x\to\exp(iux/\sqrt{\#B_n})-1$. Those functions belong to the set $\mathcal{F}(b_2,b_3)$, with

$$b_2=\left(\frac{u}{\sqrt{\#B_n}}\right)^2 \quad\text{and}\quad b_3=\left(\frac{|u|}{\sqrt{\#B_n}}\right)^3.$$

We obtain, noting by ϕ_n the characteristic function of the normalized sum $S_n/\sqrt{\#B_n}$,

$$\begin{aligned}&\left|\phi_n(u)-1+\frac{\mathrm{Var}\,S_n}{|B_n|}u^2\int_0^1 t\phi_n(tu)dt\right|\le C(\delta,M,d,u)\left\{\#B_n\,e^{-\delta m_n}+\frac{m_n^d}{\sqrt{\#B_n}}\right.\\ &\left.+\frac{m_n^d}{\sqrt{\#B_n}}\left(\sum_{k=3m_n}^{\infty}k^de^{-\delta(k-2m_n)}\right)^{1/2}+\frac{m_n^d}{\sqrt{\#B_n}}\left(\sum_{k=1}^{3m_n}e^{-\delta k}k^d\right)^{1/2}\right\}\\ &\le C(\delta,M,d,u)\left\{\#B_n\,e^{-\delta m_n}+\frac{m_n^d}{\sqrt{\#B_n}}+\frac{m_n^{3d/2}}{\sqrt{\#B_n}}e^{-\delta m_n/2}\right\}.\end{aligned}$$

For a suitable choice of the sequence m_n (for example we can take $m_n=\frac{2}{\delta}\log\#B_n$), the right hand side of the last bound tends to 0 an n goes to infinity:

$$\lim_{n\to\infty}\left|\phi_n(u)-1+\frac{\mathrm{Var}\,S_n}{\#B_n}u^2\int_0^1 t\phi_n(tu)dt\right|=0. \quad (7.3.28)$$

In order to finish the proof of Proposition 7.4, we need the following lemma.

Lemma 7.2. *Let σ^2 be a positive real number. Let (X_n) be a sequence of real valued random variables such that $\sup_{n\in\mathbb{N}} \mathbb{E}X_n^2 < +\infty$. Let ϕ_n be the characteristic function of X_n. Suppose that for any $u \in \mathbb{R}$,*

$$\lim_{n\to\infty} \left| \phi_n(u) - 1 + \sigma^2 \int_0^u t\phi_n(t)dt \right| = 0. \tag{7.3.29}$$

Then, for any $u \in \mathbb{R}$,

$$\lim_{n\to\infty} \phi_n(u) = e^{-\frac{u^2\sigma^2}{2}}.$$

Proof. Lemma 7.2 is a variant of Lemma 2 in Bolthausen (1982) [24], which is an adaptation of Stein's method. Markov inequality implies that the sequence $(\mu_n)_{n\in\mathbb{N}}$ of the laws of (X_n) is tight with the condition $\sup_{n\in\mathbb{N}} \mathbb{E}X_n^2 < \infty$. Theorem 25.10 in Billingsley (1995) [21] proves the existence of a subsequence μ_{n_k} and a probability measure μ such that μ_{n_k} converges weakly to μ as k tends to infinity. Let ϕ denote the characteristic function of μ. We deduce from (7.3.29) that, for any $u \in \mathbb{R}$,

$$\phi(u) - 1 + \sigma^2 \int_0^u t\phi(t)dt = 0,$$

or equivalently, for any $u \in \mathbb{R}$,

$$\phi'(u) + \sigma^2 u\phi(u) = 0.$$

We obtain integrating the last equation

$$\phi(u) = \exp(-\frac{\sigma^2 u^2}{2}),$$

for any $u \subset \mathbb{R}$. The proof of Lemma 7.2 is complete by the use of Theorem 25.10 in Billingsley (1995) [21] and its corollary. □

The proof of Proposition 7.4 follows by (7.3.14), (7.3.28) and Lemma 7.2. □

7.4 Conditional central limit theorem (causal)

Let S_n be the partial sums of a triangular array with stationary rows. In this section, we give necessary and sufficient conditions for S_n to satisfies the *conditional central limit theorem.* These conditions imply the weak convergence of $n^{-1/2}S_n$ to a mixture of normal distribution, but they also imply the stable convergence in the sense of Rényi (1963) [156] (see section 7.5.1). The main result of this section (Theorem 7.5) is due to Dedecker and Merlevède (2002) [44].

Definition 7.1. *Let $(\Omega, \mathcal{A}, \mathbb{P})$ be a probability space, and $T : \Omega \mapsto \Omega$ be a bijective bimeasurable transformation preserving the probability $\mathbb{P}$. An element A is said to be invariant if $T(A) = A$. We denote by $\mathcal{I}$ the σ-algebra of all invariant sets. The probability $\mathbb{P}$ is ergodic if each element of $\mathcal{I}$ has measure 0 or 1. Finally, let $\mathcal{H}$ be the space of continuous real functions φ such that $x \mapsto |(1+x^2)^{-1}\varphi(x)|$ is bounded.*

Theorem 7.5. *For each positive integer n, let $\mathcal{M}_{0,n}$ be a σ-algebra of $\mathcal{A}$ satisfying $\mathcal{M}_{0,n} \subseteq T^{-1}(\mathcal{M}_{0,n})$. Define the nondecreasing filtration $(\mathcal{M}_{i,n})_{i\in\mathbb{Z}}$ by $\mathcal{M}_{i,n} = T^{-i}(\mathcal{M}_{0,n})$ and $\mathcal{M}_{i,\inf} = \sigma\left(\bigcup_{n=1}^{\infty} \bigcap_{k=n}^{\infty} \mathcal{M}_{i,k}\right)$. Let $X_{0,n}$ be a $\mathcal{M}_{0,n}$-measurable and square integrable random variable and define the sequence $(X_{i,n})_{i\in\mathbb{Z}}$ by $X_{i,n} = X_{0,n} \circ T^i$. Finally, for any t in $[0,1]$, write $S_n(t) = X_{1,n} + \cdots + X_{[nt],n}$. Suppose that $n^{-1/2}X_{0,n}$ converges in probability to zero as n tends infinity. The following statements are equivalent:*

S1 *There exists a nonnegative $\mathcal{M}_{0,\inf}$-measurable random variable η such that, for any φ in $\mathcal{H}$, any t in $[0,1]$ and any positive integer k,*

$$\mathbf{S1}(\varphi): \quad \lim_{n\to\infty} \left\| \mathbb{E}\left(\varphi(n^{-1/2}S_n(t)) - \int \varphi(x\sqrt{t\eta})g(x)dx \,\middle|\, \mathcal{M}_{k,n} \right) \right\|_1 = 0$$

where g is the distribution of a standard normal.

S2 *(a)*

$$\lim_{t\to 0} \limsup_{n\to\infty} \mathbb{E}\left(\frac{S_n^2(t)}{nt} \left(1 \wedge \frac{|S_n(t)|}{\sqrt{n}} \right) \right) = 0\,.$$

(b) $\displaystyle\lim_{t\to 0} \limsup_{n\to\infty} \frac{1}{t\sqrt{n}} \|\mathbb{E}(S_n(t)|\mathcal{M}_{0,n})\|_1 = 0.$

(c) There exists a nonnegative $\mathcal{M}_{0,\inf}$-measurable random variable η such that,

$$\lim_{t\to 0} \limsup_{n\to\infty} \left\| \mathbb{E}\left(\frac{S_n^2(t)}{nt} - \eta \middle| \mathcal{M}_{0,n} \right) \right\|_1 = 0\,.$$

Moreover the random variable η satisfies $\eta = \eta \circ T$ almost surely.

We now give the proof of this theorem. The fact that **S1** implies **S2** is obvious. In the next sections, we focus on the consequences of condition **S2**. We start with some preliminary results.

7.4.1 Definitions and preliminary lemmas

Definition 7.2. *Let μ be a signed measure on a metric space $(S, \mathcal{B}(S))$. Denote by $|\mu|$ the total variation measure of μ, and by $\|\mu\| = |\mu|(S)$ its norm. We say that a family Π of signed measures on $(S, \mathcal{B}(S))$ is tight if for every positive ϵ*

there exists a compact set K such that $|\mu|(K^c) < \epsilon$ for any μ in Π. Denote by $\mathcal{C}(S)$ the set of continuous and bounded functions from S to $\mathbb{R}$. We say that a sequence of signed measures $(\mu_n)_{n>0}$ converges weakly to a signed measure μ if for any φ in $\mathcal{C}(S)$, $\mu_n(\varphi)$ tends to $\mu(\varphi)$ as n tends to infinity.

Lemma 7.3. *Let $(\mu_n)_{n>0}$ be a sequence of signed measure on $(\mathbb{R}^d, \mathcal{B}(\mathbb{R}^d))$, and set $\hat{\mu}_n(t) = \mu_n(\exp(i < t, \cdot >))$. Assume that the sequence $(\mu_n)_{n>0}$ is tight and that $\sup_{n>0} \|\mu_n\| < \infty$. The following statements are equivalent*

1. *the sequence $(\mu_n)_{n>0}$ converges weakly to the null measure.*

2. *for any t in $\mathbb{R}^d$, $\hat{\mu}_n(t)$ tends to zero as n tends to infinity.*

Proof of Lemma 7.3. $1 \Rightarrow 2$ is obvious. It remains to prove that $2 \Rightarrow 1$. We proceed in 3 steps.

Step 1. Let $\mathcal{D}(\mathbb{R}^d)$ be the space of functions from $\mathbb{R}^d$ to $\mathbb{C}$ which are infinitely derivable with compact support. Let φ be any element of $\mathcal{D}(\mathbb{R}^d)$ and set $\bar{\varphi}(t) = \hat{\varphi}(-t)$. From Plancherel equality, we have $\mu_n(\varphi) = (2\pi)^{-d}\hat{\mu}_n(\bar{\varphi})$. The function $\bar{\varphi}$ being infinitely derivable and fast decreasing, it belongs to $\mathbb{L}^1(\lambda)$. Since $|\hat{\mu}_n|$ converges to zero everywhere and is bounded by $\sup_{n>0} \|\mu_n\|$, the dominated convergence theorem implies that $\hat{\mu}_n(\bar{\varphi})$ tends to zero as n tends to infinity. Consequently, for any φ in $\mathcal{D}(\mathbb{R}^d)$, $\mu_n(\varphi)$ converges to zero as n tends to infinity.

Step 2. Let φ be any function from $\mathbb{R}^d$ to $\mathbb{R}$, continuous and with compact support. For any positive ϵ, there exists φ_ϵ in $\mathcal{D}(\mathbb{R}^d)$ such that $\|\varphi - \varphi_\epsilon\|_\infty \leq \epsilon$. Since furthermore $\sup_{n>0} \|\mu_n\|$ is finite, we infer from Step 1 that $\mu_n(\varphi)$ tends to zero as n tends to infinity.

Step 3. For any positive integer k, let f_k be a positive and continuous function from $\mathbb{R}^d$ to $\mathbb{R}$ satisfying: $\|f_k\|_\infty \leq 1$, $f(x) = 1$ for any x in $[-k, k]^d$, $f(x) = 0$ for any x in $\left([-k-1, k+1]^d\right)^c$.
For any continuous bounded function φ, write

$$|\mu_n(\varphi)| \leq |\mu_n(\varphi f_k)| + \|\varphi\|_\infty |\mu_n|(([-k,k]^d)^c)\,.$$

From Step 2 the first term on right hand tends to zero as n tends to infinity. Since the sequence $(\mu_n)_{n>0}$ is tight, the second term on right hand is as small as we wish by choosing k large enough. This completes the proof of 1. □

Definition 7.3. *Define the set $R(\mathcal{M}_{k,n})$ of Rademacher $\mathcal{M}_{k,n}$-measurable random variables: $R(\mathcal{M}_{k,n}) = \{2\mathbf{1}_A - 1 \,/\, A \in \mathcal{M}_{k,n}\}$. Recall that g is the $\mathcal{N}(0,1)$-distribution. For the random variable η introduced in Theorem 7.5 and any bounded random variable Z, let*

1. *$\nu_n[Z]$ be the image measure of $Z.\mathbb{P}$ by the variable $n^{-1/2}S_n(t)$.*

2. *$\nu[Z]$ be the image measure of $g.\lambda \otimes Z.\mathbb{P}$ by the variable ϕ from $\mathbb{R} \otimes \Omega$ to $\mathbb{R}$ defined by $\phi(x,\omega) = x\sqrt{t\eta(\omega)}$.*

Lemma 7.4. *Let $\mu_n[Z_n] = \nu_n[Z_n] - \nu[Z_n]$. For any φ in $\mathcal{H}$, the statement* **S1**(φ) *is equivalent to:*

S3(φ)*: for any Z_n in $R(\mathcal{M}_{k,n})$ the sequence $\mu_n[Z_n](\varphi)$ tends to zero as n tends to infinity.*

Proof of Lemma 7.4. For Z_n in $R(\mathcal{M}_{k,n})$ and φ in $\mathcal{H}$, we have

$$\begin{aligned} |\mu_n[Z_n](\varphi)| &= \Big|\mathbb{E}\Big(Z_n\Big(\varphi(n^{-1/2}S_n(t)) - \int \varphi(x\sqrt{t\eta})g(x)dx\Big)\Big)\Big| \\ &\leq \Big\|\mathbb{E}\Big(\varphi(n^{-1/2}S_n(t)) - \int \varphi(x\sqrt{t\eta})g(x)dx\,\Big|\mathcal{M}_{k,n}\Big)\Big\|_1 . \end{aligned}$$

Consequently **S1**(φ) implies **S3**(φ). Now to prove that **S3**(φ) implies **S1**(φ), choose

$$A(n,\varphi) = \Big\{\mathbb{E}\Big(\varphi(n^{-1/2}S_n(t)) - \int \varphi(x\sqrt{t\eta})g(x)dx\,\Big|\mathcal{M}_{k,n}\Big) \geq 0\Big\},$$

and $Z_n^\varphi = 2\mathbf{1}_{A(n,\varphi)} - 1$. Obviously

$$\mu_n[Z_n^\varphi](\varphi) = \Big\|\mathbb{E}\Big(\varphi(n^{-1/2}S_n(t)) - \int \varphi(x\sqrt{t\eta})g(x)dx\,\Big|\mathcal{M}_{k,n}\Big)\Big\|_1 ,$$

and **S3**(φ) implies **S1**(φ). □

7.4.2 Invariance of the conditional variance

We first prove that if **S2** holds, the random variables η satisfies $\eta = \eta \circ T$ almost surely (or equivalently that η is measurable with respect to the $\mathbb{P}$-completion of $\mathcal{I}$). From **S2**(c) and both the facts that $(X_{i,n})_{i\in\mathbb{Z}}$ is strictly stationary and $\mathcal{M}_{0,n} \subseteq \mathcal{M}_{1,n}$, we have for any t in $]0,1]$,

$$\lim_{n\to\infty} \Big\|\mathbb{E}\Big(\eta \circ T - \frac{S_n^2(t)\circ T}{nt}\Big|\mathcal{M}_{0,n}\Big)\Big\|_1 = 0 . \tag{7.4.1}$$

On the other hand, defining $\psi(x) = x^2(1-(1\wedge|x|))$ and using the fact that T preserves $\mathbb{P}$, we have

$$\begin{aligned} \Big\|\frac{S_n^2(t)}{nt} - \frac{S_n^2(t)\circ T}{nt}\Big\|_1 &\leq 2\Big\|\frac{S_n^2(t)}{nt}\Big(1 \wedge \frac{|S_n(t)|}{\sqrt{n}}\Big)\Big\|_1 \\ &\quad + \frac{1}{t}\Big\|\psi\Big(\frac{S_n(t)}{\sqrt{n}}\Big) - \psi\Big(\frac{S_n(t)\circ T}{\sqrt{n}}\Big)\Big\|_1 . \end{aligned} \tag{7.4.2}$$

To control the second term on right hand, note that the function ψ is 3-lipschitz and bounded by 1. It follows that for each positive ϵ,

$$\Big\|\psi\Big(\frac{S_n(t)}{\sqrt{n}}\Big) - \psi\Big(\frac{S_n(t)\circ T}{\sqrt{n}}\Big)\Big\|_1 \le 3\epsilon + 2\mathbb{P}(|X_{0,n} - X_{[nt],n}| > \sqrt{n}\epsilon)\,.$$

Using that $n^{-1/2}X_{0,n}$ converges in probability to 0, we derive that

$$\lim_{n\to\infty}\Big\|\psi\Big(\frac{S_n(t)}{\sqrt{n}}\Big) - \psi\Big(\frac{S_n(t)\circ T}{\sqrt{n}}\Big)\Big\|_1 = 0\,,$$

and the second term on right hand in (7.4.2) tends to 0 as n tends to infinity. This fact together with inequality (7.4.2) and Condition $\mathbf{S2}(a)$ yield

$$\lim_{t\to 0}\limsup_{n\to\infty}\Big\|\frac{S_n^2(t)}{nt} - \frac{S_n^2(t)\circ T}{nt}\Big\|_1 = 0\,,$$

which together with $\mathbf{S2}(c)$ imply that

$$\lim_{t\to 0}\limsup_{n\to\infty}\Big\|\mathbb{E}\Big(\eta - \frac{S_n^2(t)\circ T}{nt}\Big|\mathcal{M}_{0,n}\Big)\Big\|_1 = 0\,. \tag{7.4.3}$$

Combining (7.4.1) and (7.4.3), it follows that $\lim_{n\to\infty}\|\mathbb{E}(\eta - \eta\circ T|\mathcal{M}_{0,n})\|_1 = 0$, which implies that

$$\lim_{n\to\infty}\Big\|\mathbb{E}\Big(\eta - \eta\circ T\Big|\bigcap_{k\ge n}\mathcal{M}_{0,k}\Big)\Big\|_1 = 0\,.$$

Applying the martingale convergence theorem, we obtain

$$\lim_{n\to\infty}\Big\|\mathbb{E}\Big(\eta - \eta\circ T\Big|\bigcap_{k\ge n}\mathcal{M}_{0,k}\Big)\Big\|_1 = \|\mathbb{E}(\eta - \eta\circ T|\mathcal{M}_{0,\mathrm{inf}})\|_1 = 0\,. \tag{7.4.4}$$

According to $\mathbf{S2}(c)$, the random variable η is $\mathcal{M}_{0,\mathrm{inf}}$-measurable. Therefore, (7.4.4) implies that $\mathbb{E}(\eta\circ T|\mathcal{M}_{0,\mathrm{inf}}) = \eta$. The fact that $\eta\circ T = \eta$ almost surely is a direct consequence of the following elementary result.

Lemma 7.5. *Let $(\Omega, \mathcal{A}, \mathbb{P})$ be a probability space, X an integrable random variable, and $\mathcal{M}$ a σ-algebra of $\mathcal{A}$. If the random variable $\mathbb{E}(X|\mathcal{M})$ has the same law as X, then $\mathbb{E}(X|\mathcal{M}) = X$ almost surely.*

Proof of Lemma 7.5. For any real number m, consider $A_1 = \{X \le m\}$, $A_2 = \{\mathbb{E}(X|\mathcal{M}) \le m\}$, $C_1 = A_1 \cap A_2^c$ and $C_2 = A_2 \cap A_1^c$. Since by assumption the random variables X and $\mathbb{E}(X|\mathcal{M})$ are identically distributed, it follows that $\mathbb{P}(A_1) = \mathbb{P}(A_2)$, $\mathbb{P}(C_1) = \mathbb{P}(C_2)$ and $\mathbb{E}(X\mathbf{1}_{A_1}) = \mathbb{E}(X\mathbf{1}_{A_2})$. This implies in particular that $\mathbb{E}((X-m)\mathbf{1}_{C_1}) = \mathbb{E}((X-m)\mathbf{1}_{C_2})$. These terms having opposite signs, they are zero. Since $X - m$ is positive on C_2, it follows that C_2 and consequently $A_1 \Delta A_2$ have probability zero (Δ denoting the symmetric difference). Now, it is easily seen that

$$\mathbb{E}((\mathbb{E}(X|\mathcal{M}))^2\mathbf{1}_{A_1}) = \mathbb{E}((\mathbb{E}(X|\mathcal{M}))^2\mathbf{1}_{A_2}) = \mathbb{E}(X^2\mathbf{1}_{A_1}) \quad \text{and} \tag{7.4.5}$$

$$\mathbb{E}(X\mathbb{E}(X|\mathcal{M})\mathbf{1}_{A_1}) = \mathbb{E}(X\mathbb{E}(X|\mathcal{M})\mathbf{1}_{A_2}) = \mathbb{E}((\mathbb{E}(X|\mathcal{M}))^2\mathbf{1}_{A_2}) \tag{7.4.6}$$

According to (7.4.5) and (7.4.6), we obtain

$$\begin{aligned} \|(X - \mathbb{E}(X|\mathcal{M}))\mathbf{1}_{A_1}\|_2^2 &= \|X\mathbf{1}_{A_1}\|_2^2 + \|\mathbb{E}(X|\mathcal{M})\mathbf{1}_{A_1}\|_2^2 - 2\mathbb{E}(X\mathbb{E}(X|\mathcal{M})\mathbf{1}_{A_1}) \\ &= 0 \end{aligned} \tag{7.4.7}$$

Since (7.4.7) is true for any real m, it follows that $X = \mathbb{E}(X|\mathcal{M})$ almost surely, and Lemma 7.5 is proved. □.

7.4.3 End of the proof

First, note that we can restrict ourselves to bounded functions of $\mathcal{H}$: if **S2** implies **S1**(h) for any continuous and bounded function h then we easily infer from **S2**(c) that $n^{-1}S_n^2(t)$ is uniformly integrable for any t in $[0,1]$, which implies that **S1** extends to the whole space $\mathcal{H}$.

Definition 7.4. *Let $B_1^3(\mathbb{R})$ be the class of three-times continuously differentiable functions from $\mathbb{R}$ to $\mathbb{R}$ such that* $\max(\|h^{(i)}\|_\infty, i \in \{0,1,2,3\}) \leq 1$.

Suppose now that **S1**(h) holds for any h in $B_1^3(\mathbb{R})$. Applying Lemma 7.4, this is equivalent to say that **S3**(h) holds for any h in $B_1^3(\mathbb{R})$, which obviously implies that **S3**(h) holds for $h_t = e^{it}$. Using that the probability $\nu_n[1]$ is tight (since it converges weakly to $\nu[1]$) and that $|\mu_n[Z_n]| \leq \nu_n[1] + \nu[1]$, we infer that $\mu_n[Z_n]$ is tight, and Lemma 7.3 implies that **S3**(h) (and therefore **S1**(h)) holds for any continuous bounded function h.
On the other hand, from the asymptotic negligibility of $n^{-1/2}X_{0,n}$ we infer that, for any positive integer k, $n^{-1/2}(S_n(t) - S_n(t)\circ T^k)$ converges in probability to zero. Consequently, since any function h belonging to $B_1^3(\mathbb{R})$ is 1-Lipschitz and bounded, we have

$$\lim_{n\to\infty} \left\| h(n^{-1/2}S_n(t)) - h(n^{-1/2}S_n(t)\circ T^k) \right\|_1 = 0\,,$$

and **S1**(h) is equivalent to

$$\lim_{n\to\infty} \left\| \mathbb{E}\Big(h(n^{-1/2}S_n(t)\circ T^k) - \int h(x\sqrt{t\eta})g(x)dx \,\Big|\mathcal{M}_{k,n}\Big) \right\|_1 = 0\,.$$

Now, since both η and $\mathbb{P}$ are invariant by T, we infer that Theorem 7.5 is a straightforward consequence of Proposition 7.6 below:

Proposition 7.6. *Let $X_{i,n}$ and $\mathcal{M}_{i,n}$ be defined as in Theorem 7.5. If* **S2** *holds, then, for any h in $B_1^3(\mathbb{R})$ and any t in $[0,1]$,*

$$\lim_{n\to\infty}\left\|\mathbb{E}\Big(h(n^{-1/2}S_n(t)) - \int h(x\sqrt{t\eta})g(x)dx\,\Big|\mathcal{M}_{0,n}\Big)\right\|_1 = 0$$

where g is the distribution of a standard normal.

Proof of Proposition 7.6. We prove the result for $S_n(1)$, the proof of the general case being unchanged. Without loss of generality, suppose that there exists a sequence $(\varepsilon_i)_{i\in\mathbb{Z}}$ of $\mathcal{N}(0,1)$-distributed and independent random variables, independent of $\mathcal{M}_{\infty,\infty} = \sigma(\bigcup_{k,n}\mathcal{M}_{k,n})$.

Definition 7.5. *Let i, p and n be three integers such that $1 \le i \le p \le n$. Set $q = [n/p]$ and define*

$$U_{i,n} = X_{iq-q+1,n} + \cdots + X_{iq,n}, \qquad V_{i,n} = \frac{1}{\sqrt{n}}(U_{1,n} + U_{2,n} + \cdots + U_{i,n})$$
$$\Delta_i = \varepsilon_{iq-q+1} + \cdots + \varepsilon_{iq}, \qquad \Gamma_i = \sqrt{\frac{\eta}{n}}(\Delta_i + \Delta_{i+1} + \cdots + \Delta_p)\,.$$

Definition 7.6. *Let g be any function from $\mathbb{R}$ to $\mathbb{R}$. For k and l in $[1,p]$ and any positive integer $n \ge p$, set $g_{k,l;n} = g(V_{k,n} + \Gamma_l)$, with the conventions $g_{k,p+1;n} = g(V_{k,n})$ and $g_{0,l;n} = g(\Gamma_l)$. Afterwards, we shall apply this notation to the successive derivatives of the function h. For brevity we shall omit the index n.*

Let $s_n = \sqrt{\eta}(\varepsilon_1 + \cdots + \varepsilon_n)$. Since $(\varepsilon_i)_{i\in\mathbb{Z}}$ is independent of $\mathcal{M}_{\infty,\infty}$, we have, integrating with respect to $(\varepsilon_i)_{i\in\mathbb{Z}}$,

$$\begin{aligned}\mathbb{E}\Big(h(n^{-1/2}S_n(1)) &- \int h(x\sqrt{\eta})g(x)dx\,\Big|\mathcal{M}_{0,n}\Big)\\ &= \mathbb{E}(h(n^{-1/2}S_n(1)) - h(V_{p,n})|\mathcal{M}_{0,n}) + \mathbb{E}(h(V_{p,n}) - h(\Gamma_1)|\mathcal{M}_{0,n})\\ &\qquad + \mathbb{E}(h(\Gamma_1) - h(n^{-1/2}s_n)|\mathcal{M}_{0,n})\,. \quad (7.4.8)\end{aligned}$$

Here, note that $|n^{-1/2}S_n(1) - V_{p,n}| \le n^{-1/2}(|X_{n-p+2,n}| + \cdots + |X_{n,n}|)$. Using the asymptotic negligibility of $n^{-1/2}X_{0,n}$, we infer that $n^{-1/2}S_n(1) - V_{p,n}$ converges in probability to zero. Since furthermore h is 1-Lipschitz and bounded, we conclude that

$$\lim_{n\to\infty}\|h(n^{-1/2}S_n(1)) - h(V_{p,n})\|_1 = 0\,, \qquad (7.4.9)$$

and the same arguments yield

$$\lim_{n\to\infty}\|h(\Gamma_1) - h(n^{-1/2}s_n)\|_1 = 0\,. \qquad (7.4.10)$$

In view of (7.4.9) and (7.4.10), it remains to control the second term in the right hand side of (7.4.8). To this end, we use Lindeberg's decomposition.

$$h(V_{p,n}) - h(\Gamma_1) = \sum_{i=1}^{p} (h_{i,i+1} - h_{i-1,i+1}) + \sum_{i=1}^{p} (h_{i-1,i+1} - h_{i-1,i}) \,. \quad (7.4.11)$$

Now, applying Taylor's integral formula we get that:

$$\begin{cases} h_{i,i+1} - h_{i-1,i+1} & = & \frac{1}{\sqrt{n}} U_{i,n} h'_{i-1,i+1} & + & \frac{1}{2n} U_{i,n}^2 h''_{i-1,i+1} & + & R_i \\ h_{i-1,i+1} - h_{i-1,i} & = & -\sqrt{\frac{\eta}{n}} \Delta_i h'_{i-1,i+1} & - & \frac{\eta}{2n} \Delta_i^2 h''_{i-1,i+1} & + & r_i \end{cases}$$

where

$$|R_i| \le \frac{U_{i,n}^2}{n} \left(1 \wedge \frac{|U_{i,n}|}{\sqrt{n}}\right) \quad \text{and} \quad |r_i| \le \frac{\eta \Delta_i^2}{n} \left(1 \wedge \frac{\sqrt{\eta}|\Delta_i|}{\sqrt{n}}\right). \quad (7.4.12)$$

Since Δ_i is centered and independent of $\sigma\left(\mathcal{M}_{\infty,\infty} \cup \sigma(h'_{i-1,i+1})\right)$, we have $\mathbb{E}(\sqrt{\eta}\Delta_i h'_{i-1,i+1} | \mathcal{M}_{0,n}) = \mathbb{E}(\Delta_i)\mathbb{E}(\sqrt{\eta} h'_{i-1,i+1} | \mathcal{M}_{0,n}) = 0$. It follows that

$$\mathbb{E}(h(V_p) - h(\Gamma_1) | \mathcal{M}_{0,n}) = D_1 + D_2 + D_3, \quad (7.4.13)$$

where

$$\begin{aligned} D_1 &= \sum_{i=1}^{p} \mathbb{E}(n^{-1/2} U_{i,n} h'_{i-1,i+1} | \mathcal{M}_{0,n}), \\ D_2 &= \frac{1}{2} \sum_{i=1}^{p} \mathbb{E}(n^{-1}(U_{i,n}^2 - \eta\Delta_i^2) h''_{i-1,i+1} | \mathcal{M}_{0,n}), \\ D_3 &= \sum_{i=1}^{p} \mathbb{E}(R_i + r_i | \mathcal{M}_{0,n}). \end{aligned}$$

Control of D_3. From (7.4.12) and the fact that T preserves $\mathbb{P}$, we get

$$\sum_{i=1}^{p} \|R_i\|_1 \le \mathbb{E}\left(\frac{S_n^2(1/p)}{n/p}\left(1 \wedge \frac{|S_n(1/p)|}{\sqrt{n}}\right)\right),$$

and $\mathbf{S2}(a)$ implies that

$$\lim_{p\to\infty} \limsup_{n\to\infty} \sum_{i=1}^{p} \|R_i\|_1 = 0 \,. \quad (7.4.14)$$

Moreover, since for $t \in [0,1]$, the sequence $(\eta/n)^{-1/2}(\varepsilon_1 + \cdots + \varepsilon_{[nt]})$ obviously satisfies $\mathbf{S2}(a)$, the same argument applies to $\sum_{i=1}^{p} \|r_i\|_1$. Finally

$$\lim_{p\to\infty} \limsup_{n\to\infty} \|D_3\|_1 = 0 \,. \tag{7.4.15}$$

Control of D_1. Denote by $\mathbb{E}_\varepsilon$ the integration with respect to the sequence $(\varepsilon_i)_{i\in\mathbb{Z}}$. Set $l(i,n) = (i-1)[n/p]$. Bearing in mind the definition of $h'_{i-1,i+1}$ and integrating with respect to $(\varepsilon_i)_{i\in\mathbb{Z}}$ we deduce that the random variable $\mathbb{E}_\varepsilon(h'_{i-1,i+1})$ is $\mathcal{M}_{l(i,n),n}$-measurable and bounded by one. Now, since the σ-algebra $\mathcal{M}_{0,n}$ is included into $\mathcal{M}_{l(i,n),n}$, we obtain

$$\|\mathbb{E}(n^{-1/2}U_{i,n}h'_{i-1,i+1}|\mathcal{M}_{0,n})\|_1 \le \|\mathbb{E}(n^{-1/2}U_{i,n}|\mathcal{M}_{l(i,n),n})\|_1 \,.$$

Using that T preserves $\mathbb{P}$, the latter equals $\|\mathbb{E}(n^{-1/2}S_n(1/p)|\mathcal{M}_{0,n})\|_1$. Consequently $\|D_1\|_1 \le pn^{-1/2}\|\mathbb{E}(S_n(1/p)|\mathcal{M}_{0,n})\|_1$ and $\mathbf{S2}(b)$ implies that

$$\lim_{p\to\infty} \limsup_{n\to\infty} D_1 = 0 \,. \tag{7.4.16}$$

Control of D_2. Integrating with respect to $(\varepsilon_i)_{i\in\mathbb{Z}}$, we have

$$\|\mathbb{E}((U_{i,n}^2 - \eta\Delta_i^2)h''_{i-1,i+1}|\mathcal{M}_{0,n})\|_1 - \|\mathbb{E}((U_{i,n}^2 - \eta[np^{-1}])\mathbb{E}_\varepsilon(h''_{i-1,i+1})|\mathcal{M}_{0,n})\|_1 \,.$$

Arguing as for the control of D_1, we have

$$\|\mathbb{E}((U_{i,n}^2 - \eta[np^{-1}])\mathbb{E}_\varepsilon(h''_{i-1,i+1})|\mathcal{M}_{0,n})\|_1 \le \|\mathbb{E}(n^{-1}U_{i,n}^2 - \eta[np^{-1}]|\mathcal{M}_{l(i,n),n})\|_1 \,.$$

Since both η and $\mathbb{P}$ are invariant by the transformation T, the latter equals $\|\mathbb{E}(S_n^2(1/p) - \eta[np^{-1}]|\mathcal{M}_{0,n})\|_1$. Consequently

$$\|D_2\|_1 \le \left\| \mathbb{E}\left(\frac{S_n^2(1/p)}{n/p} - \eta\frac{[n/p]}{n/p} \Big| \mathcal{M}_{0,n} \right) \right\|_1$$

and $\mathbf{S2}$(c) implies that

$$\lim_{p\to\infty} \limsup_{n\to\infty} \|D_2\|_1 = 0 \,. \tag{7.4.17}$$

End of the proof of Proposition 7.6. From (7.4.15), (7.4.16) and (7.4.17) we infer that, for any h in $B_1^3(\mathbb{R})$,

$$\lim_{p\to\infty} \limsup_{n\to\infty} \|D_1 + D_2 + D_3\|_1 = 0 \,.$$

This fact together with (7.4.8), (7.4.9), (7.4.10) and (7.4.13) imply Proposition 7.6. □

7.5 Applications

7.5.1 Stable convergence

Let $(\Omega, \mathcal{A}, \mathbb{P})$ be a probability space. A sequence $(X_n)_{n>0}$ of integrable random variables is said to converge weakly in $\mathbb{L}^1$ to X if for any bounded random variable Y, we have

$$\lim_{n\to\infty} \mathbb{E}(ZX_n) = \mathbb{E}(ZX).$$

Let $\mathcal{X}$ be some polish space. A random probability μ on $\mathcal{X}$ is a function from $\mathcal{B}(\mathcal{X}) \times \Omega$ such that $\mu(.,\omega)$ is a probability measure for any ω in Ω and $\mu(B,.)$ is $\mathcal{A}$-measurable for any B in $\mathcal{B}(\mathcal{X})$.

Let μ be random probability on $\mathcal{X}$. We say that a sequence $(X_n)_{n>0}$ with values in a Polish space $\mathcal{X}$ converges stably with respect to μ if the sequence $(\varphi(X_n))_{n>0}$ converges weakly in $\mathbb{L}^1$ to $\mu(\varphi)$ for any continuous bounded function φ. The equivalence of this definition with that of Rényi (1963) [156] is proved in Aldous and Eagleson (1978) [1]. Note that, if $(X_n)_{n>0}$ converges stably with respect to μ, then it converges in distribution to the probability $\nu(A) = \int \mu(A,\omega)\mathbb{P}(d\omega)$. Here is a one consequence of the stablility.

Lemma 7.6. *Let $(X_n)_{n>0}$ and $(Y_n)_{n>0}$ be two sequences of random variables with values in two polish spaces $\mathcal{X}$ and $\mathcal{Y}$. If $(X_n)_{n>0}$ converges stably with respect to μ and $(Y_n)_{n>0}$ converges in probability to Y then (X_n, Y_n) converges in distribution to the probability $\nu(A \times B) = \int \mu(A,\omega)\mathbf{1}_{Y(\omega)\in B}\mathbb{P}(d\omega)$.*

Proof of Lemma 7.6. Let f and g be two bounded Lipschitz functions. We have

$$|\mathbb{E}(f(X_n)g(Y_n)) - \nu(f\otimes g))| = |\mathbb{E}(f(X_n)g(Y_n)) - \mathbb{E}(\mu(f)g(Y))|.$$

Consequently

$$|\mathbb{E}(f(X_n)g(Y_n)) - \nu(f\otimes g))| \le |\mathbb{E}(f(X_n)g(Y)) - \mathbb{E}(\mu(f)g(Y))| + \|g(Y_n) - g(Y)\|_1.$$

The first term on right hand tends to zero by the weak-$\mathbb{L}^1$ convergence and the second term tends to zero because Y_n converges in probability to Y and g is bounded Lipschitz. □

The next Corollary shows that if the CCLT holds then $(n^{-1/2}S_n(t))_{n>0}$ converges stably.

Corollary 7.2. *Let $X_{i,n}$, $\mathcal{M}_{i,n}$, $S_n(t)$ be as in Theorem 7.5. Suppose that the sequence $(\mathcal{M}_{0,n})_{n\ge1}$ is nondecreasing. If Condition* **S2** *is satisfied, then the sequence $(n^{-1/2}S_n(t))_{n>0}$ converges stably with respect to the random probability defined by $\mu(\varphi) = \int \varphi(x\sqrt{t\eta})g(x)dx$.*

Proof of Corollary 7.2. We shall prove that if **S2** holds, then, for any bounded random variable Z, any t in $[0,1]$ and any φ in $\mathcal{H}$,

$$\lim_{n\to\infty} \mathbb{E}\left(Z\varphi(n^{-1/2}S_n(t))\right) = \mathbb{E}\Big(Z \int \varphi(x\sqrt{t\eta})g(x)dx\Big)\,. \tag{7.5.1}$$

Since $n^{-1}S_n^2(t)$ is uniformly integrable, we only need to prove (7.5.1) for continuous bounded functions. Recall that $\mathcal{M}_{\infty,\infty} = \sigma(\bigcup_{k,n}\mathcal{M}_{k,n})$. Since both $S_n(t)$ and η are $\mathcal{M}_{\infty,\infty}$-measurable, we can and do suppose that so is Z.
Set $Z_{k,n} = \mathbb{E}(Z|\mathcal{M}_{k,n})$, and use the decomposition

$$\mathbb{E}\left(Z\varphi(n^{-1/2}S_n(t))\right) - \mathbb{E}\Big(Z\int \varphi(x\sqrt{t\eta})g(x)dx\Big) = T_1 + T_2 + T_3\,,$$

where

$$\begin{aligned} T_1 &= \mathbb{E}\left((Z - Z_{k,n})\varphi(n^{-1/2}S_n(t))\right) \\ T_2 &= \mathbb{E}\Big(Z_{k,n}\Big(\varphi(n^{-1/2}S_n(t)) - \int \varphi(x\sqrt{t\eta})g(x)dx\Big)\Big) \\ T_3 &= \mathbb{E}\Big((Z_{k,n} - Z)\int \varphi(x\sqrt{t\eta})g(x)dx\Big)\,. \end{aligned}$$

By assumption, the array $\mathcal{M}_{k,n}$ is nondecreasing in k and n. Since the random variable Z is $\mathcal{M}_{\infty,\infty}$-measurable, the martingale convergence theorem implies that $\lim_{k\to\infty}\lim_{n\to\infty}\|Z_{k,n} - Z\|_1 = 0$. Consequently,

$$\lim_{k\to\infty}\limsup_{n\to\infty}|T_1| = \lim_{k\to\infty}\limsup_{n\to\infty}|T_3| = 0\,.$$

On the other hand, Theorem 7.5 implies that T_2 tends to zero as n tends to infinity, which completes the proof of Corollary 7.2. □

We end this section with an application of the stable convergence to random normalization. The proof is straightforward, using Lemma 7.6 and Corollary 7.2.

Corollary 7.3. *Let $X_{i,n}$, $\mathcal{M}_{i,n}$, $S_n(t)$ be as in Theorem 7.5. Suppose that the sequence $(\mathcal{M}_{0,n})_{n\geq 1}$ is nondecreasing and that $\mathbb{P}(\eta > 0) = 1$. If Condition* **S2** *is satisfied, and if $(\eta_n)_{n>0}$ converges in probability to η then, for any t in $]0,1]$,*

$$\frac{n^{-1/2}S_n(t)}{\sqrt{t\eta_n \vee n^{-1}}} \quad \textit{converges in distribution to } \mathcal{N}(0,1)\,.$$

For more about stable convergence, we refer to the book by Castaing *et al.* (2004) [35].

7.5.2 Sufficient conditions for stationary sequences

For strictly stationary sequences, Theorem 7.5 writes as follows.

Theorem 7.6. *Let $\mathcal{M}_0$ be a σ-algebra of $\mathcal{A}$ satisfying $\mathcal{M}_0 \subseteq T^{-1}(\mathcal{M}_0)$ and define the nondecreasing filtration $(\mathcal{M}_i)_{i\in\mathbb{Z}}$ by $\mathcal{M}_i = T^{-i}(\mathcal{M}_0)$. Let X_0 be a $\mathcal{M}_0$-measurable, square integrable and centered random variable. Define the sequence $(X_i)_{i\in\mathbb{Z}}$ by $X_i = X_0 \circ T^i$, and $S_n = X_1 + \cdots + X_n$. The following statements are equivalent:*

S1 *There exists a nonnegative $\mathcal{M}_0$-measurable random variable η such that, for any φ in $\mathcal{H}$ and any positive integer k,*

$$\lim_{n\to\infty} \left\| \mathbb{E}\Big(\varphi(n^{-1/2}S_n) - \int \varphi(x\sqrt{\eta})g(x)dx \,\Big|\mathcal{M}_k\Big)\right\|_1 = 0$$

where g is the distribution of a standard normal.

S2 *(a) the sequence $(n^{-1}S_n^2)_{n>0}$ is uniformly integrable.*

(b) the sequence $\|\mathbb{E}(n^{-1/2}S_n|\mathcal{M}_0)\|_1$ tends to 0 as n tends to infinity.

(c) there exists a nonnegative $\mathcal{M}_0$-measurable random variable η such that $\|\mathbb{E}(n^{-1}S_n^2 - \eta|\mathcal{M}_0)\|_1$ tends to 0 as n tends to infinity.

Moreover the random variable η satisfies $\eta = \eta \circ T$ almost surely.

In Proposition 7.7 and 7.8, we give two conditions implying **S2**. For the usual central limit theorem, Proposition 7.7 is due to Gordin (1969) [97], Corollary 7.4 is due to Heyde (1974) [103] and Proposition 7.8 is due to Dedecker and Rio (2000) [50].

Proposition 7.7. *Let $(\mathcal{M}_i)_{i\in\mathbb{Z}}$ and $(X_i)_{i\in\mathbb{Z}}$ be as in Theorem 7.6. Let H_i be the Hilbert space of $\mathcal{M}_i$-measurable, centered and square integrable functions. For all integer j less than i, denote by $H_i \ominus H_j$, the orthogonal of H_j into H_i. Assume that there exists a random variable m in $H_0 \ominus H_{-1}$ such that*

$$\lim_{n\to\infty} \frac{1}{\sqrt{n}}\left\|\sum_{i=1}^{n} X_0 \circ T^i - m \circ T^i\right\|_2 = 0\,, \tag{7.5.2}$$

*then **S2** (hence **S1**) holds.*

Corollary 7.4. *Let $(\mathcal{M}_i)_{i\in\mathbb{Z}}$ and $(X_i)_{i\in\mathbb{Z}}$ be as in Theorem 7.6, and define H_i as in Proposition 7.7. Let P_i be the projection operator on $H_i \ominus H_{i-1}$: for any function f in $\mathbb{L}^2$, $P_i(f) = \mathbb{E}(f|\mathcal{M}_i) - \mathbb{E}(f|\mathcal{M}_{i-1})$. If*

$$\sum_{i=0}^{n} P_0(X_i) \text{ converges in } \mathbb{L}^2 \text{ to } m \text{ and } \lim_{n\to\infty} n^{-1/2}\|S_n\|_2 = \|m\|_2\,, \tag{7.5.3}$$

*then (7.5.2) (hence **S1**) holds.*

Proof of Proposition 7.7. Let $W_n = m\circ T + \cdots + m\circ T^n$. Since $(m\circ T^i)_{i\in\mathbb{Z}}$ is a stationary sequence of martingale differences with respect to the filtration $(\mathcal{M}_i)_{i\in\mathbb{Z}}$, it satisfies **S2**. More precisely, $n^{-1}\mathbb{E}(W_n^2|\mathcal{M}_0)$ converges to $\eta = \mathbb{E}(m^2|\mathcal{I})$ in $\mathbb{L}^1$. Now, we shall use (7.5.2) to see that the sequence $(X_i)_{i\in\mathbb{Z}}$ also satisfies **S2**.

Proof of S2(b). From (7.5.2) it is clear that $n^{-1/2}\|\mathbb{E}(S_n|\mathcal{M}_0)\|_2$ tends to zero as n tends to infinity.

Proof of **S2**(c). To see that $n^{-1}\mathbb{E}(S_n^2|\mathcal{M}_0)$ converges to η in $\mathbb{L}^1$, write

$$\begin{aligned}\frac{1}{n}\left\|\mathbb{E}(S_n^2 - W_n^2|\mathcal{M}_0)\right\|_1 &\le \frac{1}{n}\left\|S_n^2 - W_n^2\right\|_1 \\ &\le \frac{\|S_n + W_n\|_2}{\sqrt{n}}\frac{\|S_n - W_n\|_2}{\sqrt{n}}. \end{aligned} \tag{7.5.4}$$

From (7.5.2) the latter tends to zero as n tends to infinity and therefore $(X_i)_{i\in\mathbb{Z}}$ satisfies **s2**(c) with $\eta = \mathbb{E}(m^2|\mathcal{I})$.

Proof of **S2**(a). Using both that $n^{-1}W_n^2$ is uniformly integrable and that the function $x \mapsto (1\wedge|x|)$ is 1-Lipschitz, we have, for any positive real M,

$$\lim_{n\to\infty}\left(\frac{W_n^2}{n}\left|\Big(1\wedge\frac{|S_n|}{M\sqrt{n}}\Big) - \Big(1\wedge\frac{|W_n|}{M\sqrt{n}}\Big)\right|\right) = 0\,. \tag{7.5.5}$$

Since $|x^2(1\wedge|y|) - z^2(1\wedge|t|)| \le |x^2 - z^2| + z^2|(1\wedge|y|) - (1\wedge|t|)|$, we infer from (7.5.4) and (7.5.5) that

$$\lim_{n\to\infty}\left\|\frac{S_n^2}{n}\Big(1\wedge\frac{|S_n|}{M\sqrt{n}}\Big) - \frac{W_n^2}{n}\Big(1\wedge\frac{|W_n|}{M\sqrt{n}}\Big)\right\|_1 = 0\,.$$

Now, the uniform integrability of $n^{-1}W_n^2$ yields

$$\lim_{M\to\infty}\limsup_{n\to\infty}\mathbb{E}\left(\frac{S_n^2}{n}\Big(1\wedge\frac{|S_n|}{M\sqrt{n}}\Big)\right) = 0\,,$$

which means exactly that $n^{-1}S_n^2$ is uniformly integrable. □

Proof of Corollary 7.4. By assumption, the random variable m belongs to $H_0\ominus H_{-1}$. It remains to check (7.5.2). Let $m_i = m\circ T^i$ and $T_n = m_1 + \cdots + m_n$. Clearly $\mathbb{E}((S_n - T_n)^2) = \mathbb{E}(S_n^2) + \mathbb{E}(T_n^2) - 2\mathbb{E}(S_nT_n)$. By assumption $n^{-1}\mathbb{E}(S_n^2)$ converges to $\|m\|_2^2 = n^{-1}\mathbb{E}(T_n^2)$. To prove (7.5.2), it suffices to prove that $n^{-1}\mathbb{E}(S_nT_n)$ converges to $\|m\|_2^2$. Now, by stationarity

$$\frac{1}{n}\mathbb{E}(S_nT_n) = \frac{1}{n}\sum_{i=1}^{n}\sum_{j=1}^{n}\mathbb{E}(X_im_j) = \sum_{k=-n}^{n}\Big(1-\frac{|k|}{n}\Big)\mathbb{E}(mX_k)\,. \tag{7.5.6}$$

By Kronecker's lemma $n^{-1}\mathbb{E}(S_nT_n)$ converges to $\sum_{k\in\mathbb{Z}}\mathbb{E}(mX_k)$ as soon as the latter converges. Now since m belongs to $H_0 \ominus H_{-1}$, it follows that $\sum_{k\in\mathbb{Z}}\mathbb{E}(mX_k) = \sum_{k\geq 0}\mathbb{E}(mP_0(X_k))$. Since $m = \sum_{k\geq 0}P_0(X_k)$ in $\mathbb{L}^2$, the result follows. □

Proposition 7.8. *Let $(\mathcal{M}_i)_{i\in\mathbb{Z}}$, $(X_i)_{i\in\mathbb{Z}}$ and S_n be as in Theorem 7.6. Consider the condition:*

$$\sum_{k=1}^{n} X_0\mathbb{E}(X_k|\mathcal{M}_0) \text{ converges in } \mathbb{L}^1 . \tag{7.5.7}$$

If (7.5.7) is satisfied, then **S2** *holds and the sequence $\mathbb{E}(X_0^2|\mathcal{I}) + 2\mathbb{E}(X_0S_n|\mathcal{I})$ converges in $\mathbb{L}^1$ to η.*

Proof of Proposition 7.8. We first prove that $\mathbb{E}(X_0^2|\mathcal{I}) + 2\mathbb{E}(X_0S_n|\mathcal{I})$ converges in $\mathbb{L}^1$. From assumption (7.5.7), the sequence $\mathbb{E}(X_0^2|\mathcal{M}_0) + 2\mathbb{E}(X_0S_n|\mathcal{M}_0)$ converges in $\mathbb{L}^1$. The result is then a consequence of part (b) of Lemma 7.7 below:

Lemma 7.7. *We have:*

(a) Both $\mathbb{E}(X_0X_k|\mathcal{I})$ and $\mathbb{E}(\mathbb{E}(X_0X_k|\mathcal{M}_0)|\mathcal{I})$ are almost surely equal to $\mathcal{M}_0$-measurable random variables.

(b) $\mathbb{E}(X_0X_k|\mathcal{I}) = \mathbb{E}(\mathbb{E}(X_0X_k|\mathcal{M}_0)|\mathcal{I})$ almost surely.

Lemma 7.7(b) is derived from Lemma 7.7(a) via the following elementary fact, whose proof is omitted.

Lemma 7.8. *Let Y be a random variable in $\mathbb{L}^1(\mathbb{P})$ and $\mathcal{U}$, $\mathcal{V}$ two σ-algebras of $(\Omega, \mathcal{A}, \mathbb{P})$. Suppose that $\mathbb{E}(Y|\mathcal{U})$ and $\mathbb{E}(\mathbb{E}(Y|\mathcal{V})|\mathcal{U})$ are $\mathcal{V}$-measurable. Then $\mathbb{E}(Y|\mathcal{U}) = \mathbb{E}(\mathbb{E}(Y|\mathcal{V})|\mathcal{U})$ almost surely.*

Proof of Lemma 7.7(a). The fact that $\mathbb{E}(\mathbb{E}(X_0X_k|\mathcal{M}_0)|\mathcal{I})$ is almost surely equal to some $\mathcal{M}_0$-measurable random variable follows from the $\mathbb{L}^1$-ergodic theorem. Indeed the variables $\mathbb{E}(X_iX_{k+i}|\mathcal{M}_0)$ are $\mathcal{M}_0$-measurable and

$$\mathbb{E}(\mathbb{E}(X_0X_k|\mathcal{M}_0)|\mathcal{I}) = \lim_{n\to\infty}\frac{1}{n}\sum_{i=1}^{n}\mathbb{E}(X_iX_{k+i}|\mathcal{M}_0) \text{ in } \mathbb{L}^1.$$

Next, from the stationarity of the sequence $(X_i)_{i\in\mathbb{Z}}$, we have

$$\Big\|\mathbb{E}(X_0X_k|\mathcal{I}) - \frac{1}{n}\sum_{i=1}^{n}X_iX_{i+k}\Big\|_1 = \Big\|\mathbb{E}(X_0X_k|\mathcal{I}) - \frac{1}{n}\sum_{i=1-(n+k)}^{-k}X_iX_{i+k}\Big\|_1 .$$

Both this equality and the $\mathbb{L}^1$-ergodic theorem imply that $\mathbb{E}(X_0X_k|\mathcal{I})$ is the limit in $\mathbb{L}^1$ of a sequence of $\mathcal{M}_0$-measurable random variables. □

Proof of **S2**(a). Let $\overline{S}_n = \max\{|S_1|, \dots, |S_n|\}$. In Chapter 8, we shall prove that $(n^{-1}(\overline{S}_n)^2)_{n>0}$ is uniformly integrable as soon as (7.5.7) holds. □

Proof of **S2**(c). For any positive integer N, we introduce

$$\Lambda_N = [(k-1)q+1, kq]^2 \cap \{(i,j) \in \mathbb{Z}^2 / \, |i-j| > N\}, \quad \text{and} \quad \bar{\Lambda}_N = [1,n]^2 - \Lambda_N .$$

and $\eta_N = \mathbb{E}(X_0^2|\mathcal{I}) + 2(\mathbb{E}(X_0X_1|\mathcal{I}) + \cdots + \mathbb{E}(X_0X_N|\mathcal{I}))$. Since η_N converges in $\mathbb{L}^1$ to η, it suffices to prove that

$$\lim_{N\to\infty} \limsup_{n\to\infty} \Big\| \eta_N - \mathbb{E}\Big(\frac{1}{n}\sum_{i=1}^n\sum_{j=1}^n X_iX_j \Big| \mathcal{M}_0\Big)\Big\|_1 = 0 . \tag{7.5.8}$$

Using the triangle inequality and the fact that $\eta_N = \mathbb{E}(\eta_N|\mathcal{M}_0)$ almost surely, we obtain

$$\Big\| \eta_N - \mathbb{E}\Big(\frac{1}{n}\sum_{i=1}^n\sum_{j=1}^n X_iX_j \Big| \mathcal{M}_0\Big)\Big\|_1 \le \Big\| \eta_N - \frac{1}{n}\sum_{\Lambda_N} X_iX_j \Big\|_1 + \frac{1}{n}\Big\| \sum_{\bar{\Lambda}_N} \mathbb{E}(X_iX_j|\mathcal{M}_0)\Big\|_1 .$$

Applying the $\mathbb{L}^1$-ergodic theorem the first term on right hand tends to 0 as n tends to infinity. To control the second term, write

$$\begin{aligned} \frac{1}{n}\Big\| \sum_{\bar{\Lambda}_N} \mathbb{E}(X_iX_j|\mathcal{M}_0)\Big\|_1 &\le \frac{2}{n}\sum_{i=1}^n \Big\| X_i \sum_{j=i+N}^n \mathbb{E}(X_j|\mathcal{M}_i)\Big\|_1 \\ &\le \frac{2}{n}\sum_{i=1}^n \Big\| X_0 \sum_{j=N}^{n-i} \mathbb{E}(X_j|\mathcal{M}_0)\Big\|_1 . \end{aligned}$$

By assumption (7.5.7) $\lim_{N\to\infty} \max_{n-N\le i\le n} \|X_0 \sum_{j=N}^{n-i} \mathbb{E}(X_j|\mathcal{M}_0)\|_1 = 0$ and consequently

$$\lim_{N\to\infty} \limsup_{n\to\infty} \frac{2}{n}\sum_{i=1}^n \Big\| X_0 \sum_{j=N}^{n-i} \mathbb{E}(X_j|\mathcal{M}_0)\Big\|_1 = 0 .$$

This completes the proof of **S2**(c). □

Proof of **S2**(b). Let $\mathcal{M}_{-\infty} = \bigcap_{i\in\mathbb{Z}} \mathcal{M}_i$ and recall that P_i has been defined in Corollary 7.4. We have the orthogonal decomposition

$$X_k = \mathbb{E}(X_k|\mathcal{M}_{-\infty}) + \sum_{i=0}^{\infty} P_{k-i}(X_k) . \tag{7.5.9}$$

Using the stationarity of $(X_i)_{i\in\mathbb{Z}}$ we infer from (7.5.9) that

$$\sum_{i=0}^{\infty} \|P_0(X_i)\|_2^2 = \sum_{i=0}^{\infty} \|P_{-i}(X_0)\|_2^2 \le \|X_0\|_2^2 \,. \tag{7.5.10}$$

Now, from the decomposition

$$\frac{X_{-1}}{\sqrt{n}}\mathbb{E}(S_n|\mathcal{M}_0) = \frac{X_{-1}}{\sqrt{n}}\mathbb{E}(S_n|\mathcal{M}_{-1}) + \frac{X_{-1}}{\sqrt{n}}\sum_{i=1}^{n} P_0(X_i)\,,$$

we infer that

$$\frac{1}{\sqrt{n}}\|X_{-1}\mathbb{E}(S_n|\mathcal{M}_0)\|_1 \le \frac{1}{\sqrt{n}}\|X_0\mathbb{E}(S_n\circ T|\mathcal{M}_0)\|_1 + \frac{\|X_0\|_2}{\sqrt{n}}\sum_{i=1}^{n}\|P_0(X_i)\|_2\,. \tag{7.5.11}$$

By (7.5.7), the first term on right hand tends to zero as n tends to infinity. On the other hand, we infer from (7.5.10) and Cauchy-Shwarz's inequality that $n^{-1/2}\sum_{i=1}^{n}\|P_0(X_i)\|_2$ vanishes as n goes to infinity, and so does the left hand term in (7.5.11). By induction, we can prove that for any positive k,

$$\lim_{n\to\infty}\frac{1}{\sqrt{n}}\|X_{-k}\mathbb{E}(S_n|\mathcal{M}_0)\|_1 = 0\,. \tag{7.5.12}$$

Now

$$\frac{1}{\sqrt{n}}\|\mathbb{E}(|X_0||\mathcal{I})\mathbb{E}(S_n|\mathcal{M}_0)\|_1 \le \frac{1}{\sqrt{n}}\left\|\mathbb{E}(S_n|\mathcal{M}_0)\Big(\mathbb{E}(|X_0||\mathcal{I}) - \frac{1}{k}\sum_{i=1}^{k}|X_{-i}|\Big)\right\|_1 + \frac{1}{\sqrt{n}}\left\|\mathbb{E}(S_n|\mathcal{M}_0)\frac{1}{k}\sum_{i=1}^{k}|X_{-i}|\right\|_1 . \tag{7.5.13}$$

From (7.5.12), the second term on right hand tends to zero as n tends to infinity. Applying first Cauchy-Schwarz's inequality and next the $\mathbb{L}^2$-ergodic theorem, we easily deduce that the first term on right hand is as small as we wish by choosing k large enough. Therefore

$$\lim_{n\to\infty}\frac{1}{\sqrt{n}}\|\mathbb{E}(|X_0||\mathcal{I})\mathbb{E}(S_n|\mathcal{M}_0)\|_1 = 0\,. \tag{7.5.14}$$

Set $A = \{\mathbf{1}_{\mathbb{E}(|X_0||\mathcal{I})>0}\}$ and $B = A^c = \{\mathbf{1}_{\mathbb{E}(|X_0||\mathcal{I})=0}\}$.
For any positive real m, we have

$$\frac{1}{\sqrt{n}}\|\mathbf{1}_A\mathbb{E}(S_n|\mathcal{M}_0)\|_1 \le \frac{1}{m\sqrt{n}}\|\mathbb{E}(|X_0||\mathcal{I})\mathbb{E}(S_n|\mathcal{M}_0)\|_1 + \frac{1}{\sqrt{n}}\|\mathbf{1}_{0<\mathbb{E}(|X_0||\mathcal{I})<m}\mathbb{E}(S_n|\mathcal{M}_0)\|_1\,. \tag{7.5.15}$$

From (7.5.14), the first term on right hand tends to zero as n tends to infinity. Letting m tend to zero we infer that the second term on right hand of (6.12) is as small as we wish. Consequently

$$\lim_{n\to\infty} \frac{1}{\sqrt{n}} \|\mathbf{1}_A \mathbb{E}(S_n|\mathcal{M}_0)\|_1 = 0\,. \tag{7.5.16}$$

On the other hand, noting that $\mathbb{E}(|X_0|\mathbf{1}_B) = 0$, we infer that X_0 is zero on the set B. Since B is invariant by T, X_k is zero on B for any k in $\mathbb{Z}$. Now arguing as in Lemma 7.7(b), we obtain $\mathbb{E}(\mathbb{E}(|S_n||\mathcal{M}_0)|\mathcal{I}) = \mathbb{E}(|S_n||\mathcal{I})$. These two facts lead to

$$\|\mathbf{1}_B \mathbb{E}(S_n|\mathcal{M}_0)\|_1 \le \mathbb{E}(\mathbf{1}_B \mathbb{E}(\mathbb{E}(|S_n||\mathcal{M}_0)|\mathcal{I})) \le \mathbb{E}(|S_n|\mathbf{1}_B) \le 0 \tag{7.5.17}$$

Collecting (7.5.16) and (7.5.17), we conclude that $n^{-1/2}\|\mathbb{E}(S_n|\mathcal{M}_0)\|_1$ tends to zero as n tends to infinity. This completes the proof. □

7.5.3 γ-dependent sequences

Applying Corollary 7.4 and Proposition 7.8, we derive sufficient conditions for the CCLT expressed in terms of the coefficients γ_2, and γ_1. Here, the coefficients are defined by

$$\gamma_1(k) = \gamma_1(\mathcal{M}_0, X_k), \quad \text{and} \quad \gamma_2(k) = \gamma_2(\mathcal{M}_0, X_k)\,.$$

Corollary 7.5. *Let $(\mathcal{M}_i)_{i\in\mathbb{Z}}$ and $(X_i)_{i\in\mathbb{Z}}$ be as in Theorem 7.6 and define $\mathcal{M}_{-\infty} = \cap_{i\in\mathbb{Z}}\mathcal{M}_i$. Define the operators P_i as in Corollary 7.4.*

1. *If $\mathbb{E}(X_0|\mathcal{M}_{-\infty}) = 0$ and $\sum_{i\ge 0}\|P_0(X_i)\|_2 < \infty$ then (7.5.3) (hence* **S1***) holds. Moreover, η is the same as in Proposition 7.8.*

2. *Assume that*

$$\sum_{k>0} \frac{\gamma_2(k)}{\sqrt{k}} < \infty\,. \tag{7.5.18}$$

Then $\mathbb{E}(X_0|\mathcal{M}_{-\infty}) = 0$ and $\sum_{i\ge 0}\|P_0(X_i)\|_2 < \infty$, and **S1** *holds.*

Proof of Corollary 7.5:

Proof of 1. Starting from (7.5.9) with $\mathbb{E}(X_k|\mathcal{M}_{-\infty}) = 0$, we obtain

$$\mathbb{E}(X_0X_k) = \sum_{i\ge 0}\mathbb{E}(P_{-i}(X_0)P_{-i}(X_k)) = \sum_{i\ge 0}\mathbb{E}(P_0(X_i)P_0(X_{i+k}))\,.$$

Hence, using Hölder's inequality,

$$\sum_{k=1}^{\infty}|\mathbb{E}(X_0X_k)| \le \sum_{i\ge 0}\|P_0(X_i)\|_2\Big(\sum_{k=1}^{\infty}\|P_0(X_{i+k})\|_2\Big) \le \Big(\sum_{i\ge 0}\|P_0(X_i)\|_2\Big)^2, \tag{7.5.19}$$

so that $\sum_{k\in\mathbb{Z}} |\mathbb{E}(X_0X_k)|$ is finite. Consequently

$$\lim_{n\to\infty} \frac{1}{n}\mathbb{E}(S_n^2) = \sum_{k=-\infty}^{\infty} \mathbb{E}(X_0X_k) = \sum_{i=0}^{\infty}\sum_{j=0}^{\infty} \mathbb{E}(P_0(X_i)P_0(X_j)). \tag{7.5.20}$$

On the other hand

$$\mathbb{E}(m^2) = \mathbb{E}\Big(\Big(\sum_{i=0}^{\infty} P_0(X_i)\Big)^2\Big) = \sum_{i=0}^{\infty}\sum_{j=0}^{\infty} \mathbb{E}(P_0(X_i)P_0(X_j)). \tag{7.5.21}$$

Combining (7.5.20) and (7.5.21), we see that (7.5.3) holds. To compute η, note that we can in fact prove a stronger result than (7.5.19), that is

$$\sum_{k=1}^{\infty} \|\mathbb{E}(X_0X_k|\mathcal{I})\|_1 \le \Big(\sum_{i\ge 0} \|P_0(X_i)\|_2\Big)^2.$$

It follows that $n^{-1}\mathbb{E}(S_n^2|\mathcal{I})$ converges in $\mathbb{L}^1$ to $\eta = \sum_{k\in\mathbb{Z}} \mathbb{E}(X_0X_k|\mathcal{I})$. □

Proof of 2. Note first that (7.5.18) implies that $\mathbb{E}(X_k|\mathcal{M}_{-\infty}) = 0$. Let $(L_k)_{k>0}$ be a sequence of positive numbers such that $\sum_{i>0}\Big(\sum_{k=1}^{i} L_k\Big)^{-1} < \infty$. Starting from (7.5.9) and using the stationarity of $(X_i)_{i\in\mathbb{Z}}$ we obtain that

$$\sum_{k>0} L_k\|\mathbb{E}(X_k|\mathcal{M}_0)\|_2^2 = \sum_{k>0} L_k \sum_{i\le 0} \|P_i(X_k)\|_2^2 = \sum_{i>0}\Big(\sum_{k=1}^{i} L_k\Big)\|P_0(X_i)\|_2^2.$$

Let $a_i = L_1 + \cdots + L_i$. Applying Hölder's inequality in ℓ^2, we obtain that

$$\begin{aligned}\sum_{i>0} \|P_0(X_i)\|_2 &\le \Big(\sum_{i>0}\frac{1}{a_i}\Big)^{1/2}\Big(\sum_{i>0} a_i\|P_0(X_i)\|_2^2\Big)^{1/2}\\ &\le \Big(\sum_{i>0}\frac{1}{a_i}\Big)^{1/2}\Big(\sum_{k>0} L_k\|\mathbb{E}(X_k|\mathcal{M}_0)\|_2^2\Big)^{1/2}\end{aligned}$$

Since (7.5.18) holds, one can take $L_k^{-1} = \sqrt{k}\|\mathbb{E}(X_k|\mathcal{M}_0)\|_2$. The result follows. □

Corollary 7.6. *Let $(\mathcal{M}_i)_{i\in\mathbb{Z}}$ and $(X_i)_{i\in\mathbb{Z}}$ be as in Theorem 7.6. Let G_{X_0} and Q_{X_0} be as in Definition 5.1. If*

$$\sum_{k=0}^{\infty} \int_0^{\gamma_1(k)} Q_{X_0}\circ G_{X_0}(u)du < \infty \tag{7.5.22}$$

then (7.5.7) (hence **S1***) holds. Moreover (7.5.22) holds as soon as*

1. $\mathbb{P}(|X_0| > x) \le (c/x)^r$ *for* $r > 2$, *and* $\sum_{i\ge 0}(\gamma_1(i))^{(r-2)/(r-1)} < \infty$.

2. $\|X_0\|_r < \infty$ *for* $r > 2$, *and* $\sum_{i\ge 0} i^{1/(r-2)}\gamma_1(i) < \infty$.

3. $\mathbb{E}(|X_0|^2 \log(1+|X_0|)) < \infty$ *and* $\gamma_1(i) = \mathcal{O}(a^i)$ *for some* $a < 1$.

Proof of Corollary 7.6. Applying Inequality (5.2.1), we obtain that

$$\sum_{k\ge 0} \|X_0 \mathbb{E}(X_k|\mathcal{M}_0)\|_1 \le \sum_{k\ge 0} \int_0^{\gamma_1(k)} Q_{|X_0|} \circ G_{|X_0|}(u)du\,,$$

so that (7.5.22) implies (7.5.7).

Proof of 1. Since $\mathbb{P}(|X| > x) \le (c/x)^r$ we easily get that

$$\int_0^x Q_X(u)du \le \frac{c(r-1)}{r} x^{(r-1)/r} \quad \text{and then} \quad G_X(u) \ge \Big(\frac{ur}{c(r-1)}\Big)^{r/(r-1)}.$$

Set $K_{c,r} = c(c - cr^{-1})^{1/(r-1)}$. We obtain the bound

$$\begin{aligned}\sum_{i\ge 0} \int_0^{\gamma_1(i)} Q_{X_0} \circ G_{X_0}(u)du &\le K_{c,r} \sum_{i\ge 0} \int_0^{\gamma_1(i)} u^{-1/(r-1)}du \\ &\le \frac{K_{c,r}(r-1)}{r-2} \sum_{i\ge 0} \gamma_{1,i}^{(r-2)/(r-1)}.\end{aligned}$$

Proof of 2. Note first that

$$\int_0^{\|X_0\|_1} Q_{X_0}^{r-1} \circ G_{X_0}(u)du = \int_0^1 Q_{X_0}^r(u)du = \mathbb{E}(|X_0|^r)\,.$$

Define next

$$\gamma_1^{-1}(u) = \sum_{i\ge 0} \mathbf{1}_{u<\gamma_1(i)} = \inf\{k \in \mathbb{N}/\, \gamma_1(k) \le u\}\,.$$

Applying Hölder's inequality, we obtain that

$$\Big(\sum_{i\ge 0} \int_0^{\gamma_1(i)} Q_{X_0} \circ G_{X_0}(u)du\Big)^{r-1} \le \|X_0\|_r^r \Big(\int_0^{\|X_0\|_1} (\gamma_1^{-1}(u))^{(r-1)/(r-2)}du\Big)^{r-2}.$$

Here note that

$$(\gamma_1^{-1}(u))^q = \sum_{j=0}^{\infty} ((j+1)^q - j^q)\mathbf{1}_{u<\gamma_1(j)} \tag{7.5.23}$$

Now, apply (7.5.23) with $q = (r-1)/(r-2)$. Noting that $(i+1)^q - i^q \le q(i+1)^{q-1}$, we infer that

$$\int_0^{\|X\|_1} (\gamma_1^{-1}(u))^{(r-1)/(r-2)} du \le q \sum_{i\ge 0} (i+1)^{1/(r-2)} \gamma_1(i).$$

Proof of 3. Let $\rho(i) = \gamma_1(i)/\|X\|_1$ and U be a random variable uniformly distributed over $[0,1]$. We have

$$\begin{aligned}\int_0^{\|X\|_1} \gamma_1^{-1}(u) Q_{X_0} \circ G_{X_0}(u) du &= \int_0^1 \rho^{-1}(u) Q_{X_0} \circ G_{X_0}(u\|X_0\|_1) du \\ &= \mathbb{E}((\rho^{-1}(U)) Q_{X_0} \circ G_{X_0}(U\|X_0\|_1)).\end{aligned}$$

Let ϕ be the function defined on $\mathbb{R}^+$ by $\phi(x) = x(\log(1+x))^{p-1}$. Denote by ϕ^* its Young's transform. Applying Young's inequality, we have that

$$\mathbb{E}((\rho^{-1}(U)) Q_{X_0} \circ G_{X_0}(U\|X_0\|_1)) \le 2\|(\rho^{-1}(U))\|_{\phi^*} \|Q_{X_0} \circ G_{X_0}(U\|X_0\|_1)\|_{\phi}$$

Here, note that $\|Q_{X_0} \circ G_{X_0}(U\|X_0\|_1)\|_\phi$ is finite as soon as

$$\int_0^{\|X\|_1} Q_{X_0} \circ G_{X_0}(u)(\log(1 + Q_{X_0} \circ G_{X_0}(u)))^{p-1} du < \infty.$$

Setting $z = G_{X_0}(u)$, we obtain the condition

$$\int_0^1 Q_{X_0}^2(u)(\log(1 + Q_{X_0}(u)))du < \infty. \tag{7.5.24}$$

Since $Q_{X_0}(U)$ has the same distribution as $|X_0|$, we infer that (7.5.24) holds as soon as $\mathbb{E}(|X_0|^2(\log(1+|X_0|)))$ is finite. It remains to control $\|(\rho^{-1}(U))\|_{\phi^*}$. Arguing as in Rio (2000) [161] page 17, we see that $\|(\rho^{-1}(U))\|_{\phi^*}$ is finite as soon as there exists $c > 0$ such that

$$\sum_{i\ge 0} \rho(i)\, \phi'^{-1}((i+1)/c) < \infty. \tag{7.5.25}$$

Since ϕ'^{-1} has the same behavior as $x \mapsto e^x$ as x goes to infinity, we can always find $c > 0$ such that (7.5.25) holds provided that $\gamma_1(i) = \mathcal{O}(a^i)$ for some $a < 1$. □

7.5.4 $\tilde{\alpha}$ and $\tilde{\phi}$-dependent sequences

In this section, the coefficients are defined by

$$\tilde{\phi}_1(k) = \tilde{\phi}(\mathcal{M}_0, X_k) \quad \text{and} \quad \tilde{\alpha}_1(k) = \tilde{\phi}(\mathcal{M}_0, X_k).$$

Let $X_n = X_0 \circ T^i$ and

$$S_n(f) = f(X_0) + \cdots + f(X_n) \quad \text{and} \quad S_{n,0}(f) = S_n(f) - n\mathbb{E}(f(X_0)).$$

Let $\mathcal{C}(p, M, \mathbb{P}_X)$ be the closed convex hull of the class of functions g which are monotonic on an open interval and 0 elsewhere, and such that $\mathbb{E}(|g(X_0)|^p) < M$.

Corollary 7.7. *Let $(\mathcal{M}_i)_{i\in\mathbb{Z}}$ be as in Theorem 7.6. Assume that, for some $p \geq 2$,*

$$f \in \mathcal{C}(p, M, \mathbb{P}_X) \quad \text{and} \quad \sum_{k\geq 0} \frac{(\tilde{\phi}_1(k))^{(p-1)/p}}{\sqrt{k}} < \infty. \tag{7.5.26}$$

Then $S_{n,0}(f)$ satisfies **S1** *with*

$$\eta = \sum_{k\in\mathbb{Z}} \mathbb{E}(f(X_0)(f(X_k) - \mathbb{E}(f(X_k)))|\mathcal{I}). \tag{7.5.27}$$

Remark 7.5. *We can apply this result to the case of uniformly expanding maps (see Section 3.3). If T is a uniformly expanding map of $[0,1]$ preserving a probability μ, and if f belongs to $\mathcal{C}(2, M, \mu)$, then $n^{-1/2}(f\circ T + \cdots + f\circ T^n - n\mu(f))$ converges weakly in the space $([0,1], \mu)$ to a Gaussian distribution with mean 0 and variance*

$$\sigma^2(f) = \mu((f - \mu(f))^2) + 2\sum_{k>0} \mu((f - \mu(f))f \circ T^k).$$

Let $\mathcal{C}(Q)$ the closed convex hull of the class of functions g which are monotonic on an open interval and 0 elsewhere, and such that $Q_{|g(X_0)|} \leq Q$, where Q is a given quantile function.

Corollary 7.8. *Let $(\mathcal{M}_i)_{i\in\mathbb{Z}}$ be as in Theorem 7.6. Assume that, for some quantile function Q,*

$$f \in \mathcal{C}(Q) \quad \text{and} \quad \sum_{k\geq 0} \int_0^{\tilde{\alpha}(k)} Q^2(u)du < \infty. \tag{7.5.28}$$

Then $S_{n,0}(f)$ satisfies **S1** *with η given by (7.5.27).*

Proof of Corollaries 7.7 and 7.8. Without loss of generality, it suffices to prove the results for $f = \sum_{i=1}^k \alpha_i g_i$, where $\sum_{i=1}^k \alpha_i = 1$ and g_i is monotonic on an interval and 0 elsewhere, and $\mathbb{E}(|g_i(X_0)|^p) < M$ (resp. $Q_{|g_i(X_0)|} \leq Q$). To prove Corollary 7.7 it suffices to see that the sequence $(f(X_i))_{i\in\mathbb{Z}}$ satisfies (7.5.18), that is

$$\sum_{k\geq 0} \frac{\|\mathbb{E}(f(X_k)|\mathcal{M}_0) - \mathbb{E}(f(X_k))\|_2}{\sqrt{k}} < \infty. \tag{7.5.29}$$

Clearly, it suffices to check (7.5.29) for a single function g_i. Note that by monotonicity $\tilde{\phi}(\mathcal{M}_0, g_i(X_k)) \leq 2\tilde{\phi}(\mathcal{M}_0, X_k)$. Let $S_1(\mathcal{M}_0)$ be the set of all $\mathcal{M}_0$-measurable random variables Z such that $\mathbb{E}(Z^2) = 1$. Clearly,

$$\|\mathbb{E}(g_i(X_k)|\mathcal{M}_0) - \mathbb{E}(g_i(X_k))\|_2 = \sup_{Z \in S_1(\mathcal{M}_0)} |\mathrm{Cov}(Z, g_i(X_k))| \tag{7.5.30}$$

Applying (5.2.7), we have, for any conjugate exponents p, q,

$$|\mathrm{Cov}(Z, (f-g)(Y_k))| \leq 4\|g_i(X_0)\|_p \|Z\|_q (\tilde{\phi}_1(k))^{1/q}.$$

and the result easily follows. In the same way, to prove Corollary 7.8, it suffices to prove that

$$\sum_{k \geq 0} \|g_i(X_0)(\mathbb{E}(g_i(X_k)|\mathcal{M}_0) - \mathbb{E}(g_i(X_0)))\|_1 < \infty,$$

which follows (5.2.5). □

7.5.5 Sufficient conditions for triangular arrays

The next condition is the natural extension of Condition (7.5.7).

$$\lim_{N\to\infty} \limsup_{n\to\infty} \sup_{N \leq m \leq n} \Big\| X_{0,n} \sum_{k=N}^{m} \mathbb{E}(X_{k,n}|\mathcal{M}_{0,n}) \Big\|_1 = 0. \tag{7.5.31}$$

If (7.5.31) is satisfied, define $R(N, X)$ and $N(X)$ as follows:

$$R(N,X) = \limsup_{n\to\infty} \sup_{N \leq m \leq n} \Big\| X_{0,n} \sum_{k=N}^{m} \mathbb{E}(X_{k,n}|\mathcal{M}_{0,n}) \Big\|_1,$$

and $N(X) = \inf\{N > 0 : R(N, X) = 0\}$ ($N(X)$ may be infinite).

Proposition 7.9. *Let $X_{i,n}$ and $\mathcal{M}_{i,n}$ be as in Theorem 7.5. Assume that (7.5.31) and* **S2***(b) are satisfied. Assume furhtermore that, for each $0 \leq k < N(X)$, there exists an $\mathcal{M}_{0,\inf}$-measurable random variable λ_k such that for any t in $]0, 1]$,*

$$\frac{1}{nt} \sum_{i=1}^{[nt]} X_{i+k,n} X_{i,n} \text{ converges in } \mathbb{L}^1 \text{ to } \lambda_k. \tag{7.5.32}$$

Then Condition **S2** *holds with $\eta = \lambda_0 + 2\sum_{k=1}^{N(X)-1} \lambda_k$.*

Remark 7.6. *Let us have a look to a particular case, for which $N(X) = 1$. Conditions (7.5.31) and (7.5.32) are satisfied if condition R1. and R2. below are fulfilled*

R1. $\lim_{n\to\infty} \sum_{k=1}^{n} \|X_{0,n}\mathbb{E}(X_{k,n}|\mathcal{M}_{0,n})\|_1 = 0.$

R2. For any t in $]0,1]$, $\frac{1}{nt}\sum_{i=1}^{[nt]} X_{i,n}^2$ *converges in* $\mathbb{L}^1$ *to* λ.

In the stationary case, these results extend on classical results for triangular arrays of martingale differences (see for instance Hall and Heyde (1980) [100], Theorem 3.2), for which Condition R1. is satisfied. This particular case is sufficient to improve on many results in the context of kernel estimators.

We conclude with a simple result for $\tilde{\phi}$-mixing arrays. Here, the coefficients are defined by

$$\tilde{\phi}_1(k) = \sup_{n>0} \tilde{\phi}(\mathcal{M}_{0,n}, X_{k,n}) .$$

Corollary 7.9. *Let* $X_{i,n}$ *and* $\mathcal{M}_{i,n}$ *be as in Theorem 7.5. If there exists two conjugate exponent* $p \le q$ *such that*

$$\sup_{n>0} \|X_{0,n}\|_p \cdot \|X_{0,n}\|_q < \infty \quad \textit{and} \quad \sum_{k=0}^{\infty} (\tilde{\phi}_1(k))^{1/p} < \infty , \tag{7.5.33}$$

then (7.5.31) holds. If furthermore, for any $k > 0$ *the sequence* $\|X_{0,n}X_{k,n}\|_1$ *converges to 0 as n tends to infinity, then* $N(X) = 1$. *If furthermore Lindeberg's condition holds:*

$$\textit{for any } \epsilon > 0, \quad \lim_{n\to\infty} \mathbb{E}(X_{0,n}^2 \mathbf{1}_{|X_{0,n}|>\epsilon\sqrt{n}}) = 0 , \tag{7.5.34}$$

then **S2** *holds as soon as* $\mathbb{E}(X_{0,n}^2)$ *converges, and* $\eta = \lim_{n\to\infty} \mathbb{E}(X_{0,n}^2)$.

Proof of Proposition 7.9.

Proof of **S2***(a).* Write first

$$\begin{aligned} \mathbb{E}\Big(\frac{S_n^2(t)}{nt}(1 \wedge \frac{|S_n(t)|}{\sqrt{n}})\Big) &\le \mathbb{E}\Big(\frac{S_n^2(t)}{nt}\mathbf{1}_{|S_n(t)|>2\sqrt{n}\epsilon}\Big) + \frac{2\epsilon}{nt}\mathbb{E}(S_n^2(t)) \\ &\le \frac{4}{nt}\mathbb{E}\left((|S_n(t)| - \epsilon)_+^2\right) + \frac{2\epsilon}{nt}\mathbb{E}(S_n^2(t)) \quad (7.5.35) \end{aligned}$$

Since (7.5.31) holds, $(nt)^{-1}\mathbb{E}(S_n^2(t))$ is bounded, so that the second term on right hand is a small as we wish. Consequently, we infer from (7.5.35) that **S2**(a) holds as soon as,

$$\text{for any positive } \epsilon, \quad \lim_{t\to 0} \limsup_{n\to\infty} \frac{1}{nt}\mathbb{E}\left((|S_n(t)| - \epsilon)_+^2\right) = 0 . \tag{7.5.36}$$

In fact, we shall prove that (7.5.36) holds with $\overline{S}_n(t) = \sup_{s\in[0,t]} |S_n(s)|$ instead of $|S_n(t)|$. Define $S_n^*(t) = \sup_{s\in[0,t]}(S_n(s))_+$ and $G(t,\epsilon,n)$ by $G(t,\epsilon,n)$

$= \{S_n^*(t) > \epsilon\}$. From Proposition 5.8, we have, for any positive integer N,

$$\frac{1}{nt}\mathbb{E}\left((S_n^*(t)-\epsilon)_+^2\right) \le 8\mathbb{E}\Big(\mathbf{1}_{G(t,\epsilon,n)}\frac{1}{nt}\sum_{k=1}^{[nt]}\sum_{i=0}^{N-1}|X_{k,n}X_{k+i,n}|\Big) + 8\sup_{N\le m\le [nt]}\Big\|X_{0,n}\sum_{k=N}^{m}\mathbb{E}(X_{k,n}|\mathcal{M}_{0,n})\Big\|_1. \quad (7.5.37)$$

Since (7.5.31) holds, the second term on right hand is as small as we wish by choosing N large enough. To control the first term, note that by Proposition 5.8 and Markov's inequality, $\lim_{t\to 0}\limsup_{n\to\infty}\mathbb{P}(G(t,\epsilon,n)) = 0$. The result follows by noting that $2|X_{k,n}X_{l,n}| \le X_{k,n}^2 + X_{l,n}^2$ and by using (7.5.32) for $k = 0$. □

Proof of **S2**(*c*). For any finite integer $0 \le N \le N(X)$, define the variable $\eta_N = \lambda_0 + 2(\lambda_1 + \cdots + \lambda_{N-1})$ and the two sets

$$\begin{aligned} \Lambda_N &= [1,[nt]]^2 \cap \{(i,j)\in\mathbb{Z}^2 / |i-j| < N\} \quad \text{and} \\ \overline{\Lambda}_N &= [1,[nt]]^2 \cap \{(i,j)\in\mathbb{Z}^2 / j-i \ge N\}, \quad \text{so that,} \end{aligned}$$

$$\Big\|\mathbb{E}\Big(\frac{S_n^2(t)}{nt} - \eta_N\Big|\mathcal{M}_{0,n}\Big)\Big\|_1 \le \Big\|\eta_N - \frac{1}{nt}\sum_{\Lambda_N}X_{i,n}X_{j,n}\Big\|_1 + \frac{2}{nt}\Big\|\sum_{\overline{\Lambda}_N}\mathbb{E}(X_{i,n}X_{j,n}|\mathcal{M}_{0,n})\Big\|_1. \quad (7.5.38)$$

Hence, it remains to show that

$$\lim_{N\to N(X)}\limsup_{n\to\infty}\frac{1}{nt}\Big\|\sum_{\overline{\Lambda}_N}\mathbb{E}(X_{i,n}X_{j,n}|\mathcal{M}_{0,n})\Big\|_1 = 0\,. \quad (7.5.39)$$

Using first the inclusion $\mathcal{M}_{0,n} \subseteq \mathcal{M}_{i,n}$ for any positive i and second the stationarity of the sequence, we obtain succesively

$$\begin{aligned} \frac{1}{nt}\Big\|\sum_{\overline{\Lambda}_N}\mathbb{E}(X_{i,n}X_{j,n}|\mathcal{M}_{0,n})\Big\|_1 &\le \frac{1}{nt}\Big\|\sum_{\overline{\Lambda}_N}X_{i,n}\mathbb{E}(X_{j,n}|\mathcal{M}_{i,n})\Big\|_1 \\ &\le \sup_{N\le m\le n}\Big\|X_{0,n}\sum_{k=N}^{m}\mathbb{E}(X_{k,n}|\mathcal{M}_{0,n})\Big\|_1. \end{aligned}$$

and (7.5.39) follows from (7.5.31). This completes the proof. □

Proof of Corollary 7.9. Using the stationarity and the inequality (5.2.7), we infer that

$$\|X_{0,n}\mathbb{E}(X_{k,n}|\mathcal{M}_{0,n})\|_1 \leq 2\|X_{0,n}\|_p\|X_{0,n}\|_q(\tilde{\phi}_1(k))^{1/p}\,, \tag{7.5.40}$$

so that (7.5.31) follows easily from (7.5.33). Moreover, if for each $k > 0$ the sequence $\|X_{0,n}X_{k,n}\|_1$ converges to 0, the fact that $N(X) = 1$ follows from (7.5.40) and the dominated convergence theorem.

To prove the last point of this corollary, it suffices, by applying Proposition (7.9), to prove that for any t in $]0,1]$,

$$\lim_{n\to\infty}\frac{1}{nt}\Big\|\sum_{k=1}^{[nt]}(X_{k,n}^2 - \mathbb{E}(X_{0,n}^2))\Big\|_1 = 0\,. \tag{7.5.41}$$

According to the condition (7.5.34), it suffices to prove that

$$\lim_{\epsilon\to 0}\limsup_{n\to\infty}\mathrm{Var}\Big(\frac{1}{nt}\sum_{k=1}^{[nt]}X_{k,n}^2\mathbf{1}_{|X_{k,n}|\leq\sqrt{n}\epsilon}\Big) = 0\,. \tag{7.5.42}$$

Setting $Y_{k,n} = X_{k,n}^2\mathbf{1}_{|X_{k,n}|\leq\sqrt{n}\epsilon}$, we have the elementary inequality

$$\mathrm{Var}\Big(\frac{1}{nt}\sum_{k=1}^{[nt]}Y_{k,n}\Big) \leq \frac{2}{nt}\sum_{k=0}^{n}|\mathrm{Cov}(Y_{0,n},Y_{k,n})|\,. \tag{7.5.43}$$

Now, bearing in mind the definition of $Y_{k,n}$ and applying (5.2.7), we have successively

$$\begin{aligned}
|\mathrm{Cov}(Y_{0,n},Y_{k,n})| &= |\mathrm{Cov}(Y_{0,n},X_{k,n}^2\mathbf{1}_{|X_{k,n}|\leq\sqrt{n}K})| \\
&\leq 2(\tilde{\phi}_1(k))^{1/p}\|X_{0,n}^2\mathbf{1}_{|X_{0,n}|\leq\sqrt{n}\epsilon}\|_p\|X_{0,n}^2\mathbf{1}_{|X_{0,n}|\leq\sqrt{n}\epsilon}\|_q \\
&\leq 2n\epsilon^2(\tilde{\phi}_1(k))^{1/p}\|X_{0,n}\|_p\|X_{0,n}\|_q\,,
\end{aligned}$$

and (7.5.42) follows from (7.5.43) and (7.5.33). □

Chapter 8

Donsker Principles

In this chapter we give sufficient conditions for the smoothed partial sum process of a sequence (resp. field) of real-valued random variables to converge to a Brownian motion (resp. Brownian sheet). In the non causal case, we show that the conditions on $\kappa(r)$ and $\lambda(r)$ (implied by η dependence) given in Theorems 7.1 and 7.2 respectively, are sufficient to obtain the weak invariance principle. In Section 8.2 we give a general result for η dependent random fields having moments of order 4. For the causal case, we present the functional version of the conditional central limit theorem (CCLT) established in Section 7.4. We shall see in Section 8.4 that the sufficient conditions for the CCLT for γ, $\tilde{\alpha}$ and $\tilde{\phi}$-dependent sequences, are also sufficient for the conditional invariance principle.

8.1 Non causal stationary sequences

Let $(X_i)_{i\in\mathbb{Z}}$ be a stationary sequence of centered and square integrable random variables, and let U_n be the Donsker line

$$U_n(t) = \frac{1}{\sqrt{n}}\sum_{i=1}^{[nt]} X_i + \frac{nt-[nt]}{\sqrt{n}} X_{[nt]+1}.$$

In this short section, we show that under the same assumptions as in Theorem 7.1 or Theorem 7.2, the weak invariance principle holds. This follows easily from the control of $\|S_n\|_{2+\delta}$ obtained at the end of Section 7.2.1.

Theorem 8.1. *Assume that $(X_i)_{i\in\mathbb{Z}}$ satisfy the assumptions of Theorem 7.1 or the assumptions of Theorem 7.2. Then the process U_n converges weakly in $(C([0,1]), \|\cdot\|_\infty)$ to a Wiener process with variance $\sigma^2 = \sum_{k\in\mathbb{Z}} \mathrm{Cov}(X_0, X_k)$.*

Proof. The finite dimensional convergence of the process U_n can be proved as in Section 7.2.1. It remains to prove the tightness. Recall that under the assumptions of Theorem 7.1 or of Theorem 7.2, there exist there exists $\delta > 0$ and $C > 0$ such that:

$$\mathbb{E}|S_p|^{2+\delta} \leq Cp^{1+\delta/2}.$$

The tightness follows then from standard arguments, which can be found in many papers. For instance, if λ denotes the Lebesgue measure, we infer from the moment inequality above that (U_n, λ) belongs to the class $\mathcal{C}(1 + \delta/2, 2 + \delta)$ defined in Bickel and Wichura (1971) [19] (see the inequality (3) of this paper). The tightness of the process $\{U_n(t), t \in [0, 1]\}$ follows by applying Theorem 3 of the above paper. □

Remark. The same result follows the same lines for the Bernoulli shift models with dependent inputs from theorem 7.3.

8.2 Non causal random fields

Here we establish the Donsker principle for non causal weak dependent sequences and random fields. A block B in $[0, 1]^d$ is a subset of $[0, 1]^d$ of the form $]s, t] = \prod_{p=1}^{d}]s_p, t_p]$. Let $(X_j)_{j \in \mathbb{Z}^d}$ be a random field and B be a block in $[0, 1]^d$. Let $nB = \{nx, x \in B\}$, and

$$S_n(B) = \frac{1}{n^{d/2}} \sum_{j \in nB \cap \mathbb{Z}^d} X_j. \tag{8.2.1}$$

For any $j \in \mathbb{Z}^d$, denote by C^j the unit with upper corner j, and define the continuous process:

$$U_n(B) = \frac{1}{n^{d/2}} \sum_{j \in \mathbb{Z}^d} \lambda(nB \cap C^j) X_j,$$

where λ denotes the Lebesgue measure on $\mathbb{R}^d$. For $t \in [0, 1]^d$, define $S_n(t) = S_n([0, t])$ and $U_n(t) = U_n([0, t])$.

Theorem 8.2. *Let $(X_j)_{j \in \mathbb{Z}^d}$ be a η-weak dependent stationary centered random field. Assume that $\mathbb{E}|X_0|^4 = M^4 \leq \infty$. If $\eta(r) \leq cr^{-a}$, with $a > 3d$, then the process U_n converges weakly in $(C([0, 1]^d), \|\cdot\|_\infty)$ to a Brownian sheet with variance $\sigma^2 = \sum_{k \in \mathbb{Z}^d} \mathrm{Cov}(X_0, X_k)$.*

8.2.1 Moment inequality

First we establish a bound for the fourth moment of a partial sum:

Lemma 8.1. *Assume that the assumptions of theorem 8.2 are satisfied, then for any block B in $[0,1]^d$, there exists $C>0$ such that:*

$$\mathbb{E}(S_n(B)^4) \le C\lambda(B)^2$$

Proof. For a finite sequence $\mathbf{k}=(k_1,\dots,k_q)$ of elements of $\mathbb{Z}^d$, define $\Pi_{\mathbf{k}} = \prod_{i=1}^q X_{k_i}$. For any integer $q\ge 1$, set:

$$A_q(n) = \sum_{\mathbf{k}\in(nB\cap\mathbb{Z}^d)^q} |E(\Pi_{\mathbf{k}})|, \tag{8.2.2}$$

then

$$|\mathbb{E}(S_n(B))^4| \le n^{-2d}A_4(n). \tag{8.2.3}$$

The gap of $\mathbf{k}$ is defined by the max of the integers r such that the sequence may be split into two non-empty subsequences $\mathbf{k}^1$ and $\mathbf{k}^2\subset\mathbb{Z}^d$ whose mutual distance equals r ($d(\mathbf{k}^1,\mathbf{k}^2)=\min\{\|i-j\|_1/\, i\in\mathbf{k}^1, j\in\mathbf{k}^2\}=r$). If the sequence is constant, its gap is 0. Define the set $G_r(q,n)=\{\mathbf{k}\in(nB)^q$ and the gap of $\mathbf{k}$ is $r\}$. Sorting the sequences of indices by their gap:

$$A_q(n) \le \sum_{k\in nB}\mathbb{E}|X_k|^q + \sum_{r=1}^{\infty}\sum_{\mathbf{k}\in G_r(q,n)} |\mathrm{Cov}(\Pi_{\mathbf{k}^1},\Pi_{\mathbf{k}^2})| \tag{8.2.4}$$

$$+ \sum_{r=1}^{\infty}\sum_{\mathbf{k}\in G_r(q,n)} |\mathbb{E}(\Pi_{\mathbf{k}^1})\,\mathbb{E}(\Pi_{\mathbf{k}^2})|. \tag{8.2.5}$$

Define $V_q(n)$ as the sum of the right hand side of (8.2.4). We get

$$A_4(n)\le V_4(n)+V_2(n)^2.$$

Denote by N the cardinality of $nB\cap\mathbb{Z}^d$. To build a sequence $\mathbf{k}$ belonging to $G_r(q,n)$, we first fix one of the N points of $nB\cap\mathbb{Z}^d$. We choose a second point on the ℓ^1-sphere of radius r centered on the first point. The third point is in a ball of radius r centered on one of the preceding points, and so on. We get

$$\begin{aligned}\#G_r(q,4) &\le N2d(2r+1)^{d-1}.2(2r+1)^d.3(2r+1)^d \le 12dN3^{3d}r^{3d-1},\\ \#G_r(q,2) &\le N2d(2r+1)^{d-1}\le 2dN3^{d-1}r^{d-1}\end{aligned}$$

and

$$\begin{aligned}V_4(n) &\le NM^4+12dN3^{3d}\sum_{r=1}^{\infty} r^{3d-1}\eta(r),\\ V_2(n) &\le NM^2+2dN3^{d-1}\sum_{r=1}^{\infty} r^{d-1}\eta(r),\end{aligned}$$

so that

$$A_4(n) \leq C(N + N^2).$$

Because N is an integer and $|N - n^d\lambda(B)| \leq 2dN/n$, Lemma 8.1 is proved. □

8.2.2 Finite dimensional convergence

To prove the convergence of the finite dimensional distributions, we note that it is sufficient to prove it for the finite dimensional distribution of $S_n(B)$, because $U_n(B) - S_n(B)$ tends to zero in probability. Considering B and C two disjoint blocks of $[0,1]^d$, we check that the joint distribution of $(S_n(B), S_n(C))$ satisfies:

$$\lim_{n\to\infty} \mathrm{Cov}(S_n(B), S_n(C)) = 0.$$

Denote b^- and b^+ the lower and upper vertex of block B. If the domains are non intersecting, for at least one coordinate (say the first), we have (say) $b_1^+ \leq c_1^-$. Then

$$\begin{aligned}
|\mathrm{Cov}(S_n(B), S_n(C))| &\leq n^{-d} \sum_{i\in nB} \sum_{j\in nC} |\mathrm{Cov}(X_i, X_j)| \\
&\leq n^{-d} \sum_{i\in nB} \left(\sum_{j\in nC, j_1 \leq n^\beta} |\mathrm{Cov}(X_i, X_j)| \right. \\
&+ \left. \sum_{j\in nC, j_1 > n^\beta} |\mathrm{Cov}(X_i, X_j)| \right) \\
&\leq n^{-d} \left(n^{\beta+d-1} \sum_{r\in\mathbb{N}} r^{d-1}\eta(r) + n^d \sum_{r>n^\beta} r^{d-1}\eta(r) \right) \\
&= o(n^{\beta-1}, n^{\beta(d-a)})
\end{aligned}$$

Taking $\beta < 1$ gives the result.

Now, let ν and μ be two reals, we show that $S_n = \nu S_n(B) + \mu(S_n(C))$ tends to a Gaussian distribution. Write

$$S_n = n^{-d/2} \sum_{i\in\{0,\ldots,n\}^d} \alpha_{i,n} X_i$$

where $\alpha_{i,n} = \nu + \mu$ if $i \in nB \cap nC$, $\alpha_{i,n} = \nu$ if $i \in nB \setminus nC$, $\alpha_{i,n} = \mu$ if $i \in nC \setminus nB$ and $\alpha_{i,n} = 0$ elsewhere. We use the Bernstein blocking technique, (1939) [13]. Let $p(n)$ and $q(n)$ be sequences of integers such that $p(n) = o(n)$ and $q(n) = o(p(n))$. Assume that the Euclidean division of n by $(p+q)$ gives a quotient k and a remainder r. Denote $\overline{j} = (j, \ldots, j)$. Define $K = \{1, \ldots, k+1\}^d$

and order K by the lexicographic order; for $i \in \{1,\dots,k\}^d$, define the blocks $P_i = [(p+q)(i-\overline{1}),\dots,(p+q)i - q\overline{1}]$. If $r>0$, for $i \in \{0,\dots,1\}^d \setminus \{\overline{0}\}$ define the blocks $P_i = [(p+q)(\overline{k}+i),\dots,(p+q)(\overline{k}+i)+(r\vee p)\overline{1}]$. Denote Q the set of indices that are not in one of the P_i. Note that the cardinality of Q is less than $d(k+1)qp^{d-1}$. For each block P_i and Q, we define the partial sums:

$$\begin{aligned} u_i &= n^{-d/2}\sum_{j\in P_i}\alpha_{j,n}X_j,\\ v &= n^{-d/2}\sum_{j\in Q}\alpha_{j,n}X_j. \end{aligned}$$

Recall lemma 11 in Doukhan and Louhichi (1999) [67].

Lemma 8.2. *Let $S_n = n^{-d/2}\sum_{j\in\{0,\dots,n\}^d}\alpha_{j,n}X_j$ be a sum of centered triangular array; set $\sigma_n^2 = \operatorname{Var} S_n$. Assume that :*

$$\lim_{n\to\infty}\frac{1}{\sigma_n^2}Ev^2 = 0. \tag{8.2.6}$$

$$\sum_{j\in K}\left|\operatorname{Cov}\left(g\left(\frac{t}{\sigma_n}\sum_{i\in K, i<j}u_i\right),\ h\left(\frac{t}{\sigma_n}u_j\right)\right)\right| \to 0,\ \textit{for all } t\in\mathbb{R}, \tag{8.2.7}$$

where h and g are either the sine or the cosine function,

$$\lim_{n\to\infty}\frac{1}{\sigma_n^2}\sum_{i\in K}E|u_i|^2\mathbf{1}_{\{|u_i|\ge\epsilon\sigma_n\}} = 0,\ \textit{for all } \epsilon>0, \tag{8.2.8}$$

$$\textit{and}\quad \lim_{n\to\infty}\frac{1}{\sigma_n^2}\sum_{i=1}^{k}E|u_i|^2 = 1. \tag{8.2.9}$$

Then S_n/σ_n converges in distribution to a Gaussian $\mathcal{N}(0,1)$-distribution.

First note that

$$\sum_{j\in\mathbb{N}^d}\operatorname{Cov}(\alpha_{0,n}X_0,\alpha_{j,n}X_j) < \infty \tag{8.2.10}$$

so that σ_n^2 tends to a constant. If this constant is zero then the limit of S_n is 0. If it is not, we check the conditions of the preceding lemma for the array $\alpha_{j,n}X_j$. To check (8.2.6), note that

$$\begin{aligned} \mathbb{E}v^2 &\le (|\nu|+|\mu|)^2 n^{-d}\sum_{i,j\in Q}|\operatorname{Cov}(X_i,X_j)| \le (|\nu|+|\mu|)^2 n^{-d}\sum_{i\in Q}\sum_{j\in Q}\eta(|j-i|)\\ &\le 2(|\nu|+|\mu|)^2\frac{d(k+1)qp^{d-1}}{n^d}\sum_{r=0}^{\infty}r^{d-1}\eta(r) = o(1). \end{aligned}$$

Consider 8.2.7. Note that $g\left(\frac{t}{\sigma_n}\sum_{i\in K, i\neq j} u_i\right)$ is a function of at most $((k+1)p)^d$ variables X_l and that its Lipschitz modulus is less than $t(|\nu|+|\mu|)/n^{d/2}\sigma_n$. Similarly $h(tu_j/\sigma_n)$ is a function of at most p^d variables and its Lipschitz modulus is less than $t(|\nu|+|\mu|)/n^{d/2}\sigma_n$. Using the weak dependence property, we get

$$\left|\mathrm{Cov}\left(g\left(\frac{t}{\sigma_n}\sum_{i\in K, i\neq j} u_i\right), h\left(\frac{t}{\sigma_n}u_j\right)\right)\right| \leq p^d((k+1)^d+1)\cdot\frac{t(|\nu|+|\mu|)}{n^{d/2}\sigma_n}\eta(q).$$

and

$$\begin{aligned}\sum_{j\in K}\left|\mathrm{Cov}\left(g\left(\frac{t}{\sigma_n}\sum_{i\in K, i\neq j} u_i\right), h\left(\frac{t}{\sigma_n}u_j\right)\right)\right| &\leq p^d(k+1)^{d+1}\cdot\frac{t(|\nu|+|\mu|)}{n^{d/2}\sigma_n}\eta(q)\\ &= \mathcal{O}(n^{3d/2}p^{-d}q^{-a}).\end{aligned}$$

Taking $p=n^{5/6}$ and $q=n^{5d/6a}$ gives a bound tending to 0.

To prove (8.2.8), it is sufficient to show that $\mathbb{E}|u_i|^4=\mathcal{O}(k^{-2d})$. But

$$\mathbb{E}\left(\frac{1}{n^{d/2}}\sum_{j\in P_i}\alpha_{j,n}X_j\right)^4 \leq \frac{(|\nu|+|\mu|)^4}{n^{2d}}\mathbb{E}\left(\sum_{j\in P_i}X_j\right)^4 \leq \frac{p^{2d}}{n^{2d}}\mathbb{E}\left(S_p([0,\bar{1}])\right)^4,$$

and we conclude with the moment inequality

$$\mathbb{E}\left(S_p([0,\bar{1}])\right)^4=\mathcal{O}(1).$$

In order to prove (8.2.9), note that (8.2.6) implies that

$$\lim_{n\to\infty}\frac{1}{\sigma_n^2}\mathrm{Var}\left(\sum_{i\in K}u_i\right)=1.$$

But

$$\begin{aligned}\left|\mathrm{Var}\left(\sum_{i\in K}u_i\right)-\sum_{i\in K}E|u_i|^2\right| &\leq 2\sum_{i\in K; i\neq j}|\mathrm{Cov}(u_i,u_j)|\\ &\leq 2(k+1)^d\sum_{j=q}^{\infty}\eta(j)\\ &= \mathcal{O}(np^{-d}q^{-a+1}).\end{aligned}$$

Taking $p=n^{5/6}$ and $q=n^{5d/6a}$ gives a bound tending to 0. □

8.2.3 Tightness

Recall that λ is the Lebesgue measure on $\mathbb{R}^d$. Applying Lemma 8.1, we infer that (S_n, λ) and (U_n, λ) belong to the class $\mathcal{C}(2,4)$ defined in Bickel and Wichura (1971) [19] (see the inequality (3) of this paper). The tightness or relative compactness of the process $\{U_n(t), t \in [0,1]^d\}$ follows by applying Theorem 3 of [19].

8.3 Conditional (causal) invariance principle

In this section we give the functional version of Theorem 7.5. Denote by $\mathcal{H}^*$ the space of continuous functions φ from $(C([0,1]), \|\cdot\|_\infty)$ to $\mathbb{R}$ such that $x \mapsto |(1+\|x\|_\infty^2)^{-1}\varphi(x)|$ is bounded.

Theorem 8.3. *Let $X_{i,n}$, $\mathcal{M}_{i,n}$ and $S_n(t)$ be as in Theorem 7.5. For any t in $[0,1]$, let $U_n(t) = S_n(t)+(nt-[nt])X_{[nt]+1,n}$ and define $\overline{S}_n(t) = \sup_{0\le s\le t}|S_n(s)|$. The following statements are equivalent:*

S1* *There exists a nonnegative $\mathcal{M}_{0,\text{inf}}$-measurable random variable η such that, for any φ in $\mathcal{H}^*$ and any positive integer k,*

$$\mathbf{S1}^*(\varphi): \quad \lim_{n\to\infty}\Big\|\mathbb{E}\Big(\varphi(n^{-1/2}U_n) - \int \varphi(x\sqrt{\eta})W(dx)\ \Big|\mathcal{M}_{k,n}\Big)\Big\|_1 = 0$$

where W is the distribution of a standard Wiener process.

S2* *Properties **S2**(b) and (c) of Theorem 7.5 hold, and (a) is replaced by:*

(a^*) the sequence $(n^{-1}(\overline{S}_n(1))^2)_{n>0}$ is uniformly integrable, and

$$\lim_{t\to 0}\limsup_{n\to\infty}\mathbb{E}\left(\frac{(\overline{S}_n(t))^2}{nt}\Big(1\wedge\frac{\overline{S}_n(t)}{\sqrt{n}}\Big)\right) = 0\,.$$

Moreover the random variable η satisfies $\eta = \eta\circ T$ almost surely.

Remark 8.1. *Let $X_{i,n}$, $\mathcal{M}_{i,n}$ and U_n be as in Theorem 8.3. Suppose that the sequence $(\mathcal{M}_{0,n})_{n\ge1}$ is nondecreasing. If Condition **S1*** is satisfied, then the sequence $(n^{-1/2}U_n)_{n>0}$ converges stably with respect to the random probability defined by $\mu(\varphi) = \int\varphi(x\sqrt{\eta})W(x)dx$. The proof of this result is the same as that of Corollary 7.2.*

We now give the proof of this theorem. Once again, it suffices to prove that **S2*** implies **S1***.

8.3.1 Preliminaries

Definition 8.1. *Recall that $R(\mathcal{M}_{k,n})$ is the set of Rademacher $\mathcal{M}_{k,n}$-measurable random variables: $R(\mathcal{M}_{k,n}) = \{2\mathbf{1}_A - 1 \,/\, A \in \mathcal{M}_{k,n}\}$. For the random variable η introduced in Theorem 8.3 and any bounded random variable Z, let*

1. *$\nu_n^*[Z]$ be the image measure of $Z \cdot \mathbb{P}$ by the process $n^{-1/2}U_n$.*

2. *$\nu^*[Z]$ be the image measure of $W \otimes Z.\mathbb{P}$ by the variable ϕ from $C([0,1]) \otimes \Omega$ to $C([0,1])$ defined by $\phi(x,\omega) = x\sqrt{\eta(\omega)}$.*

We need the functional analogue of Lemma 7.4 (the proof is unchanged).

Lemma 8.3. *Let $\mu_n^*[Z_n] = \nu_n^*[Z_n] - \nu^*[Z_n]$. For any φ in $\mathcal{H}^*$ the statement* **S1**$^*(\varphi)$ *is equivalent to:*

S3$^*(\varphi)$*: for any Z_n in $R(\mathcal{M}_{k,n})$ the sequence $\mu_n^*[Z_n](\varphi)$ tends to zero as n tends to infinity.*

Suppose that **S1**$^*(\varphi)$ holds for any bounded function φ of $\mathcal{H}^*$. Since the sequence $(n^{-1}(S_n^*(1))^2)_{n>0}$ is uniformly integrable, **S1**$^*(\varphi)$ obviously extends to the whole space $\mathcal{H}^*$. Consequently, we can restrict ourselves to the space of continuous bounded functions from $C([0,1])$ to $\mathbb{R}$. According to Lemma 8.3, the proof of Theorem 8.3 will be complete if we show that, for any Z_n in $R(\mathcal{M}_{k,n})$, the sequence $\mu^*[Z_n]$ converges weakly to the null measure as n tends to infinity.

Definition 8.2. *For $0 \le t_1 < \cdots < t_d \le 1$, define the functions $\pi_{t_1 \dots t_d}$ and $Q_{t_1 \dots t_d}$ from $C([0,1])$ to $\mathbb{R}^d$ by the equalities $\pi_{t_1 \dots t_d}(x) = (x(t_1), \dots, x(t_d))$ and $Q_{t_1 \dots t_d}(x) = (x(t_1), x(t_2) - x(t_1), \dots, x(t_d) - x(t_{d-1}))$. For any signed measure μ on $(C([0,1]), \mathcal{B}(C([0,1])))$ and any function f from $C([0,1])$ to $\mathbb{R}^d$, denote by μf^{-1} the image measure of μ by f.*

Let μ and ν be two signed measures on $(C([0,1]), \mathcal{B}(C([0,1])))$. Recall that if $\mu\pi^{-1}_{t_1 \dots t_d} = \nu\pi^{-1}_{t_1 \dots t_d}$ for any positive integer d and any d-tuple such that $0 \le t_1 < \cdots < t_d \le 1$, then $\mu = \nu$. Consequently Theorem 8.3 is a straightforward consequence of the two following items

1. relative compactness: for any Z_n in $R(\mathcal{M}_{k,n})$, the family $(\mu_n^*[Z_n])_{n>0}$ is relatively compact with respect to the topology of weak convergence.

2. finite dimensional convergence: for any positive integer d, any d-tuple $0 \le t_1 < \cdots < t_d \le 1$ and any Z_n in $R(\mathcal{M}_{k,n})$ the sequence $\mu^*[Z_n]\pi^{-1}_{t_1 \dots t_d}$ converges weakly to the null measure as n tends to infinity.

8.3.2 Finite dimensional convergence

Clearly it is equivalent to take $Q_{t_1\dots t_d}$ instead of $\pi_{t_1\dots t_d}$ in item 2. The following lemma shows that finite dimensional convergence is a consequence of Condition **S2** of Theorem 7.5. The stronger condition **S2*** is only required for tightness.

Lemma 8.4. *For any a in $\mathbb{R}^d$ define f_a from $\mathbb{R}^d$ to $\mathbb{R}$ by $f_a(x) =< a, x >$. If* **S2** *holds then, for any a in $\mathbb{R}^d$, any d-tuple $0 \le t_1 < \cdots < t_d \le 1$ and any Z_n in $R(\mathcal{M}_{k,n})$, the sequence $\mu_n^*[Z_n](f_a \circ Q_{t_1\dots t_d})^{-1}$ converges weakly to the null measure.*

Write $\mu_n^*[Z_n](f_a \circ Q_{t_1\dots t_d})^{-1}(\exp(i\cdot)) = \mu_n^*[Z_n]Q_{t_1\dots t_d}^{-1}(\exp(i < a, \cdot >))$. According to Lemma 8.4, the latter converges to zero as n tends to infinity. Taking $Z_n = 1$, we infer that the probability measure $\nu_n^*[1]Q_{t_1\dots t_d}^{-1}$ converges weakly to the probability measure $\nu^*[1]Q_{t_1\dots t_d}^{-1}$ and hence is tight. Since $|\mu^*[Z_n]Q_{t_1\dots t_d}^{-1}| \le \nu_n^*[1]Q_{t_1\dots t_d}^{-1} + \nu^*[1]Q_{t_1\dots t_d}^{-1}$, the sequence $(\mu^*[Z_n]Q_{t_1\dots t_d}^{-1})_{n>0}$ is tight. Consequently we can apply Lemma 7.3 to conclude that $\mu^*[Z_n]Q_{t_1\dots t_d}^{-1}$ converges weakly to the null measure. □

Proof of Lemma 8.4. According to Lemma 8.3, we have to prove the property **S1**$^*(\varphi \circ f_a \circ Q_{t_1\dots t_d})$ for any continuous bounded function φ. Arguing as in Section 7.4.3, we can restrict ourselves to the class of function $B_1^3(\mathbb{R})$. Let h be any element of $B_1^3(\mathbb{R})$ and write

$$
\begin{aligned}
&h \circ f_a \circ Q_{t_1\dots t_d}(n^{-1/2}U_n) - \int h \circ f_a \circ Q_{t_1\dots t_d}(x\sqrt{\eta})W(dx) \\
&\quad = \sum_{\ell-1}^{d} h_\ell\Big(a_\ell\Big(\frac{U_n(t_\ell) - U_n(t_{\ell-1})}{\sqrt{n}}\Big)\Big) - \int h_\ell(a_\ell x\sqrt{(t_\ell - t_{\ell-1})\eta})\, g(x)\, dx\,,
\end{aligned}
$$

where the random variable $h_\ell(x)$ is equal to

$$
\int h\Big(\sum_{i=1}^{\ell-1} ai\Big(\frac{U_n(t_i) - U_n(t_{i-1})}{\sqrt{n}}\Big) + x + \sum_{i=\ell+1}^{d} a_i x_i \sqrt{(t_i - t_{i-1})\eta}\Big) \prod_{i=\ell+1}^{d} g(x_i)dx_i\,.
$$

Note that for any ω in Ω, the random function h_ℓ belongs to $B_1^3(\mathbb{R})$. To complete the proof of Lemma 8.4, it suffices to see that, for any positive integers k and ℓ, the sequence

$$
\Big\|\mathbb{E}\Big(h_\ell\Big(a_l\Big(\frac{U_n(t_\ell) - U_n(t_{\ell-1})}{\sqrt{n}}\Big)\Big) - \int h_\ell\Big(a_\ell x\sqrt{(t_\ell - t_{\ell-1})\eta}\Big)\, g(x)\, dx\Big|\mathcal{M}_{k,n}\Big)\Big\|_1 \tag{8.3.1}
$$

tends to zero as n tends to infinity. Since h_ℓ is 1-Lipshitz and bounded, we infer from the asymptotic negligibility of $n^{-1/2}X_{0,n}$ that

$$\lim_{n\to\infty}\Big\|h_\ell\Big(a_\ell\Big(\frac{U_n(t_\ell)-U_n(t_{\ell-1})}{\sqrt{n}}\Big)\Big)-h_\ell\Big(a_\ell\Big(\frac{S_n(t_\ell-t_{\ell-1})\circ T^{[nt_{\ell-1}]+1}}{\sqrt{n}}\Big)\Big)\Big\|_1=0. \tag{8.3.2}$$

Denote by g_ℓ the random function $g_\ell = h_\ell\circ T^{-[nt_{\ell-1}]-1}$. Combining (8.3.1), (8.3.2) and the fact that $\mathcal{M}_{k-1-[nt_{\ell-1}],n}\subseteq \mathcal{M}_{k,n}$, we infer that it suffices to prove that

$$\lim_{n\to\infty}\Big\|\mathbb{E}\Big(g_\ell(a_\ell n^{-1/2}S_n(u))-\int g_\ell(a_\ell x\sqrt{u\eta})\,g(x)\,dx\Big|\mathcal{M}_{k,n}\Big)\Big\|_1=0\,. \tag{8.3.3}$$

Since the random functions g_ℓ is $\mathcal{M}_{0,n}$-measurable (8.3.3) can be proved exactly as property **S1** of Theorem 7.5 (see Section 7.4.3). This completes the proof of Lemma 8.4. □

8.3.3 Relative compactness

In this section, we shall prove that the sequence $(\mu_n^*[Z_n])_{n>0}$ is relatively compact with respect to the topology of weak convergence. That is, for any increasing function f from $\mathbb{N}$ to $\mathbb{N}$, there exists an increasing function g with value in $f(\mathbb{N})$ and a signed mesure μ on $(C([0,1]),\mathcal{B}(C([0,1])))$ such that $(\mu_{g(n)}^*[Z_{g(n)}])_{n>0}$ converges weakly to μ.

Let Z_n^+ (resp. Z_n^-) be the positive (resp. negative) part of Z_n, and write

$$\mu_n^*[Z_n]=\mu_n^*[Z_n^+]-\mu_n^*[Z_n^-]=\nu_n^*[Z_n^+]-\nu_n^*[Z_n^-]-\nu^*[Z_n^+]+\nu^*[Z_n^-]\,,$$

where $\nu_n^*[Z]$ and $\nu^*[Z]$ are defined in 1. and 2. of Definition 8.1. Obviously, it is enough to prove that each sequence of finite positive measures $(\nu_n^*[Z_n^+])_{n>0}$, $(\nu_n^*[Z_n^-])_{n>0}$, $(\nu^*[Z_n^+])_{n>0}$ and $(\nu^*[Z_n^-])_{n>0}$ is relatively compact. We prove the result for the sequence $(\nu_n^*[Z_n^+])_{n>0}$, the other cases being similar.

Let f be any increasing function from $\mathbb{N}$ to $\mathbb{N}$. Choose an increasing function l with value in $f(\mathbb{N})$ such that

$$\lim_{n\to\infty}\mathbb{E}(Z_{l(n)}^+)=\liminf_{n\to\infty}\mathbb{E}(Z_{f(n)}^+)\,.$$

We must sort out two cases:

1. If $\mathbb{E}(Z_{l(n)}^+)$ converges to zero as n tends to infinity, then, taking $g=l$, the sequence $(\nu_{g(n)}^*[Z_{g(n)}^+])_{n>0}$ converges weakly to the null measure.

2. If $\mathbb{E}(Z_{l(n)}^+)$ converges to a positive real number as n tends to infinity, we introduce, for n large enough, the probability measure p_n defined by p_n

$= (\mathbb{E}(Z^+_{l(n)}))^{-1}\nu^*_{l(n)}[Z^+_{l(n)}]$. Obviously if $(p_n)_{n>0}$ is relatively compact with respect to the topology of weak convergence, then there exists an increasing function g with value in $l(\mathbb{N})$ (and hence in $f(\mathbb{N})$) and a measure ν such that $(\nu^*_{g(n)}[Z^+_{g(n)}])_{n>0}$ converges weakly to ν. Since $(p_n)_{n>0}$ is a family of probability measures, relative compactness is equivalent to tightness. Here we apply Theorem 8.2 in Billingsley (1968) [20]: to derive the tightness of the sequence $(p_n)_{n>0}$ it is enough to show that, for each positive ϵ,

$$\lim_{\delta\to 0}\limsup_{n\to\infty} p_n(x / w(x,\delta) \geq \epsilon) = 0\,, \tag{8.3.4}$$

where $w(x,\delta)$ is the modulus of continuity of the function x. According to the definition of p_n, we have

$$p_n(x/\,w(x,\delta)\geq\epsilon) = \frac{1}{\mathbb{E}(Z^+_{l(n)})} Z^+_{l(n)}\cdot\mathbb{P}\left(w\Big(\frac{U_{l(n)}}{\sqrt{l(n)}},\delta\Big)\geq\epsilon\right).$$

Since both $\mathbb{E}(Z^+_{l(n)})$ converges to a positive number and $Z^+_{l(n)}$ is bounded by one, we infer that (8.3.4) holds if

$$\lim_{\delta\to 0}\limsup_{n\to\infty}\mathbb{P}\left(w\Big(\frac{U_{l(n)}}{\sqrt{l(n)}},\delta\Big)\geq\epsilon\right) = 0\,. \tag{8.3.5}$$

From Theorem 8.3 and inequality (8.16) in Billingsley (1968) [20], it suffices to prove that, for any positive ϵ,

$$\lim_{\delta\to 0}\limsup_{n\to\infty}\frac{1}{\delta}\mathbb{P}\left(\frac{\overline{S}_{l(n)}(\delta)}{\sqrt{l(n)\delta}}\geq\frac{\epsilon}{\sqrt{\delta}}\right) = 0\,. \tag{8.3.6}$$

We conclude by noting that (8.3.6) follows straightforwardly from $\mathbf{S2}(a^*)$ and Markov's inequality.

Conclusion. In both cases there exists an increasing function g with value in $f(\mathbb{N})$ and a measure ν such that $(\nu^*_{g(n)}[Z^+_{g(n)}])_{n>0}$ converges weakly to ν. Since this is true for any increasing function f with value in $\mathbb{N}$, we conclude that the sequence $(\nu^*_n[Z^+_n])_{n>0}$ is relatively compact with respect to the topology of weak convergence. Of course, the same arguments apply to the sequences $(\nu^*_n[Z^-_n])_{n>0}$, $(\nu^*[Z^+_n])_{n>0}$ and $(\nu^*[Z^-_n])_{n>0}$, which implies the relative compactness of the sequence $(\mu^*_n[Z_n])_{n>0}$. □

8.4 Applications

8.4.1 Sufficient conditions for stationary sequences

For strictly stationary sequences, Theorem 8.3 writes as follows.

Theorem 8.4. *Let $(\mathcal{M}_i)_{i\in\mathbb{Z}}$ and $(X_i)_{i\in\mathbb{Z}}$ be as in Theorem 7.6. Define $S_n = X_1 + \cdots + X_n$ and $U_n(t) = S_{[nt]} + (nt - [nt])X_{[nt]+1}$. The following statements are equivalent:*

S1* *There exists a nonnegative $\mathcal{M}_0$-measurable random variable η such that, for any φ in $\mathcal{H}^*$ and any positive integer k,*

$$\lim_{n\to\infty}\Big\|\mathbb{E}\Big(\varphi(n^{-1/2}U_n) - \int \varphi(x\sqrt{\eta})W(dx)\ \Big|\mathcal{M}_k\Big)\Big\|_1 = 0$$

where W is the distribution of a standard Wiener process.

S2* *Properties* **S2***(b) and (c) of Theorem 7.6 hold, and (a) is replaced by:*

(a^*) the sequence $(n^{-1}(\max_{1\le i\le n}|S_i|)^2)_{n>0}$ is uniformly integrable.

Under the conditions of Proposition 7.8, **S1*** holds:

Proposition 8.1. *Let $(\mathcal{M}_i)_{i\in\mathbb{Z}}$ and $(X_i)_{i\in\mathbb{Z}}$ be as in Theorem 7.6 and define $\mathcal{M}_{-\infty} = \cap_{i\in\mathbb{Z}}\mathcal{M}_i$. Define the operators P_i as in Corollary 7.4.*

1. *If $\mathbb{E}(X_0|\mathcal{M}_{-\infty}) = 0$ and $\sum_{i\ge 0}\|P_0(X_i)\|_2 < \infty$ then* **S1*** *holds. Moreover, η is the same as in Proposition 7.8.*

2. *If (7.5.7) is satisfied, then* **S1*** *holds and η is the same as in Proposition 7.8.*

Remark 8.2. *As in Chapter 7, we deduce from Proposition 8.1 sufficient conditions for the functional CCLT in terms of the coefficients γ_1, γ_2, $\tilde\alpha_1$ and $\tilde\phi_1$. More precisely,* **S1*** *holds if either (7.5.18), (7.5.22), (7.5.26) or (7.5.28) is satisfied.*

Proof of Proposition 8.1. In view of Corollary 7.5 and Proposition 7.8, it is enough to prove that **S2**(a^*) holds.

Proof of item 1. According to Proposition 5.9, for any two sequences of nonnegative numbers $(a_m)_{m\ge0}$ and $(b_m)_{m\ge0}$ such that $K = \sum_{m\ge0} a_m^{-1}$ is finite and $\sum_{m\ge0} b_m = 1$, we have

$$\frac{1}{n}\mathbb{E}\left((S_n^* - M\sqrt{n})_+^2\right) \le 4K\sum_{m=0}^{\infty} a_m\mathbb{E}\Big(\frac{1}{n}\sum_{k=1}^{n} P_{k-m}^2(X_k)\mathbf{1}_{\Gamma(m,n,b_mM\sqrt{n})}\Big), \tag{8.4.1}$$

where $\Gamma(m,n,\lambda) = \Big(0 \vee \max_{1\le k\le n}\big\{\sum_{\ell=1}^{k} P_{\ell-m}(X_\ell)\big\} > \lambda\Big)$. Here, we take $b_m = 2^{-m-1}$ and $a_m = (\|P_0(Y_{m,i})\|_2 + (m+1)^{-2})^{-1}$. By assumption, $\sum a_m^{-1}$ is finite.

Since for all $m \geq 0$

$$a_m \mathbb{E}\Big(\frac{1}{n}\sum_{k=1}^{n} P_{k-m}^2(X_k)\mathbf{1}_{\Gamma(m,n,b_m M\sqrt{n})}\Big) \leq \frac{\|P_0(X_m)\|_2^2}{\|P_0(X_m)\|_2 + (m+1)^2} \leq \|P_0(X_m)\|_2 \,,$$

we infer from (8.4.1) that for any $\epsilon > 0$, there exists $N(\epsilon)$ such that

$$\frac{1}{n}\mathbb{E}\left((S_n^* - M\sqrt{n})_+^2\right) \leq \epsilon + 4K \sum_{m=0}^{N(\epsilon)} a_m \mathbb{E}\Big(\frac{1}{n}\sum_{k=1}^{n} P_{k-m}^2(X_k)\mathbf{1}_{\Gamma(m,n,b_m M\sqrt{n})}\Big) . \tag{8.4.2}$$

Now by Doob's maximal inequality

$$\mathbb{P}\big(\Gamma(m,n,b_m M\sqrt{n})\big) \leq \frac{4\sum_{k=1}^{n}\|P_{k-m}(X_k)\|_2^2}{b_m^2 M^2 n} = \frac{4\|P_0(X_m)\|_2^2}{b_m^2 M^2} \,,$$

and consequently

$$\lim_{M\to\infty}\sup_{n>0}\mathbb{P}\big(\Gamma(m,n,b_m M\sqrt{n})\big) = 0 \,. \tag{8.4.3}$$

Since $n^{-1}\sum_{k=1}^{n} P_{k-m}^2(X_k)$ converges in $\mathbb{L}^1$ (apply the ergodic theorem), we infer from (8.4.3) that

$$\lim_{M\to\infty}\limsup_{n\to\infty}\mathbb{E}\Big(\frac{1}{n}\sum_{k=1}^{n} P_{k-m}^2(X_k)\mathbf{1}_{\Gamma(m,n,b_m M\sqrt{n})}\Big) = 0 \,. \tag{8.4.4}$$

Combining (8.4.2) and (8.4.4), we conclude that

$$\lim_{M\to\infty}\limsup_{n\to\infty}\frac{1}{n}\mathbb{E}\left((S_n^* - M\sqrt{n})_+^2\right) = 0 \,. \tag{8.4.5}$$

Of course, the same arguments apply to the sequence $(-X_k)_{k\in\mathbb{Z}}$ so that (8.4.4) holds for $\max_{1\leq k\leq n}|S_k|$ instead of S_n^*. This completes the proof. □

Proof of item 2. Let $A_k(\lambda) = \{\max_{1\leq i\leq k}|S_i| > \lambda\}$. From Proposition 5.8 applied to the sequences $(X_i)_{i\in\mathbb{Z}}$ and $(-X_i)_{i\in\mathbb{Z}}$ we get that

$$\mathbb{E}\Big(\Big(\max_{1\leq i\leq n}|S_i| - \lambda\Big)_+^2\Big) \leq 8\sum_{k=1}^{n}\Big(\mathbb{E}(X_k^2\mathbf{1}_{A_k(\lambda)}) + 2\|\mathbf{1}_{A_k(\lambda)}X_k\mathbb{E}(S_n - S_k \mid \mathcal{F}_k)\|_1\Big). \tag{8.4.6}$$

By assumption the sequence $(X_k^2)_{k>0}$ and the array $(X_k\mathbb{E}(S_n - S_k \mid \mathcal{F}_k))_{1\leq k\leq n}$ are uniformly integrable. It follows that the $\mathbb{L}^1$-norms of the above random variables are each bounded by some positive constant K. Hence, from (8.4.6) with $\lambda = 0$ we get that $\mathbb{E}((\max_{1\leq i\leq n}|S_i|)^2) \leq 24Kn$. It follows that

$$\mathbb{P}(A_k(M\sqrt{n})) \leq (nM^2)^{-1}\mathbb{E}\Big(\Big(\max_{1\leq i\leq n}|S_i|\Big)^2\Big) \leq 24KM^{-2}. \tag{8.4.7}$$

From the inequality (8.4.7) and the uniform integrability of both $(X_k^2)_{k>0}$ and $(X_k\mathbb{E}(S_n - S_k \mid \mathcal{F}_k))_{1\le k\le n}$ we infer that

$$\lim_{M\to\infty} \limsup_{n\to\infty} n^{-1}\mathbb{E}\Big(\Big(\max_{1\le i\le n} |S_i| - M\sqrt{n}\Big)_+^2\Big) = 0\,.$$

This completes the proof. □

8.4.2 Sufficient conditions for triangular arrays

Under the conditions of Proposition 7.9, **S1*** holds:

Proposition 8.2. *Let $X_{i,n}$ and $\mathcal{M}_{i,n}$ be as in Proposition 7.9. If (7.5.31) and (7.5.32) hold, then* **S1*** *holds with the same η as in Proposition 7.9.*

Proof of Proposition 8.2. In view of Proposition 7.9, it is enough to prove that **S2**(a^*) holds. In fact, this follows from the inequality (7.5.37). □.

Chapter 9

Law of the iterated logarithm (LIL)

In this chapter, we derive laws of the iterated logarithm. We first give a bounded law of the iterated logarithm in a non causal setting. We then focus on τ-dependent sequences for which we derive a causal strong invariance principle. The main tool to prove it is the Fuk-Nagaev type inequality given in Theorem 5.3 of Chapter 5.

9.1 Bounded LIL under a non causal condition

In this section, we derive a bounded law of the iterated logarithm under a non causal condition detailed in the assumptions of Theorem 4.5 of Chapter 4. We get the following theorem:

Theorem 9.1. *Suppose that $(X_n)_{n\in\mathbb{Z}}$ is a stationary process satisfying the assumptions of Theorem 4.5 of Chapter 4. If $\sigma_n^2 = \mathrm{Var}\,(\sum_{i=1}^n X_i)$, assume that $\sigma^2 = \lim_{n\to\infty} \sigma_n^2/n > 0$. Then we have*

$$\limsup_{n\to\infty} \frac{1}{\sigma\sqrt{2n\log\log n}} |S_n| \le 1 \qquad a.s. \tag{9.1.1}$$

Proof of Theorem 9.1. Let $a > 1$. We define the subsequence $(n_k)_{k\in\mathbb{Z}}$ as $n_k = [a^k]$. We obtain from Theorem 4.5 that, for any $n_k \le n < n_{k+1}$ and any fixed c,

$$\begin{aligned}\mathbb{P}\left(\frac{|S_n|}{\sqrt{2n\sigma^2}} > c\sqrt{\log\log n_k}\right) &\leq 2\,\exp\left(-c^2\log\log n_k\frac{\sigma^2 n}{\sigma_n^2}(1+o(1))\right)\\ &= 2\,\exp\left(-c^2\log\log n_k(1+o(1))\right)\\ &= \mathcal{O}\left(k^{-c^2(1+o(1))}\right).\end{aligned}$$

This implies by the maximal inequality given in Theorem 2.2 in Móricz *et al.*(1982) [133] that

$$\mathbb{P}\left(\max_{n_k\leq n<n_{k+1}}\frac{|S_n|}{\sqrt{2n\sigma^2}} > c\sqrt{\log\log n_k}\right) \leq C\,k^{-c'}, \tag{9.1.2}$$

where $c' < c^2$ can be chosen arbitrarily close to c^2 and C is an appropriate finite constant (see the remark following the proof of Theorem 2.2 in Móricz *et al.*(1982) [133]). Since $\lim_{k\to\infty}\max_{n_k\leq n<n_{k+1}}\frac{\log\log n}{\log\log n_k} = 1$, we conclude from (9.1.2) by the Borel-Cantelli lemma that for any $c > 1$,

$$\limsup_{n\to\infty}\frac{1}{\sigma\sqrt{2n\log\log n}}|S_n| \leq c \qquad \text{a.s.}$$

This implies (9.1.1). □

9.2 Causal strong invariance principle

In this section, we present a strong invariance principle for partial sums of $\tau_{1,\infty}-$dependent sequences. Let $(X_n)_{n\in\mathbb{Z}}$ be a stationary sequence of zero-mean square integrable real valued random variables. Let $\mathcal{M}_i = \sigma(X_j, j \leq i)$. Define

$$S_n = X_1 + \cdots + X_n \quad \text{and} \quad S_n(t) = S_{[nt]} + (nt - [nt])X_{[nt]+1}.$$

We assume that $\sigma_n^2/n = \operatorname{Var}(S_n)/n$ converges to some constant σ^2 as n tends to infinity (this will always be true for any of the conditions we shall use hereafter). For $\sigma > 0$, we study the almost sure behavior of the partial sum process

$$\left\{\sigma^{-1}(2n\log\log n)^{-1/2}S_n(t)\;\Big/\;t\in[0,1]\right\}. \tag{9.2.1}$$

Before stating the main result, let us recall existing results in the i.i.d. case or in other frames of dependence.

Let $\mathcal{S}$ be the subset of $C([0,1])$ consisting of all absolutely continuous functions with respect to the Lebesgue measure such that

$$h(0) = 0 \text{ and } \int_0^1 (h'(t))^2dt \leq 1.$$

In 1964, Strassen [180] proved that if the sequence $(X_n)_{n\in\mathbb{Z}}$ is i.i.d. then the process defined in (9.2.1) is relatively compact with a.s. limit set $\mathcal{S}$. This result is known as the functional law of the iterated logarithm (FLIL for short). Heyde and Scott (1973) [104] extended the FLIL to the case where $\mathbb{E}(X_1|\mathcal{M}_0) = 0$ and the sequence is ergodic. Starting from this result and from a coboundary decomposition due to Gordin (1969) [97], Heyde (1975) [105] proved that the FLIL holds if $\mathbb{E}(S_n|\mathcal{M}_0)$ converges in $\mathbb{L}_2$ and the sequence is ergodic. Heyde's condition holds as soon as

$$\sum_{k=1}^{\infty} k \int_0^{\gamma_1(k)/2} Q \circ G(u)\, du < \infty, \tag{9.2.2}$$

where the functions $Q = Q_{|X_0|}$ and $G = G_{|X_0|}$ have been defined in Chapter 5 and $\gamma_1(k) = \|\mathbb{E}(X_k|\mathcal{M}_0)\|_1$ is the coefficient introduced in Section 2.2.4 of Chapter 2.

Other types of dependence have been soon considered for the FLIL (see for instance the review paper by Philipp (1986) [150]). For ρ and ϕ-mixing sequences, a strong invariance principle is given in Shao (1993) [174]. The case of strongly (α-)mixing sequences has been considered by Oodaira and Yoshihara (1971) [137], Dehling and Philipp (1982) [54] , and Bradley (1983) [29] among others. In 1995, Rio [159] proved a FLIL (and even a strong invariance principle) for the process defined in (9.2.1) as soon as the DMR (Doukhan, Massart and Rio (1994) [70]) condition (9.2.3) is satisfied

$$\sum_{k=1}^{\infty} \int_0^{2\,\alpha_X(k)} Q^2(u)\, du < \infty, \tag{9.2.3}$$

where $\alpha_X(k)$ has been defined in Section 1.2 of Chapter 1.

Considering Corollary 7.6 which gives the central limit theorem for γ-dependent sequences, we think that a reasonable condition for the FLIL is condition (9.2.2) without the k in front of the integral. Actually, we can only prove this conjecture with $\tau_{1,\infty}(k)$ instead of $\gamma_1(k)$, that is the FLIL holds as soon as

$$\sum_{k=1}^{\infty} \int_0^{\tau_{1,\infty}(k)/2} Q \circ G(u)\, du < \infty. \tag{9.2.4}$$

Theorem 9.2. *Let $(X_n)_{n\in\mathbb{Z}}$ be a strictly stationary sequence of centered and square integrable random variables satisfying (9.2.4). Then σ_n^2/n converges to σ^2, and there exists a sequence $(Y_n)_{n\in\mathbb{N}}$ of independent $\mathcal{N}(0,\sigma^2)$-distributed random variables (possibly degenerate) such that*

$$\sum_{i=1}^{n}(X_i - Y_i) = o\left(\sqrt{n \log\log n}\right) \quad a.s.$$

Such a result is known as a strong invariance principle. If $\sigma > 0$, Theorem 9.2 and Strassen's FLIL for the Brownian motion yield the FLIL for the process (9.2.1).

As in Corollary 7.6, we obtain simple sufficient conditions for the FLIL to hold:

Corollary 9.1. *Let $(X_n)_{n\in\mathbb{Z}}$ be a strictly stationary sequence of centered and square integrable random variables. Any of the following conditions implies (9.2.4) and hence the FLIL.*

1. $\mathbb{P}(|X_0| > x) \le (c/x)^r$ *for some* $r > 2$, *and* $\sum_{i\ge 0} (\tau_{1,\infty}(i))^{(r-2)/(r-1)} < \infty$.
2. $\|X_0\|_r < \infty$ *for some* $r > 2$, *and* $\sum_{i\ge 0} i^{1/(r-2)}\tau_{1,\infty}(i) < \infty$.
3. $\mathbb{E}(|X_0|^2 \log(1 + |X_0|)) < \infty$ *and* $\tau_{1,\infty}(i) = \mathcal{O}(a^i)$ *for some* $a < 1$.

Condition (9.2.4) is essentially optimal as shown in Corollary 9.2 below, derived from the examples given in Doukhan, Massart and Rio (1994) [70]:

Corollary 9.2. *For any $r > 2$, there is stationary Markov chain $(X_n)_{n\in\mathbb{Z}}$ such that*

1. $\mathbb{E}(X_0) = 0$ *and, for any nonnegative real* x, $\mathbb{P}(|X_0| > x) = \min(1, x^{-r})$.
2. *The sequence* $(\tau_{1,\infty}(i))_{i\ge 0}$ *satisfies* $\sup_{i\ge 0} i^{(r-1)/(r-2)}\tau_{1,\infty}(i) < \infty$.
3. $\limsup_{n\to\infty} (n \log\log n)^{-1/2}|S_n| = +\infty$ *almost surely.*

This corollary follows easily from Proposition 3 in Doukhan, Massart and Rio (1994) [69]. Let us now write the proof of Theorem 9.2.

Proof of Theorem 9.2. We first need to give some precise notations.

Notations 9.1. *Define the set*

$$\Psi = \left\{\psi \Big/ \mathbb{N} \to \mathbb{N}, \psi \text{ increasing}, \ \frac{\psi(n)}{n} \to_{n\to\infty} \infty, \ \psi(n) = o(n\sqrt{LLn})\right\}.$$

If ψ is some function of Ψ, let

$$M_1 = 0 \text{ and } M_n = \sum_{k=1}^{n-1}(\psi(k) + k), \qquad \text{for } n \ge 2\,.$$

For $n \ge 1$, define the random variables

$$U_n = \sum_{i=M_n+1}^{M_n+\psi(n)} X_i\,, \quad V_n = \sum_{i=M_{n+1}+1-n}^{M_{n+1}} X_i\,, \quad \text{and}$$

$$U_n' = \sum_{i=M_n+1}^{M_{n+1}} |X_i| \,.$$

If $Lx = \max(1, \log x)$, *define the truncated random variables*

$$\overline{U}_n = \max\left(\min\left(U_n, \frac{n}{\sqrt{LLn}}\right), \frac{-n}{\sqrt{LLn}}\right).$$

Theorem 9.2 is a consequence of the following Proposition

Proposition 9.1. *Let* $(X_n)_{n\in\mathbb{Z}}$ *be a strictly stationary sequence of centered and square integrable random variables satisfying condition (9.2.4). Then* σ_n^2/n *converges to* σ^2 *and there exist a function* $\psi \in \Psi$ *and a sequence* $(W_n)_{n\in\mathbb{N}}$ *of independent* $\mathcal{N}\left(0, \psi(n)\sigma^2\right)$*-distributed random variables (possibly degenerate) such that*

(a) $\displaystyle\sum_{i=1}^{n}(W_i - \overline{U}_i) = o\left(\sqrt{M_n LLn}\right)$ *a.s.*

(b) $\displaystyle\sum_{n=1}^{\infty} \frac{\mathbb{E}(|U_n - \overline{U}_n|)}{n\sqrt{LLn}} < \infty$

(c) $U_n' = o\left(n\sqrt{LLn}\right)$ *a.s.*

Proof of Proposition 9.1. It is adapted from the proof of Proposition 2 in Rio (1995) [159].

Proof of (b). Note first that

$$\mathbb{E}|U_n - \overline{U}_n| = \mathbb{E}\left(\left(|U_n| - \frac{n}{\sqrt{LLn}}\right)_+\right) \text{ so that}$$

$$\mathbb{E}|U_n - \overline{U}_n| = \int_{\frac{n}{\sqrt{LLn}}}^{+\infty} \mathbb{P}(|U_n| > t)dt \,. \tag{9.2.5}$$

In the following we write Q instead of $Q_{|X_0|}$. Since U_n is distributed as $S_{\psi(n)}$, we infer from Theorem 5.3 that

$$\mathbb{P}(|U_n| > t) \le 4\left(1 + \frac{t^2}{25\, r\, s_{\psi(n)}^2}\right)^{-\frac{r}{2}} + \frac{20\,\psi(n)}{t}\int_0^{S\left(\frac{t}{5r}\right)} Q(u)du. \tag{9.2.6}$$

Consider the two terms

$$A_{1,n} = \frac{4}{n\sqrt{LLn}}\int_{\frac{n}{\sqrt{LLn}}}^{+\infty}\left(1 + \frac{t^2}{25\, r\, s_{\psi(n)}^2}\right)^{-\frac{r}{2}} dt \,,$$

$$A_{2,n} = \frac{20\,\psi(n)}{n\sqrt{LLn}} \int_{\frac{n}{\sqrt{LLn}}}^{+\infty} \frac{1}{t} \int_0^{S\left(\frac{t}{5r}\right)} Q(u)du\,dt\,.$$

From (9.2.5) and (9.2.6), we infer that

$$\frac{\mathbb{E}|U_n - \overline{U}_n|}{n\sqrt{LLn}} \leq A_{1,n} + A_{2,n}\,. \tag{9.2.7}$$

Study of $A_{1,n}$. Since the sequence $(X_n)_{n\in\mathbb{N}}$ satisfies (9.2.4), $s^2_{\psi(n)}/\psi(n)$ converges to some positive constant. Let C_r denote some constant depending only on r which may vary from line to line. We have that

$$A_{1,n} \leq \frac{4}{n\sqrt{LLn}} \int_{\frac{n}{\sqrt{LLn}}}^{+\infty} \frac{t^{-r}}{C_r\, s^{-r}_{\psi(n)}} dt \leq C_r\, s^r_{\psi(n)}\, \frac{n^{-r}}{LLn^{1-\frac{r}{2}}}.$$

We infer that $A_{1,n} = \mathcal{O}(\psi(n)^{r/2} n^{-r} LLn^{(r-2)/2})$ as n tends to infinity. Since $\psi \in \Psi$ and $r > 2$, we infer that $\sum_{n\geq 1} A_{1,n}$ is finite.

Study of $A_{2,n}$. We use the elementary result: if $(a_i)_{i\geq 1}$ is a sequence of positive numbers, then there exists a sequence of positive numbers $(b_i)_{i\geq 1}$ such that $b_i \to \infty$ and $\sum_{i\geq 1} a_i b_i < \infty$ if and only if $\sum_{i\geq 1} a_i < \infty$ (note that $b_n^2 = \left(\sum_{i=n}^{\infty} a_i\right)^{-1}$ works). Consequently $\sum_{n\geq 1} A_{2,n}$ is finite for some $\psi \in \Psi$ if and only if

$$\sum_{n\geq 1} \frac{1}{\sqrt{LLn}} \int_{\frac{n}{\sqrt{LLn}}}^{+\infty} \frac{1}{t} \int_0^{S\left(\frac{t}{5r}\right)} Q(u)du\,dt < +\infty. \tag{9.2.8}$$

Recall that $S = R^{-1}$, with the notations of Theorem 5.3. To prove (9.2.8), write

$$\begin{aligned}
\int_{\frac{n}{\sqrt{LLn}}}^{+\infty} \frac{1}{t} \int_0^{S\left(\frac{t}{5r}\right)} Q(u)du\,dt &= \int_{\frac{n}{\sqrt{LLn}}}^{+\infty} \frac{1}{t} \int_0^1 1_{R(u)\geq \frac{t}{5r}}\, Q(u)du\,dt \\
&= \int_0^1 Q(u) \int_{\frac{n}{\sqrt{LLn}}}^{5\,r\,R(u)} \frac{1}{t} dt du \\
&= \int_0^1 Q(u)\, \log \frac{5\,r\,R(u)}{\frac{n}{\sqrt{LLn}}}\, 1_{R(u)\geq \frac{n}{5\,r\,\sqrt{LLn}}} du.
\end{aligned}$$

Consequently (9.2.8) holds if and only if

$$\int_0^1 Q(u) \sum_{n\geq 1} \frac{1}{\sqrt{LLn}} \log \frac{5\,r\,R(u)}{\frac{n}{\sqrt{LLn}}}\, 1_{\left\{R(u)\geq \frac{n}{5\,r\,\sqrt{LLn}}\right\}} du < +\infty. \tag{9.2.9}$$

To see that (9.2.9) holds, we shall prove the following result: if f is any increasing function such that $f(0) = 0$ and $f(1) = 1$, then for any positive R we have that

$$\sum_{n\geq 1} \log\left(\frac{R}{f(n)}\right)(f(n) - f(n-1))\, 1_{f(n)\leq R} \leq (R-1)\vee 0 \leq R. \qquad (9.2.10)$$

Applying this result to $f(x) = x(LLx)^{-1/2}$ and $R = 5rR(u)$, and noting that $(LLn)^{-1/2} \leq C\,(f(n) - f(n-1))$ for some constant $C > 1$, we infer that

$$\int_0^1 Q(u) \sum_{n\geq 1} \frac{1}{\sqrt{LLn}} \log \frac{5\,r\,R(u)}{\frac{n}{\sqrt{LLn}}}\, 1_{(R(u)\geq \frac{n}{5\,r\,\sqrt{LLn}})} du \leq 5Cr \int_0^1 Q(u)R(u)du\,,$$

which is finite as soon as (9.2.4) holds.
It remains to prove (9.2.10). If $R \leq 1$ the result is clear. Now, for $R > 1$, let x_R be the largest integer such that $f(x_R) \leq R$ and write $R^* = f(x_R)$. Note first that

$$\sum_{n\geq 1} (\log R)\,(f(n) - f(n-1))\, 1_{f(n)\leq R} \leq R^* \log R. \qquad (9.2.11)$$

On the other hand, we have that

$$\sum_{n\geq 1} \log(f(n))\,(f(n) - f(n-1))\, 1_{f(n)\leq R} = \sum_{n=1}^{x_R} \log(f(n))\,(f(n) - f(n-1))\,.$$

It follows that

$$\sum_{n\geq 1} \log(f(n))\,(f(n) - f(n-1)) \geq \int_1^{R^*} \log x\, dx = R^* \log R^* - R^* + 1. \qquad (9.2.12)$$

Using (9.2.11) and (9.2.12) we get that

$$\sum_{n\geq 1} \log\left(\frac{R}{f(n)}\right)(f(n) - f(n-1))\, 1_{f(n)\leq R} \leq R^* - 1 + R^*(\log R - \log R^*). \qquad (9.2.13)$$

Using Taylor's inequality, we have that $R^*(\log R - \log R^*) \leq R - R^*$ and (9.2.10) follows. The proof of (b) is complete.

Proof of (c). Let $T_n = \sum_{i=M_n+1}^{M_{n+1}} (|X_i| - \mathbb{E}|X_i|)$. We easily see that

$$U'_n = (\psi(n) + n)\,\mathbb{E}|X_1| + T_n. \qquad (9.2.14)$$

By definition of Ψ, we have $\psi(n) = o\left(n\sqrt{LLn}\right)$. Here note that

$$T_n \leq \frac{n}{\sqrt{LLn}} + 0 \vee \left(T_n - \frac{n}{\sqrt{LLn}}\right). \qquad (9.2.15)$$

Using same arguments as for the proof of (b), we obtain that

$$\sum_{n\geq 1} \frac{\mathbb{E}\left(0 \vee \left(T_n - \frac{n}{\sqrt{LLn}}\right)\right)}{n\sqrt{LLn}} < +\infty\,, \text{ so that}$$

$$\sum_{n\geq 1} \frac{0 \vee \left(T_n - \frac{n}{\sqrt{LLn}}\right)}{n\sqrt{LLn}} < +\infty \text{ a.s.}$$

Consequently $\max(0, T_n - n(LLn)^{-1/2}) = o(n\sqrt{LLn})$ almost surely, and the result follows from (9.2.14) and (9.2.15).

Proof of (a). In the following, $(\delta_n)_{n\geq 1}$ and $(\eta_n)_{n\geq 1}$ denote independent sequences of independent random variables with uniform distribution over $[0,1]$, independent of $(X_n)_{n\geq 1}$. Since $\overline{U}_n$ is a 1-Lipschitz function of U_i, $\tau(\sigma(U_i, i \leq n-1), \overline{U}_n) \leq \psi(n)\tau(n)$. Using Lemma 5.2 and arguing as in the proof of Theorem 5.2, we get the existence of a sequence $(\overline{U}_n^*)_{n\geq 1}$ of independent random variables with the same distribution as the random variables $\overline{U}_n$ such that $\overline{U}_n^*$ is a measurable function of $(\overline{U}_l, \delta_l)_{l\leq n}$ and

$$\mathbb{E}\left(|\overline{U}_n - \overline{U}_n^*|\right) \leq \psi(n)\,\tau(n).$$

Since (9.2.4) holds, we have that

$$\sum_{n\geq 1} \frac{\mathbb{E}\left|\overline{U}_n - \overline{U}_n^*\right|}{\sqrt{M_n\, LLn}} < \infty \quad \text{so that} \quad \sum_{n\geq 1} \frac{|\overline{U}_n - \overline{U}_n^*|}{\sqrt{M_n\, LLn}} < +\infty \text{ a.s.}$$

Applying Kronecker's lemma, we obtain that

$$\sum_{i=1}^{n} (\overline{U}_i - \overline{U}_i^*) = o\left(\sqrt{M_n\, LLn}\right) \text{ a.s.} \tag{9.2.16}$$

We infer from (9.2.4) and from Dedecker and Doukhan (2003) [43] that

$$(\psi(n))^{-1} \operatorname{Var} U_n \to_{n\to\infty} \sigma^2 \qquad \text{and} \qquad (\psi(n))^{-1/2}\, U_n \xrightarrow[n\to\infty]{\mathcal{D}} \mathcal{N}\left(0, \sigma^2\right).$$

Hence the sequence $(U_n^2/\psi(n))_{n\geq 1}$ is uniformly integrable (Theorem 5.4. in Billingsley (1968) [20]). Consequently, since the random variables $\overline{U}_n^*$ have the same distribution as the random variables $\overline{U}_n$, we deduce from the above limit results, from Strassen's representation theorem (see Dudley (1968) [79]), and from Skorohod's lemma (1976) [179] that one can construct some sequence

$(W_n)_{n\geq 1}$ of $\sigma(\overline{U}^*_n, \eta_n)$-measurable random variables with respective distribution $\mathcal{N}\left(0, \psi(n)\,\sigma^2\right)$ such that

$$\mathbb{E}\left(\left(\overline{U}^*_n - W_n\right)^2\right) = o\left(\psi(n)\right) \text{ as } n \to +\infty, \tag{9.2.17}$$

which is exactly equation (5.17) of the proof of Proposition 2(c) in Rio (1995) [159]. The end of the proof is the same as that of Rio.

Proof of Theorem 9.2. By Skohorod's lemma (1976) [179], there exists a sequence $(Y_i)_{i\geq 1}$ of independent $\mathcal{N}\left(0, \sigma^2\right)$-distributed random variables satisfying $W_n = \sum_{i=M_n+1}^{M_n+\psi(n)} Y_i$ for all positive n. Define the random variable

$$V'_n = \sum_{i=M_{n+1}+1-n}^{M_{n+1}} Y_i.$$

Let $n(k) := \sup\{n \geq 0 / \, M_n \leq k\}$, and note that by definition of M_n we have $n(k) = o(\sqrt{k})$. Applying Proposition 9.1(c) we see that

$$\left|\sum_{i=1}^{k} X_i - \sum_{i=1}^{n(k)} (U_i + V_i)\right| \leq U'_{n(k)} = o\left(\sqrt{k\,LLk}\right) \;\; a.s. \tag{9.2.18}$$

From (5.26) in Rio (1995) [159], we infer that

$$\sum_{i=1}^{n(k)} V_i = o\left(\sqrt{k\,LLk}\right) \;\; a.s. \quad \text{and} \quad \sum_{i=1}^{n(k)} V'_i = o\left(\sqrt{k\,LLk}\right) \;\; a.s. \tag{9.2.19}$$

Gathering (9.2.18), (9.2.19) and Proposition 9.1(a) and (b), we obtain that

$$\sum_{i=1}^{k} X_i - \sum_{i=1}^{n(k)} (W_i + V'_i) = o\left(\sqrt{k\,LLk}\right) \;\; a.s. \tag{9.2.20}$$

Clearly $\sum_{i=1}^{k} Y_i - \sum_{i=1}^{n(k)} (W_i + V'_i)$ is normally distributed with variance smaller than $\psi(n(k)) + n(k)$. Since $n(k) = o(\sqrt{k})$ we have that $\psi(n(k)) + n(k) = o(\sqrt{kLLk})$ by definition of ψ. An elementary calculation on Gaussian random variables shows that

$$\sum_{i=1}^{k} Y_i - \sum_{i=1}^{n(k)} (W_i + V'_i) = o\left(\sqrt{k\,LLk}\right) \;\; a.s. \tag{9.2.21}$$

Theorem 9.2 follows from (9.2.20) and (9.2.21).

Chapter 10

The Empirical process

In this chapter, we prove central limit theorems for the empirical distribution function of weakly dependent stationary sequences (or fields). Except in the last section (Section 10.6), where the oscillations of the empirical distribution of weakly dependent random fields are studied, all the results are mainly based on the tightness criterion given in Proposition 4.2. In Section 10.1, we give a sufficient condition for the tightness, based on the control of the covariances between indicators of half lines. In Section 10.2 we prove an empirical central limit theorem for η-dependent sequences by assuming an exponential decay of the coefficients. In Sections 10.3 and 10.4, we give sufficient conditions in terms of the coefficients $\tilde{\alpha}$, $\tilde{\beta}$, $\tilde{\phi}$, θ and τ. In Section 10.5 we present the applications of such results to the empirical copula process.

Definition 10.1 (Empirical process). *We recall the definition of the empirical process for the different cases considered:*

- **Stationary real valued or multivariate sequence**: *Let $(X_n)_{n\in\mathbb{Z}}$ be a sequence of $\mathbb{R}^d$ valued random variables. Define in $\mathbb{R}^d$ the partial order by $s \le t$ if and only if $s_i \le t_i$ for $i = 1, \ldots, d$. The empirical process of X is defined by*

$$F_n(t) = \frac{1}{n}\sum_{i=1}^{n} \mathbf{1}_{\{X_i \le t\}}. \tag{10.0.1}$$

- **Stationary real valued random field**: *For N in $\mathbb{N}$, let B_N be the closed ball of radius N for the ℓ^∞-norm on $T = \mathbb{Z}^d$ and $n = \#B_N = (2N+1)^d$. Let $(X_t)_{t\in T}$ be a real valued random field. We define the empirical process*

$$F_n(t) = \frac{1}{n}\sum_{k\in B_N} \mathbf{1}_{\{X_k \le t\}}. \tag{10.0.2}$$

In any case, if the variables are identically distributed with common distribution function F, we define the normalized empirical process by

$$U_n(t) = \sqrt{n}\,(F_n(t) - F(t))\,.$$

We need to define the limit processes in the following central limit theorems. $(\mathbf{B}(t))_{t\in[0,1]^d}$ is a zero-mean Gaussian process with covariance function

$$\Gamma(t,s) = \sum_{k\in T} \mathrm{Cov}(\mathbf{1}_{X_0\le t}, \mathbf{1}_{X_k\le s}). \tag{10.0.3}$$

where $T = \mathbb{Z}$ in the case of random sequences and $T = \mathbb{Z}^d$ in the case of random fields.

10.1 A simple condition for the tightness

We consider a stationary real valued sequence $(X_n)_{n\in\mathbb{Z}}$ with continuous common repartition function F. We assume without loss of generality that the marginal distribution of this sequence is the uniform law on $[0,1]$.
Assume that the sequence $(X_n)_{n\in\mathbb{Z}}$ satisfies the following weak dependence condition:
Let $\mathcal{F} = \{x \mapsto \mathbf{1}_{s<x\le t}/ \text{ for } s,t \in [0,1]\}$. We assume that for any $m \in \{1,2,3\}$ and any $0 \le t_1 \le t_2 \le t_3 \le t_4$,

$$\sup_{f\in\mathcal{F}} \left| \mathrm{Cov}\left(\prod_{i=1}^{m} f(X_{t_i}), \prod_{i=m+1}^{4} f(X_{t_i}) \right) \right| \le \varepsilon(r), \tag{10.1.1}$$

where $r = t_{m+1} - t_m$ and $\varepsilon(r)$ does only depend on r (in this case a weak dependence condition holds for a class of functions $\mathbb{R}^u \to \mathbb{R}$ working only with the values $u = 1$, 2 or 3).

Proposition 10.1. *Let $(X_n)_{n\in\mathbb{Z}}$ be a real valued stationary sequence fulfilling (10.1.1) with*

$$\varepsilon(r) = \mathcal{O}(r^{-5/2-\nu}), \quad \textit{for some} \quad \nu > 0. \tag{10.1.2}$$

Then the process U_n is tight in $\mathcal{D}([0,1])$.

Proof of Proposition 10.1. The moment inequality (4.3.17), together with conditions (10.1.1) and (10.1.2), yields the existence of a positive constant C such that for any s, t in $[0,1]$

$$\begin{aligned}\|U_n(t)-U_n(s)\|_4 &\le C\Big\{\Big(\sum_{r=0}^{n-1} r^{-a}\wedge|t-s|\Big)^{1/2}+\Big(\frac{1}{n}\sum_{r=0}^{n-1}(r+1)^2\varepsilon(r)\Big)^{1/4}\Big\}\\ &\le C\Big\{\Big(\sum_{r\ge|t-s|^{-1/a}} r^{-a}\Big)^{1/2}\\ &\quad+\Big(\sum_{r<|t-s|^{-1/a}}|t-s|\Big)^{1/2}+n^{\frac{2-a}{4}}\Big\}\\ &\le C\{|t-s|^{\frac{a-1}{2a}}+n^{\frac{2-a}{4}}\}.\end{aligned}$$

The last bound together with the tightness criterion given in Proposition 4.2 proves that the sequence $\{U_n(t), t\in[0,1]\}$ is tight.

Remark 10.1. *Stationary associated sequences (see Section 1.4) satisfy the requirement of Proposition 10.1 if*

$$\sup_{|k|\ge r}\sup_{x,y\in\mathbb{R}}\operatorname{Cov}(\mathbf{1}_{X_0>x},\mathbf{1}_{X_k>y})=\mathcal{O}(r^{-5/2-\nu}).$$

Using the following inequality :

$$\sup_{|k|\ge r}\sup_{x,y\in\mathbb{R}}\operatorname{Cov}(\mathbf{1}_{X_0>x},\mathbf{1}_{X_k>y})\le C\sup_{|k|\ge r}\operatorname{Cov}^{1/3}(X_0,X_k),$$

for an universal constant C, Yu (1993) [195] proves the tightness under the condition $\operatorname{Cov}(X_0,X_r)=\mathcal{O}(r^{-a})$, for $a>15/2$. However in this case, the paper by Louhichi (2000) [124] proves the tightness under the condition $\operatorname{Cov}(X_0,X_r)=\mathcal{O}(r^{-a})$, for $a>4$.

10.2 η-dependent sequences

In this section, $\mathbf{Y}$ is a stationary η-dependent sequence in $[0,1]^d$ with uniform marginal distributions.

Theorem 10.1. *Assume that $(\mathbf{Y}_i)_{i\in\mathbb{Z}}$ is a stationary η-dependent zero-mean process in $[0,1]^d$ with uniform marginal distributions. Assume that there exist some constants $C>0$ and $a>4d+2$ such that $\eta(r)\le Cr^{-a}$. Then the process U_n converges in distribution in $\mathcal{D}([0,1]^d)$ to the Gaussian process $\mathbf{B}$.*

The following lemma is used to prove Theorem 10.1.

Lemma 10.1. *Assume that $(Y_i)_{i\in\mathbb{Z}^p}$ is a stationary multivariate random field with value in $[0,1]^d$ and that the density of marginal distribution of the vectors*

Y_i are bounded by C_Y. For $s \le t$ in $[0,1]^d$, denote $g_{t,s}(x) = \mathbf{1}\{x \le t\} - \mathbf{1}\{x \le s\}$. Let $\mathbf{i} = (i_1, \dots, i_u)$ in $(\mathbb{Z}^p)^u$ and $\mathbf{j} = (j_1, \dots, j_v)$ in $(\mathbb{Z}^p)^v$ be two sets of indices that are r-distant in $\mathbb{L}^1$-distance. Let G and H be two bounded Lipschitz functions on $\mathbb{R}^u$ and $\mathbb{R}^v$ respectively. Denote $Y_{\mathbf{i}} = (Y_{i_1}, \dots, Y_{i_u})$. Then

$$|\mathrm{Cov}\left(G(g_{t,s}(Y_{\mathbf{i}})), H(g_{t,s}(Y_{\mathbf{j}}))\right)| \le \psi(G,H)\epsilon_r, \tag{10.2.1}$$

where, setting $\phi(G,H) = d_G\|H\|_\infty \mathrm{Lip}\,(G)$, we define

- *if Y is η-dependent, $\epsilon_r = \eta_r^{1/2}$*

$$\psi(G,H) = 4(C_Y d)^{1/2}(\phi(G,H) + \phi(H,G)), \tag{10.2.2}$$

- *if Y is κ-dependent, $\epsilon_r = \kappa_r^{1/3}$*

$$\psi(G,H) = 2(4(C_Y d))^{2/3}(\phi(G,H) + \phi(H,G))^{2/3}(\phi(G,H)\phi(H,G))^{1/3} \tag{10.2.3}$$

Proof of lemma 10.1 For $\delta \ge 0$, define the δ-approximations of $\mathbf{1}_{\{x \ge t\}}$ by:

$$h_{\delta,t}(x) = \prod_{p=1}^{d}\left(\frac{(x^{(p)} - t^{(p)} + \delta)}{\delta}\mathbf{1}_{\{t^{(p)}-\delta < x^{(p)} < t^{(p)}\}} + \mathbf{1}_{\{x^{(p)} \ge t^{(p)}\}}\right).$$

Define $g_{\delta,t,s} = h_{\delta,t} - h_{\delta,s}$. Then its Lipschitz modulus is equal to δ^{-1}, where the distance in $\mathbb{R}^d$ is $d_1(x,y) = \sum_{p=1}^d |x^{(p)} - y^{(p)}|$ and $\mathbb{E}\,|g_{s,t}(Y_0) - g_{t,s,\delta}(Y_0)| \le 2dC_Y\delta$ because the density of the variable Y_i is bounded by C_Y and the two functions are equal except on $2d$ blocks of width δ.
Define $G_0(Y_{\mathbf{i}}) = G\left(g_{t,s}(Y_{\mathbf{i}})\right)$ and $G_\delta(Y_{\mathbf{i}}) = G\left(g_{t,s,\delta}(Y_{\mathbf{i}})\right)$.

$$\begin{aligned}
&|\mathrm{Cov}(G_0(Y_{\mathbf{i}}), H_0(Y_{\mathbf{j}})) - \mathrm{Cov}(G_\delta(Y_{\mathbf{i}}), H_\delta(Y_{\mathbf{j}}))| \\
&\qquad \le |\mathbb{E}\left(G_0(Y_{\mathbf{i}})H_0(Y_{\mathbf{j}})\right) - \mathbb{E}\left(G_\delta(Y_{\mathbf{i}})H_\delta(Y_{\mathbf{j}})\right)| \\
&\qquad\qquad + |\mathbb{E}\left(G_0(Y_{\mathbf{i}})\right)\mathbb{E}\left(H_0(Y_{\mathbf{j}})\right) - \mathbb{E}\left(G_\delta(Y_{\mathbf{i}})\right)\mathbb{E}\left(H_\delta(Y_{\mathbf{j}})\right)|
\end{aligned}$$

After substitution of the variables one by one, the first term of the right hand side is bounded by:

$$(u\|H\|_\infty \mathrm{Lip}\,(G) + v\|G\|_\infty \mathrm{Lip}\,(H))\mathbb{E}\,|g_{t,s}(Y_0) - g_{t,s,\delta}(Y_0)|\,.$$

so that:

$$|\mathbb{E}\left(G_0(Y_{\mathbf{i}})H_0(Y_{\mathbf{j}})\right) - \mathbb{E}\left(G_\delta(Y_{\mathbf{i}})H_\delta(Y_{\mathbf{j}})\right)| \le 2C_Y d\delta(\phi(G,H) + \phi(H,G)).$$

The bound of the second term is the same.
Now if Y is η-dependent:

$$|\mathrm{Cov}\left(G_\delta(Y_{\mathbf{i}}), H_\delta(Y_{\mathbf{j}})\right)| \le (\phi(G,H) + \phi(H,G))\frac{\eta_r}{\delta}.$$

Hence

$$|\mathrm{Cov}\,(G_0(Y_{\mathbf{i}}), H_0(Y_{\mathbf{j}})\,| \le (\phi(G,H) + \phi(H,G))\left(4C_Y d\delta + \frac{\eta_r}{\delta}\right)$$

If Y is κ-dependent:

$$|\mathrm{Cov}\,(G_\delta(Y_{\mathbf{i}}), H_\delta(Y_{\mathbf{j}}))\,| \le (\phi(G,H)\phi(H,G))\frac{\kappa_r}{\delta^2}.$$

Hence

$$|\mathrm{Cov}\,(G_0(Y_{\mathbf{i}}), H_0(Y_{\mathbf{j}}))\,| \le 4C_Y d(\phi(G,H) + \phi(H,G))\delta + \phi(G,H)\phi(H,G)\frac{\kappa_r}{\delta^2}$$

Choosing the optimal δ, relation (10.2.1) is proved. □
We apply this lemma for the case of products of indicator functions, namely for $s \le t$ in $\mathbb{R}^d$, for any $\mathbf{k}$ multi-index of $\mathbb{Z}^u$, we define

$$\Pi_{\mathbf{k}} = \prod_{j=1}^{u} g_{t,s}(Y_{k_j}) - F(t) + F(s).$$

We note that $\Pi_{\mathbf{k}} = G(g_{t,s}(Y_{\mathbf{k}}))$ where the function G is defined by $G(x_1,\dots x_u) = \prod_{j=1}^{u}(x_j - c)$ with $c = F(t) - F(s)$. Here $\|G\|_\infty = \mathrm{Lip}\,(G) = 1$.

Corollary 10.1. *Assume that $(\mathbf{Y}_i)_{i\in\mathbb{Z}}$ is a stationary η-dependent zero-mean process in $[0,1]^d$ with uniform marginal distributions, then for any sequences $\mathbf{i} = (i_1,\dots,i_u)$ and $\mathbf{j} = (j_1,\dots,j_v)$ such that $r \le j_1 - i_u$:*

$$\left|\mathrm{Cov}\left(\Pi_{\mathbf{i}}, \Pi_{\mathbf{j}}\right)\right| \le (u+v)\epsilon(r)\,, \tag{10.2.4}$$

with $\epsilon(r) = 4\sqrt{d\eta(r)}$.

Next we prove a Rosenthal type inequality.

Proposition 10.2. *Assume that $\mathbf{Y}$ is a stationary η-dependent zero-mean process in $[0,1]^d$ with uniform marginal distributions. Assume moreover that condition (10.2.4) is satisfied with $\epsilon(r) = Cr^{-a}$. Then, for $l < (a+1)/2$ and (s,t) such that $\mathbb{E}|x_0(s,t)| < C$, we have*

$$\mathbb{E}(U_n(t) - U_n(s))^{2l} \le \frac{(4l-2)!3^{2l}}{(2l-1)!}\left(\left(k_l\,\frac{2}{3}\left(\frac{\mathbb{E}|x_0(s,t)|}{C}\right)^{1-1/a}\right)^l + (2l)!\frac{n^{1-l}k_l}{3}\left(\frac{\mathbb{E}|x_0(s,t)|}{C}\right)^{1+(1-2l)/a}\right), \tag{10.2.5}$$

where $k_l = \left(C + \frac{C2^a}{a-2l+1}\right)$.

Proof of Proposition 10.2. Let $s \le t$ be in $\mathbb{R}^d$. As $|x_i(s,t)| \le 1$, we get for any sequences $\mathbf{i} \in \mathbb{Z}^u$, $\mathbf{j} \in \mathbb{Z}^v$,

$$\left|\mathrm{Cov}\left(\Pi_{\mathbf{i}}, \Pi_{\mathbf{j}}\right)\right| \le 2\,\mathbb{E}|x_0(s,t)|. \tag{10.2.6}$$

For any integer $q \ge 1$, set

$$A_q(n) = \sum_{\mathbf{k}\in\{1,\ldots,n\}^q} \left|\mathbb{E}\left(\Pi_{\mathbf{k}}\right)\right|, \tag{10.2.7}$$

then

$$\mathbb{E}(U_n(s) - U_n(t))^{2l} \le (2l)!n^{-l}A_{2l}(n). \tag{10.2.8}$$

Let $q \ge 2$.
For a finite sequence $\mathbf{k} = (k_1, \ldots, k_q)$ of elements of $\mathbb{Z}$, let $(k_{(1)}, \ldots, k_{(q)})$ be the same sequence ordered from the smaller to the larger. The gap $r(\mathbf{k})$ in the sequence is defined as the maximum of the integers $k_{(i+1)} - k_{(i)}$, $i = 1, \ldots, q-1$. Choose any index $j < q$ such that $k_{(j+1)} - k_{(j)} = r$, and define the two non-empty subsequences $\mathbf{k}^1 = (k_{(1)}, \ldots, k_{(j)})$ and $\mathbf{k}^2 = (k_{(j+1)}, \ldots, k_{(q)})$. Define $G_r(q,n) = \{\mathbf{k} \in \{1,\ldots,n\}^q \ / \ r(\mathbf{k}) = r\}$. Sorting the sequences of indices by their gaps, we get

$$\begin{aligned} A_q(n) \ \le\ & \sum_{k=1}^{n} \mathbb{E}|x_0(s,t)|^q \\ & + \sum_{r=1}^{n-1} \sum_{\mathbf{k}\in G_r(q,n)} \left|\mathrm{Cov}\left(\Pi_{\mathbf{k}^1}, \Pi_{\mathbf{k}^2}\right)\right| && (10.2.9)\\ & + \sum_{r=1}^{n-1} \sum_{\mathbf{k}\in G_r(q,n)} \left|\mathbb{E}\left(\Pi_{\mathbf{k}^1}\right)\mathbb{E}\left(\Pi_{\mathbf{k}^2}\right)\right|. && (10.2.10) \end{aligned}$$

Define

$$B_q(n) = \sum_{k=1}^{n} \mathbb{E}|x_0(s,t)|^q + \sum_{r=1}^{n-1} \sum_{\mathbf{k}\in G_r(q,n)} \left|\mathrm{Cov}\left(\Pi_{\mathbf{k}^1}, \Pi_{\mathbf{k}^2}\right)\right|.$$

In order to prove that the expression (10.2.10) is bounded by the product $\sum_m A_m(n)A_{q-m}(n)$ we make a first summation over the $\mathbf{k}$'s with $\#\mathbf{k}^1 = m$. Hence

$$A_q(n) \le B_q(n) + \sum_{m=1}^{q-1} A_m(n)A_{q-m}(n). \tag{10.2.11}$$

Now we give a bound of $B_q(n)$. To build a sequence $\mathbf{k}$ belonging to $G_r(q,n)$, we first fix one of the n points of $\{1,\ldots,n\}$. We choose a second point among the two points that are at distance r from the first point. The i-th point lies in an interval of radius r centered at one of the $i-1$ preceding points. Thus for $r \in \mathbb{N}^*$, we have

$$\#G_r(q,n) \le n \cdot 2 \cdot 2(2r+1)\cdots(q-1)(2r+1) \le 2n(q-1)!(3r)^{q-2} .$$

We use condition (10.2.4) and condition (10.2.6) to deduce:

$$\begin{aligned} B_q(n) &\le n\,\mathbb{E}|x_0(s,t)| + 2n(q-1)!\sum_{r=1}^{n-1}(3r)^{q-2}\min(q\epsilon(r), 2\,\mathbb{E}|x_0(s,t)|) \\ &\le n\,\mathbb{E}|x_0(s,t)| + n\,3^{q-1}\,q!\left(\sum_{r=1}^{n-1} r^{q-2}\min(\epsilon(r),\mathbb{E}|x_0(s,t)|)\right) . \end{aligned}$$

Denote by R the integer such that $R < (\mathbb{E}|x_0(s,t)|/C)^{-1/a} \le R+1$. For any $2 \le q \le 2l$:

$$\begin{aligned} B_q(n) &\le n\,\mathbb{E}|x_0(s,t)| + 3^{q-1}nq!\left(\mathbb{E}|x_0(s,t)|\sum_{r=1}^{R-1} r^{q-2} + C\sum_{r=R}^{\infty} r^{q-2-a}\right) \\ &\le 3^{q-1}nq!\left(\frac{\mathbb{E}|x_0(s,t)|}{q-1}R^{q-1} + \frac{C}{a-q+1}R^{q-1-a}\right) \\ &\le 3^{q-1}nq!(\mathbb{E}|x_0(s,t)|/C)^{-(q-1)/a}\left(\frac{\mathbb{E}|x_0(s,t)|}{q-1} + \frac{C}{a-q+1}R^{-a}\right). \end{aligned}$$

But $R \ge 1$, so that $(\mathbb{E}|x_0(s,t)|/C)^{-1/a} \le 2R$, and

$$B_q(n) \le 3^{q-1}nq!(\mathbb{E}|x_0(s,t)|/C)^{1-(q-1)/a}\left(C + \frac{C2^a}{a-2l+1}\right) .$$

We find that:

$$B_q(n) \le \left(3\left(\frac{\mathbb{E}|x_0(s,t)|}{C}\right)^{-1/a}\right)^q \frac{n\,k_l}{3}\left(\frac{\mathbb{E}|x_0(s,t)|}{C}\right)^{1+1/a} q! \qquad (10.2.12)$$

so $B_q(n)$ is bounded by a function M^qV_q that satisfies condition (4.3.24) and gives

$$\begin{aligned} A_{2l}(n) &\le \frac{(4l-2)!}{(2l)!(2l-1)!}\left(3\left(\left(\frac{\mathbb{E}|x_0(s,t)|}{C}\right)^{-1/a}\right)^{2l}\right. \\ &\quad \left(\left(n\,k_l\,\frac{2}{3}\left(\frac{\mathbb{E}|x_0(s,t)|}{C}\right)^{1+1/a}\right)^l + (2l)!\frac{n\,k_l}{3}\left(\frac{\mathbb{E}|x_0(s,t)|}{C}\right)^{1+1/a}\right) \\ &\le \frac{(4l-2)!3^{2l}}{(2l)!(2l-1)!}\left(\left(2\frac{n\,k_l}{3}\left(\frac{\mathbb{E}|x_0(s,t)|}{C}\right)^{1-1/a}\right)^l\right. \\ &\quad \left. +(2l)!\frac{n\,k_l}{3}\left(\frac{\mathbb{E}|x_0(s,t)|}{C}\right)^{1+(1-2l)/a}\right), \end{aligned}$$

and (10.2.5) is proved. □

Proof of theorem 10.1.

CLT for the finite dimensional distributions of U_n. Let $(s_1, \ldots, s_m)$ be a fixed sequence of elements in $[0,1]^d$. Denote $\mathbf{B}_n$ the vector-valued process

$$\mathbf{B}_n = (U_n(s_1), \ldots, U_n(s_m)).$$

To prove a CLT for the vector $\mathbf{B}_n$ is equivalent to prove the Gaussian convergence for any linear combination of its coordinates. Let $(\alpha_1, \ldots, \alpha_m)$ be a real vector such that $\sum_{j=1}^m \alpha_j \neq 0$.
Define $Z_i = \sum_{j=1}^m \alpha_j (\mathbf{1}\{\mathbf{Y}_i \leq s_j\} - P(\mathbf{Y}_i \leq s_j))$. Define also

$$\mathbf{S}_n = \frac{1}{\sqrt{n}} \sum_{1 \leq i \leq n} Z_i = \sum_{1 \leq j \leq m} \alpha_j U_n(s_j).$$

We use the Bernstein blocking technique, as in chapter 8. Let $p(n)$ and $q(n)$ be sequences of integers such that $p(n) = o(n)$ and $q(n) = o(p(n))$. Assume that the Euclidean division of n by $(p+q)$ gives a quotient k and a remainder r. For $i = 1, \ldots, k$, we define the interval $P_i = \{(p+q)(i-1)+1, \ldots, (p+q)i - q\}$ and if $r \neq 0$, $P_{k+1} = \{(p+q)k+1, \ldots, (p+q)k + r \vee p\}$. Q the set of indices that are not in one of the P_i. Note that the cardinal of Q is less than $(k+1)q$. For each block P_i $(1 \leq i \leq k+1)$ and Q, we define the partial sums:

$$u_i = \frac{1}{\sqrt{n}} \sum_{j \in P_i} Z_j, \qquad v = \frac{1}{\sqrt{n}} \sum_{j \in Q} Z_j.$$

We use lemma 8.2. We check the conditions for the sequence Z_j. To check (8.2.6), note that $\sigma_n^2 \leq \frac{1}{n} \sum_{r=0}^{n-1} \epsilon(r)$ and that $\mathbb{E} v^2 \leq \frac{(k+1)q}{n} \sum_{r=0}^{n-1} \epsilon(r)$.
Let us check (8.2.7). Using (10.2.4) with $\operatorname{Lip} G = \operatorname{Lip} H = t \max \alpha_j / \sqrt{n} \sigma_n$, $d_G = mpk$ and $d_H = mp$, we get

$$\left| \operatorname{Cov}\left(g\left(\frac{t}{\sigma_n} \sum_{i=1}^{j-1} u_i \right), h\left(\frac{t}{\sigma_n} u_j \right) \right) \right| \leq mp\,(k+1) \frac{t \sum_j \alpha_j}{\sigma_n} \epsilon(q).$$

and

$$\sum_{j=2}^{k+1} \left| \operatorname{Cov}\left(g\left(\frac{t}{\sigma_n} \sum_{i=1}^{j-1} u_i \right), h\left(\frac{t}{\sigma_n} u_j \right) \right) \right| \leq mp\,(k+1)^2 \frac{t \sum_j \alpha_j}{\sigma_n} \epsilon(q) = \mathcal{O}(n^{3/2} p^{-1} q^{-a}).$$

Taking $p = n^{5/6}$ and $q = n^{5/6a}$ gives a bound tending to 0.
To prove (8.2.8), it is sufficient to show that $\mathbb{E}|u_i|^4 = \mathcal{O}(k^{-2})$. But

$$\left(\sum_{j\in P_i} Z_j\right)^4 = \frac{p^2}{n^2}\left(\sum_{i=1}^m \alpha_i B_p(s_i)\right)^4 \leq \frac{p^2}{n^2} m^3 \sum_{i=1}^m \alpha_i^4 \left(B_p(s_i) - B_p(0)\right)^4,$$

and we conclude by applying Proposition 10.2 for $l = 2$ to the couples $(0, s_i)$.
In order to prove (8.2.9), note that (8.2.6) implies that

$$\lim_{n\to\infty} \frac{1}{\sigma_n^2} \mathrm{Var}\left(\sum_{i=1}^{k+1} u_i\right) = 1.$$

But

$$\begin{aligned}
\left|\mathrm{Var}\left(\sum_{i=1}^{k+1} u_i\right) - \sum_{i=1}^{k+1} E|u_i|^2\right| &\leq 2 \sum_{1\leq i\neq j\leq k} |\mathrm{Cov}(u_i, u_j)| \\
&\leq \frac{4(k+1)p}{n} \sum_{j=q}^{\infty} \epsilon(j) = \mathcal{O}(q^{-a+1}) = o(1).
\end{aligned}$$

Taking $p = n^{5/6}$ and $q = n^{5/6a}$ gives a bound tending to 0. □

Tightness of U_n. We use the criteria of Proposition 4.2. Define

$$\mathcal{F} = \{g_{s,t}/g_{s,t}(x) = \mathbf{1}\{x \leq t\} - \mathbf{1}\{x \leq s\} - F(t) + F(s); s, t \in [0,1]^d\}.$$

By definition $U_n(t) - U_n(s) = Z_n(g_{s,t})$ and $\|g_{s,t}\|_{P_Y,1} = \mathbb{E}|x_0(s,t)|$. Recalling that $a > 2d+1$, from the Rosenthal inequality (10.2.5) for $l = d+1$ we get

$$\|Z_n(g_{s,t})\|_p \leq C(\|g_{o,t}\|_{P_Y,1}^{1/r} + n^{1/q-1/2}),$$

with $p = q = 2l = 2d+2$, $r = 2a/(a-1)$. For the class of functions considered, the covering number $\mathcal{N}_{P_Y,1}(x, \mathcal{F}))$ is $\mathcal{O}(x^{-d})$ so that

$$\int_0^1 x^{(1-r)/r}(\mathcal{N}_{P_Y,1}(x,\mathcal{F}))^{1/p} dx \leq C \int_0^1 x^{-\frac{1}{2}\left(\frac{a+1}{a}+\frac{d}{d+1}\right)} dx.$$

Because $a > d+1$, the exponent is greater than -1, and the integral is finite. As $p = q = 2d+2$, the last condition holds directly. □

10.3 $\tilde{\alpha}$, $\tilde{\beta}$ and $\tilde{\phi}$-dependent sequences

Consider the three following conditions:
$(\mathcal{C}_1)$ There exists $\varepsilon > 0$ such that $\tilde{\alpha}_2(k) = \mathcal{O}\left(k^{-7/3-\varepsilon}\right)$ if $d = 1$, and $\tilde{\alpha}_2(k)$

$= \mathcal{O}\left(k^{-2d-\varepsilon}\right)$ if $d > 1$.
$(\mathcal{C}_2)$ There exists $\varepsilon > 0$ such that $\tilde{\beta}_2(k) = \mathcal{O}\left(k^{-2d-\varepsilon}\right)$.
$(\mathcal{C}_3)$ There exists $\varepsilon > 0$ such that $\tilde{\phi}_2(k) = \mathcal{O}\left(k^{-1-\varepsilon}\right)$.

Theorem 10.2. *If one of the conditions $(\mathcal{C}_i)$, $i = 1, \dots, 3$ holds, then the process U_n converges in distribution in $\mathcal{D}(\mathbb{R}^d)$ to the Gaussian process* $\mathbf{B}$.

Remark 10.2. *For $d = 1$, the condition $(\mathcal{C}_1)$ is better than the condition $\alpha_X(k) = \mathcal{O}(k^{-1-\sqrt{2}-\epsilon})$ given in Shao and Yu (1996) [175] for strongly mixing sequences (recall that $\alpha_X(k)$ has been defined in Section 1.2 of Chapter 1). In fact, in Theorem 7.3 of his book, Rio (2000) [161] has shown that the rate $\alpha_X(k) = \mathcal{O}(k^{-1-\epsilon})$ is sufficient for the weak convergence of the d-dimensional distribution function.*

Proof of Theorem 10.2. We keep the notations of Chapter 4. The finite dimensional convergence of U_n can be proved as before. Let us prove the tightness of U_n. Let $\mathcal{F} = \{x \mapsto \mathbf{1}_{x \le t}\,,\, t \in \mathbb{R}^d\}$, and let $\mathcal{G} = \{f - h\,,\, f, h \in \mathcal{F}\}$. We have to prove that the process $\{Z_n(f), f \in \mathcal{F}\}$ is asymptotically tight, that is there exists a semi metric ρ on $\mathcal{F}$ such that $(\mathcal{F}, \rho)$ is totally bounded, and, for every $\epsilon > 0$,

$$\lim_{\delta \to 0} \limsup_{n \to \infty} \mathbb{P}\Big(\sup_{\rho(f,g) \le \delta,\, f,g \in \mathcal{F}} |Z_n(f) - Z_n(g)| > \epsilon \Big) = 0\,. \tag{10.3.1}$$

Since $\mathcal{N}_{Q,1}(x, \mathcal{F}) = \mathcal{O}(x^{-d})$ for any finite measure Q on $\mathbb{R}^d$, the set $(\mathcal{F}, \|\cdot\|_{Q,1})$ is totally bounded. Consequently, the property (10.3.1) follows from (4.5.2) by applying Markov's inequality.
Let us prove that condition $(\mathcal{C}_1)$ implies (4.5.2). For any s, t in $\mathbb{R}^d$, let $f_{s,t}(x) = \mathbf{1}_{x \le t} - \mathbf{1}_{x \le s}$ and $\tilde{f}_{s,t}(x) = f_{s,t}(x) - \int f_{s,t}(x) P(dx)$. With proposition 5.6 for $(\tilde{f}_{s,t}(X_i))_{i \in \mathbb{Z}}$ for any $p \ge 1$, the quantity $\|Z_n(\tilde{f}_{s,t})\|_p$ is bounded by

$$\sqrt{pV_\infty} + n^{1/3-1/2}\Big(3p^2(\|\tilde{f}_{s,t}(X_0)^3\|_{p/3} + M_1(p) + M_2(p) + M_3(p))\Big)^{1/3}, \tag{10.3.2}$$

where V_∞, $M_1(p)$, $M_2(p)$ and $M_3(p)$ are defined in Proposition 5.6. Let P be the law of X_0. Use inequality (5.2.5), then for any $r > 2$, $g \in \mathcal{G}$,

$$\|Z_n(g)\|_p \le 4(p\,\|g\|_{P,1})^{1/r}\Big(\sum_{k \ge 0} (\tilde{\alpha}_1(k))^{(r-2)/r}\Big)\Big) + n^{1/3-1/2}\Big(3p^2\Big(1 + 10\sum_{k=1}^{+\infty} k(\tilde{\alpha}_2(k))^{3/p}\Big)\Big)^{1/3}. \tag{10.3.3}$$

We then apply Proposition 4.2 with $q = 3$. Recall that $\mathcal{N}_{P,1}(x, \mathcal{F}) = \mathcal{O}(x^{-d})$. If $d = 1$, we can take $r = 7/2$ and $p > 7/2$ such that (4.5.2) holds under C_1. If

$d > 1$ we can take $r = 3$ and $p > 3d$ such that (4.5.2) holds under $(\mathcal{C}_1)$.
Let us prove that condition $(\mathcal{C}_2)$ implies (4.5.2). Define the measure Q on $\mathbb{R}^d$ by

$$Q(dx) = B(x)\,P(dx) = \left(1 + 4\sum_{k=1}^{+\infty} b_k(x)\right) P(dx), \tag{10.3.4}$$

where $b_k(x)$ is the function from $\mathbb{R}^d$ to $[0,1]$ such that $b(\sigma(X_0), X_k) = b_k(X_0)$ and P is the law of X_0. Note that Q is finite as soon as $\sum_{k=1}^{+\infty} \beta_1(k)$ is finite. Applying the inequality (5.2.6), we obtain that for any g in $\mathcal{G}$,

$$\|Z_n(g)\|_p \le (p\,\|g\|_{Q,1})^{1/2} + n^{1/3-1/2}\left(3p^2\left(1 + 10\sum_{k=1}^{+\infty} k\left(\tilde{\beta}_2(k)\right)^{3/p}\right)\right)^{1/3}.$$

We then apply Proposition 4.2 with $r = 2$ and $q = 3$. Since $\mathcal{N}_{Q,1}(x, \mathcal{F}) = \mathcal{O}(x^{-d})$, we can take $p > 3d$ such that (4.5.2) holds under $(\mathcal{C}_2)$.

Let us prove that condition $(\mathcal{C}_3)$ implies (4.5.2). Applying Proposition 5.7 to the sequence $(\tilde{f}_{s,t}(X_i))_{i\in\mathbb{Z}}$, we obtain, for any $p \ge 1$,

$$\begin{aligned}\|Z_n(\tilde{f}_{s,t})\|_p &\le (p(V_\infty + 2M_0(p)))^{1/2}\\ &\quad + n^{1/3-1/2}\Big(3p^2\Big(\|\tilde{f}(X_0)^3\|_{p/3} + \tilde{M}_1(p) + \tilde{M}_2(p) + M_3(p)\Big)\Big)^{1/3},\end{aligned}$$

where V_∞, $M_0(p)$, $\tilde{M}_1(p)$, $\tilde{M}_2(p)$ and $M_3(p)$ are defined in Proposition 5.7. Applying the inequality (5.2.7), we get that for any g in $\mathcal{G}$,

$$\begin{aligned}\|Z_n(g)\|_p &\le (p\|g\|_{Q,1})^{1/2} + \Big(2p\sum_{k=N}^{+\infty}\tilde{\phi}_2(k)\Big)^{1/2}\\ &+ n^{1/3-1/2}\Big(3p^2\Big(1 + 2\sum_{k=1}^{N} k\tilde{\phi}_2(k) + 4\sum_{k=1}^{\infty}\tilde{\phi}_2(k)(k\wedge N) + 4\sum_{k=1}^{+\infty}\tilde{\phi}_2(k)\Big)\Big)^{1/3}.\end{aligned} \tag{10.3.5}$$

We take now $N = n^\alpha$ with $\alpha = 1/(2+\varepsilon)$. If $(\mathcal{C}_3)$ holds, we infer from (10.3.5) that there exists some positive constant C such that, for any g in $\mathcal{G}$,

$$\|Z_n(g)\|_p \le C\|g\|_{Q,1}^{1/2} + Cn^{-\varepsilon/(4+2\varepsilon)}.$$

To conclude we apply Proposition 4.2 with $r = 2$ and $q = 2 + \varepsilon$. Since $\mathcal{N}_{Q,1}(x, \mathcal{F}) = \mathcal{O}(x^{-d})$, (4.5.2) holds under $(\mathcal{C}_3)$. □

10.4 θ and τ-dependent sequences

Consider the three following conditions:

($\mathcal{C}_4$) Each component of X_1 has a bounded density and there exists $\varepsilon > 0$ such that $\theta_{1,2}(k) = \mathcal{O}\left(k^{-14/3-\varepsilon}\right)$ if $d = 1$, and $\theta_{1,2}(k) = \mathcal{O}\left(k^{-4d-\varepsilon}\right)$ if $d > 1$.

($\mathcal{C}_5$) Each component of X_1 has a bounded density and there exists $\varepsilon > 0$ such that $\tau_{1,2}(k) = \mathcal{O}\left(k^{-4d-\varepsilon}\right)$.

($\mathcal{C}_6$) Each component of X_1 has a bounded density and there exists $\varepsilon > 0$ such that $\tau_{\infty,2}(k) = \mathcal{O}\left(k^{-2-\varepsilon}\right)$.

Theorem 10.3. *If one of the conditions ($\mathcal{C}_i$), $i = 4, \ldots, 6$ holds, then the process U_n converges in distribution in $\mathcal{D}(\mathbb{R}^d)$ to the Gaussian process* $\mathbf{B}$.

Proof of Theorem 10.3. The result is immediate by using Theorem 10.2 and Lemma 5.1. □

10.5 Empirical copula processes

Copulas describe the dependence structure between some random vectors. They have been introduced a long time ago by Sklar (1959) [178] and have been rediscovered recently, especially for their applications in finance and biostatistics. Briefly, a d-dimensional copula is a distribution function on $[0,1]^d$, whose marginal distributions are uniform and that summarizes the dependence "structure" independently of the specification of the marginal distributions.

To be specific, consider a random vector $\mathbf{X} = (X_1, \ldots, X_d)$ in $\mathbb{R}^d$, whose joint distribution function is F and whose marginal distribution functions are denoted by F_j, $j = 1, \ldots, d$. Then there exists a unique copula C defined on the product of the values taken by the r.v. $F_j(X_j)$, such that

$$C(F_1(x_1), \ldots, F_d(x_d)) = F(x_1, \ldots, x_d),$$

for any $\mathbf{x} = (x_1, \ldots, x_d) \in \mathbb{R}^d$. C is called the copula associated with $\mathbf{X}$. When F is continuous, it is defined on $[0,1]^d$, with an obvious extension to $\overline{\mathbb{R}}^d$. When F is discontinuous, there are several choices to expand C on the whole $[0,1]^d$ (see Nelsen (1999) [134] for a complete theory).

The natural empirical counterpart of C is the so-called empirical copula, defined by

$$C_n(\mathbf{u}) = F_n(F_{n,1}^{-1}(u_1), \ldots, F_{n,d}^{-1}(u_d)),$$

for every $u_1, \ldots, u_d$ in $[0,1]$, where F_n denotes the empirical process as in Definition 10.0.1 and $F_{n,i}$ the empirical process of the i-th marginal distribution. We use the usual "generalized inverse" notations, for every $j = 1, \ldots, d$, $F_j^{-1}(u) = \inf\{t \,/\, F_j(t) \geq u\}$.

Empirical copulas have been introduced by Deheuvels (1979, 1981a, 1981b), [51],[52],[53] in an i.i.d. framework. This author studied the consistency of C_n

and the limiting behavior of $n^{1/2}(C_n - C)$ under the strong assumption of independence between margins. Fermanian *et al.*(2002) [86] proved some functional CLT for this empirical copula process in a more general framework and provide some extensions. Note that the results of [86] are available under the sup-norm and outer expectations assumptions, as in van der Vaart and Wellner (1996) [183].

Assume that the process $(\mathbf{Y}_i)_{i\in\mathbb{Z}}$, $\mathbf{Y} = (F_1(X_1), \ldots, F_d(X_d))$ is weakly dependent. Note that the covariance structure of the limit process $\mathbf{B}$ depends not only on the copula C (via the term associated with $i = 0$ *e.g.*), but also on the joint law between $\mathbf{X}_0$ and $\mathbf{X}_i$, for every i. This is different from the i.i.d. case, where $\mathbf{B}$ becomes a Brownian bridge whose covariance structure is a function of C only. Actually, the covariances of $\mathbf{B}$ depend here on every successive copulas of the random vectors $(\mathbf{X}_0, \mathbf{X}_i)$. We can state:

Theorem 10.4. *If $(\mathbf{Y}_i)_{i\in\mathbb{Z}}$ is weakly dependent and if the empirical process of $(\mathbf{Y}_i)_{i\in\mathbb{Z}}$ converges in distribution in $\mathcal{D}([0,1]^d)$ to a Gaussian process $\mathbf{B}$, if C has some continuous first partial derivatives, then the process $n^{1/2}(C_n - C)$ tends weakly to a Gaussian process $\mathbf{G}$ in $\mathcal{D}([0,1]^d)$. Moreover, this process has continuous sample paths and can be written as*

$$\mathbf{G}(\mathbf{u}) = \mathbf{B}(\mathbf{u}) - \sum_{j=1}^{d} \partial_j C(\mathbf{u})\mathbf{B}(u_1, \ldots, u_{j-1}, 0, u_{j+1}, \ldots, u_d), \qquad (10.5.1)$$

for every $\mathbf{u} \in [0,1]^d$.

Note that the covariance structure of $n^{1/2}(C_n - C)$ is involved, because of both (10.0.3) and (10.5.1).

Proof of theorem 10.4. The proof is directly adapted from Fermanian *et al.*(2002) [86]. Briefly, we can assume that the law of $\mathbf{X}$ is compactly supported on $[0,1]^d$, eventually by working with $\mathbf{Y} = (F_1(X_1), \ldots, F_d(X_d))$. Indeed, it can be proved that the empirical copulas associated with $\mathbf{Y}$ and $\mathbf{X}$ are equal on all the points $(i_1/n, \ldots, i_d/n)$, $i_1, \ldots, i_d$ in $\{0, \ldots, n\}$ (lemma 3 in [86]), thus on $[0,1]^d$ as a whole.

Consider the usual norm on $l^\infty([0,1])$ and the Skohorod metric δ on $D([0,1]^d$. Define the mappings

$$\phi_1 : \begin{cases} (D([0,1]^d), \delta) \to (D([0,1]^d), \delta) \times (D([0,1]), \delta)^{\otimes d} \\ F \mapsto (F, F_1, \ldots, F_d) \end{cases}$$

$$\phi_2 : \begin{cases} (D([0,1]^d), \delta) \times (D([0,1]), \delta)^{\otimes d} \to (l^\infty([0,1]^d)) \times (l^\infty([0,1]))^{\otimes d} \\ (F, F_1, \ldots, F_d) \mapsto (F, F_1^{-1}, \ldots, F_d^{-1}) \end{cases}$$

$$\phi_3 : \begin{cases} (l^\infty([0,1]^d) \times l^\infty([0,1]))^{\otimes d} \to (l^\infty([0,1]^d)) \\ (F, G_1, \ldots, G_d) \mapsto F(G_1, \ldots, G_d) . \end{cases}$$

Clearly, ϕ_1 is Hadamard-differentiable because it is linear. Moreover, ϕ_2 is Hadamard-differentiable tangentially to the corresponding product of continuous functions by applying theorem 3.9.23 in van der Vaart and Wellner (1996) [183]. Note that, for any function $h \in C([0,1])$, the convergence of a sequence h_n towards h in $(D([0,1]), \delta)$ is equivalent to the convergence in $(D([0,1]), \|\cdot\|_\infty)$. Thus, working with the Skorohod metric is not an hurdle here. At last, ϕ_3 is Hadamard-differentiable by applying theorem 3.9.27 in [183]. Thus, the chain rule applies : $\phi = \phi_3 \circ \phi_2 \circ \phi_1$ is Hadamard-differentiable tangentially to $C([0,1]^d)$. The result follows by applying the functional Δ-method to the empirical process of $\mathbf{Y}$ and to the function ϕ (see theorem 3.9.4 in [183]). □

10.6 Random fields

In this section, we give rates of convergence in the central limit theorem for a η- or κ-weak dependent field ξ. Assume that the density of the variable ξ_i is bounded by C_ξ. Following lemma 4.1, we get $(\mathcal{I}, c)$-weak dependence with

- for η-weak dependence, $\epsilon(r) = \sqrt{\eta(r)}$ and $c(d_f, d_g) = 2\sqrt{8C_\xi}(d_f + d_g)$.
- for κ-weak dependence: $\epsilon(r) = (\kappa(r))^{\frac{1}{3}}$ and $c(d_f, d_g) = 2(8C_\xi)^{\frac{2}{3}}(d_f+d_g)^{\frac{4}{3}}$.

For the sake of simplicity, we shall assume that the process takes its values in $[0,1]$. This may be achieved by using the quantile transform. Let $(S, |\cdot|_S)$ be the space of càdlàg functions $D([0,1])$ with the Skorohod metric. Let π denote the Prohorov distance between distribution functions on $(S, |\cdot|_S)$. If X and Y are two processes on S, we also denote $\pi(X, Y)$ the Prohorov distance between their distributions.

Central limit theorem for the empirical process

Let U_n be the normalized version of the empirical process defined by (10.0.2) with respect to the closed ball B_N of radius N and cardinality $n = \#B_N = (2N+1)^d$. Denote $g_{s,t}(u) = \mathbf{1}_{s<u\le t} - F(t) + F(s)$ the interval counting functions. Denote $x_i(t) = g_{0,t}(\xi_i)$.

Theorem 10.5. *Assume that $(\xi_n)_{n\in T}$ is a centered and η- or κ-dependent process. Assume that the density of the variable ξ_i is bounded by C_ξ. Let $X(t)$ be the centered normal process with variance $\Sigma_{s,t} = \sum_{j\in T} \mathrm{Cov}(x_0(s), x_j(t))$. Assume that there exist $C > 0$ and $b > 0$ such that $\epsilon(r) \le Ce^{-br}$, then*

$$\pi(U_n, X) = \mathcal{O}\left(n^{-\alpha_0}\log(n)^{-\beta_0}\right),$$

where

$$\begin{aligned}\alpha_0 &= \frac{1}{8d+24},\\ \beta_0 &= \frac{10d^2+39d+28}{8d+24}.\end{aligned}$$

The proof is based on the well known result on the Prohorov distance:

Lemma 10.2. *Let δ be a positive real and D be a finite subset $\{x_1, \ldots, x_m\} \subset [0,1]$, such that every x in $[0,1]$ is in a δ-neighborhood of some x_i. Let X and Y be two distributions on S. Define the δ-oscillation of X*

$$w_X(\delta) = \sup_{\|x-y\|<\delta} \left(\|X(x) - X(y)\|\right),$$

and $\varepsilon_X(\delta) = \inf\{\varepsilon \in \mathbb{R} \,/\, \mathbb{P}(w_X(\delta) > \varepsilon) \le \varepsilon\}$. The Prohorov distance between X and Y is bounded by:

$$\pi(X,Y) \le \varepsilon_X(\delta) + \varepsilon_Y(\delta) + \pi(X_D, Y_D),$$

where X_D is the finite dimensional distribution of X on the subset D.

We need to compute bounds for the δ-oscillations and distance between laws. These are based on moment inequalities for the process U_n.

Proposition 10.3. *Assume that $\epsilon(r) \le Ce^{-br}$, with $b > 0$. For (s,t) such that $|t-s| < C_\xi e^{-4b}/C$ and $|t-s| < Ce^{3b}/C_\xi$:*

$$\begin{aligned}\mathbb{E}(U_n(t) - U_n(s))^{2l} \le{}& \frac{(4l-2)!}{(2l-1)!}\left(6(1 \vee b^{-2})\log\left(1/|t-s|\right)\right)^{2dl}\\ &\times \left(\left(2(2d)!C_\xi|t-s|\right)^l + (2l)!(2ld)!n^{1-l}C_\xi|t-s|\right). \quad (10.6.1)\end{aligned}$$

Note that $U_n(t) - U_n(s) = \frac{1}{\sqrt{n}}\sum_{k\in B_N} g_{s,t}(\xi_k)$. The proposition is a consequence of the bound of the covariance of quantities depending on the functions $g_{s,t}(\xi_k)$.

Proof of proposition 10.3. We adapt the proof of Proposition 10.2 to the series $(g_{s,t}(\xi_k))_{k\in B_N}$. For a sequence $\mathbf{k} = (k_1, \ldots, k_q)$ of elements of T, define $\xi_{\mathbf{k}} = (\xi_{k_1}, \ldots, \xi_{k_q})$ and when s and t are fixed, $\Pi_{\mathbf{k}} = \prod_{i=1}^q g_{s,t}(\xi_{k_i})$. For any integer $q \ge 1$, set:

$$A_q(N) = \sum_{\mathbf{k}\in B_N^q} \left|E\left(\Pi_{\mathbf{k}}\right)\right|, \quad (10.6.2)$$

then,

$$|\mathbb{E}(U_n(s) - U_n(t))^{2l}| \le (2l)!n^{-l}A_{2l}(N). \quad (10.6.3)$$

Let $q \geq 2$. For a finite sequence $\mathbf{k} = (k_1, \ldots, k_q)$ of elements of T, the gap is defined by the max of the integers r such that the sequence may be split into two non-empty subsequences $\mathbf{k}^1$ and $\mathbf{k}^2 \subset \mathbb{Z}^d$ whose mutual distance equals r ($d(\mathbf{k}^1, \mathbf{k}^2) = \min\{\|i-j\|_1 / i \in \mathbf{k}^1, j \in \mathbf{k}^2\} = r$). If the sequence is constant, its gap is 0. Define the set $G_r(q,N) = \{\mathbf{k} \in B_N^q$ and the gap of $\mathbf{k}$ is $r\}$. Sorting the sequences of indices by their gap:

$$\begin{aligned} A_q(N) &\leq \sum_{k_1 \in B_N} \mathbb{E}|g_{s,t}(\xi_{k_1})|^q + \sum_{r=1}^{2N} \sum_{\mathbf{k} \in G_r(q,N)} \left|\mathrm{Cov}\left(\Pi_{\mathbf{k}^1}, \Pi_{\mathbf{k}^2}\right)\right| \\ &+ \sum_{r=1}^{2N} \sum_{\mathbf{k} \in G_r(q,N)} \left|\mathbb{E}\left(\Pi_{\mathbf{k}^1}\right) \mathbb{E}\left(\Pi_{\mathbf{k}^2}\right)\right|. \end{aligned} \tag{10.6.4}$$

Define

$$B_q(N) = \sum_{k_1 \in B_N} \mathbb{E}|g_{s,t}(\xi_{k_1})|^q + \sum_{r=1}^{2N} \sum_{\mathbf{k} \in G_r(q,N)} \left|\mathrm{Cov}\left(\Pi_{\mathbf{k}^1}, \Pi_{\mathbf{k}^2}\right)\right|.$$

We get

$$A_q(N) \leq B_q(N) + \sum_{m=1}^{q-1} A_m(N) A_{q-m}(N).$$

To build a sequence $\mathbf{k}$ belonging to $G_r(q,N)$, we first fix one of the n points of B_N. We choose a second point on the ℓ^1-sphere of radius r centered on the first point. The third point is in a ball of radius r centered on one of the preceding points, and so on... Thus

$$\#G_r(q,N) \leq n \cdot 2d(2r+1)^{d-1} \cdot 2(2r+1)^d \cdots (q-1)(2r+1)^d \leq ndq!(3r)^{d(q-1)-1}.$$

We use Lemma 4.1 to deduce:

$$B_q(N) \leq n\Big(C_\xi |t-s| + dq! \sum_{r=1}^{2N} (3r)^{d(q-1)-1} \min(\epsilon(r), C_\xi |t-s|)\Big).$$

Let R be an integer to be chosen later.

$$B_q(N) \leq nd3^{d(q-1)} q! \Big(C_\xi |t-s| \sum_{r=0}^{R-1} r^{d(q-1)-1} + C \sum_{r=R}^{\infty} r^{d(q-1)-1} e^{-br}\Big).$$

Comparing summations with integrals we get

$$\begin{aligned} B_q(N) &\leq n(3(1 \vee b^{-1}))^{d(q-1)} q! (d(q-1))! \times \\ &\quad \times R^{d(q-1)} C_\xi |t-s| \left(1 + \frac{Ce^{4b}}{C_\xi |t-s|} e^{-b(R+1)}\right). \end{aligned}$$

Choose R as the integer part of $\frac{1}{b}\log\left(Ce^{4b}/C_\xi|t-s|\right)$ and assume that $(s,t)\in T$ are such that $|t-s|\le e^{-4b}C_\xi/C$ and $|t-s|\le e^{3b}C/C_\xi$. Then $R\ge 1$ and

$$B_q(N)\le\left(6(1\vee b^{-2})\log\left(1/|t-s|\right)\right)^{dq} nC_\xi|t-s|\,q!\,(dq)!, \qquad (10.6.5)$$

so that $B_q(N)$ is bounded by a function M^qV_q that satisfies condition (4.3.24). Then

$$A_{2l}(N)\le\frac{(4l-2)!}{(2l)!(2l-1)!}\left(6(1\vee b^{-2})\log\left(1/|t-s|\right)\right)^{2dl}$$
$$\left(\left(2(2d)!nC_\xi|t-s|\right)^l+(2l)!(2ld)!nC_\xi|t-s|\right),$$

and (10.6.1) is proved. □

Oscillations of the empirical process

Using Proposition 10.3, we give a bound for the modulus of continuity of U_n.

Proposition 10.4. *If $\epsilon(r)\le Ce^{-br}$ with $b>0$, then for $\delta\ge 1/n$:*

$$\varepsilon_{U_n}(\delta)\le K_1(C_\xi,b,d)\delta^{1/2}\log^{d+1}(1/\delta), \qquad (10.6.6)$$

where $K_1(C_\xi,b,d)=8\left(6(1\vee b^{-2})\right)^d\left(2(2d)!C_\xi\right)^{\frac{1}{2}}$.

Proof of proposition 10.4. We show that exponential moment of $U_n(t)-U_n(s)$ are finite and use Stroock's method to find the oscillation.

Lemma 10.3. *Let $f(u)=|u|^{1/2}\log^d(1/u)$. Assume that $\epsilon(r)\le Ce^{-br}$ with $b>0$, and that $\delta\ge 1/n$. Then there exists a constant c_0 such that for every $c<c_0$ and every (s,t) such that $|t-s|\le\delta$:*

$$\mathbb{E}\Big(\exp\Big(c\frac{|U_n(t)-U_n(s)|}{f(t-s)}\Big)\Big)\le B(c)<+\infty.$$

Proof of lemma 10.3 Using the moment inequality (10.6.1) and Stirling's formula:

$$\mathbb{E}\left(\frac{|U_n(t)-U_n(s)|}{f(t,s)}\right)^{2p}\le p^{2p}\left(8e^2C_\xi(2d)!\left(6(1\vee b^{-2})\right)^{2d}\right)^p.$$

$$\begin{aligned}\mathbb{E}\left(\exp\left(c\frac{|U_n(t)-U_n(s)|}{f(t-s)}\right)\right) &\le \sum_{k=0}^{\infty}\frac{c^k}{k!}\left(\mathbb{E}\left(\frac{|U_n(t)-U_n(s)|}{f(t,s)}\right)^{2k}\right)^{\frac{1}{2}}\\ &\le \sum_{k=0}^{\infty}\frac{c^kk^k}{k!}\left(8e^2C_\xi(2d)!\left(6(1\vee b^{-2})\right)^{2d}\right)^{\frac{k}{2}}.\end{aligned}$$

For $c_0 = \left(6(1 \vee b^{-2})\right)^{-d} \left(8e^2 C_\xi (2d)!\right)^{-\frac{1}{2}}$, the lemma is true.
We apply a lemma of Garsia (1965) [90]. Let $c = c_0/2$, $B(c)$ as in lemma 10.3, $\psi(u) = e^{cu} - 1$ and $0 < \delta < e^{-2d}$. The lemma says that if

$$Y(\omega) = \int_0^\delta \int_0^\delta \psi\left(\frac{|U_n(t,\omega) - U_n(s,\omega)|}{f(|t-s|)}\right) dsdt < \infty,$$

then $|U_n(t,\omega) - U_n(s,\omega)| \leq 8\psi^{-1}(4Y(\omega)/\delta^2) f(\delta)$.
Using Markov inequality and the fact that $\mathbb{E}(Y) \leq B(c)\delta^2$:

$$\mathbb{P}(|U_n(t) - U_n(s)| \geq \lambda) \leq \mathbb{P}\left(Y \geq \frac{\delta^2}{4}\psi\left(\frac{\lambda}{8f(\delta)}\right)\right) \leq \frac{4B(c)}{\psi\left(\frac{\lambda}{8f(\delta)}\right)}.$$

For $\lambda = (4/c) f(\delta) \log(1/\delta)$, $\mathbb{P}(|U_n(t) - U_n(s)| \geq \lambda) \leq 4B(c)\delta^{1/2}/(1-\delta^{1/2})$. For δ sufficiently small, this term is less than λ, so that $\varepsilon_{U_n}(\delta) \leq K_1 \delta^{1/2} \log^d(1/\delta)$ with $K_1 = 4/c$. □

Oscillations of the limit process

Proposition 10.5. *If $\epsilon(r) \leq Ce^{-br}$ with $b > 0$, then etting $K_3(C_\xi, b, d) = (C_\xi d!(2d+7))^{1/2} 2^d \left(1 \vee b^{-1}\right)^d$,*

$$\varepsilon_X(\delta) \leq K_3(C_\xi, b, d)\delta^{1/2} \log^{(d+1)/2}(1/\delta). \tag{10.6.7}$$

Proof of proposition 10.5. We use a chaining argument to bound the δ-oscillations of the non-stationary Gaussian limit process X. Define a semi-metric $\bar{\rho}$ on $[0,1]$ by $\bar{\rho}(s,t)^2 = \operatorname{Var}(X(t) - X(s))$. As a Gaussian process, X satisfies the exponential inequality:

$$\mathbb{P}(|X(t) - X(s)| \geq \lambda \bar{\rho}(s,t)) \leq \exp\{-\lambda^2/2\}.$$

By definition, $\bar{\rho}(s,t)^2 = \sum_{j \in T} \operatorname{Cov}(x_0(t) - x_0(s), x_j(t) - x_j(s))$. Using the Cauchy-Schwarz inequality:

$$|\operatorname{Cov}(x_0(t) - x_0(s), x_r(t) - x_r(s))| \leq \operatorname{Var}(x_0(t) - x_0(s)) \leq C_\xi |t-s|,$$

and by definition of $\epsilon(r)$,

$$|\operatorname{Cov}(x_0(t) - x_0(s), x_r(t) - x_r(s))| \leq \epsilon(r).$$

For a metric ν over $[0,1]$, we denote $N_\nu(\varepsilon)$ the covering number (see Pollard (1984) [151], p.143). It is the cardinality of the smallest set S of points of $[0,1]$, so that for any t, $\nu(t,S) \leq \varepsilon$. The corresponding covering integral is

$$J_\nu(\varepsilon) = \int_0^\varepsilon \left(2\log\left(N_\nu(u)^2/u\right)\right)^{1/2} du.$$

Computing as in (10.6.5), it is easy to show that

$$\bar{\rho}(s,t)^2 = \sum_{r=0}^{\infty}(2r+1)^{d-1}\epsilon(r) \wedge C_\xi|t-s| \leq K^2|t-s|\log^d(1/|t-s|).$$

where $K = 2\sqrt{C_\xi d!}(1 \vee b^{-1})^d$. Let $\varepsilon > 0$. The inclusion of the balls of the two metrics implies that $N_{\bar{\rho}}(K\varepsilon^{1/2}\log^{d/2}(1/\varepsilon)) \leq N_{|\cdot|}(\varepsilon)$. Thus $N_{\bar{\rho}}(u) \leq 2^{d+1}(K/u)^2\log^{d+1}(K/u) \leq 2^d(K/u)^{d+2}$. The corresponding covering integral is:

$$\begin{aligned} J_{\bar{\rho}}(\delta) &\leq \int_0^\delta \left(4\log\left(2^{d+1}(K/u)^{d+3}\right) + 2\log(1/u)\right)^{1/2} du \\ &\leq 2\delta\log^{1/2}\left(2^{d+1}K^{d+3}\right) + (4d+14)^{1/2}\log^{-1/2}(1/\delta)\int_0^\delta \log(u)du \\ &\leq 2\delta\log^{1/2}\left(2^{d+1}K^{d+3}\right) + (4d+14)^{1/2}\delta\log^{1/2}(1/\delta). \end{aligned}$$

Taking $\varepsilon = K\delta^{1/2}\log^{d/2}(1/\delta)$:

$$\mathbb{P}\left(\max_{|t-s|\leq\delta}|X(t)-X(s)| \geq 26J_{\bar{\rho}}(K\delta^{1/2}\log^{d/2}(1/\delta))\right) \leq K\delta^{1/2}\log^{d/2}(1/\delta).$$

For a sufficiently small δ,

$$\varepsilon_X(\delta) \leq J_{\bar{\rho}}(K\delta^{1/2}\log^{d/2}(1/\delta)) \leq K(2d+7)^{1/2}\delta^{1/2}\log^{(d+1)/2}(1/\delta).$$

Distance between the finite dimensional laws

Let $m \in \mathbb{N}$. Let $D = \{t_1, \ldots, t_m\} \subset [0,1]$. We denote z^i the m-vector $(x_i(t_1), \ldots, x_i(t_m))$. Define the partial sum s_n:

$$s_n = \frac{1}{\sqrt{n}}\sum_{i\in B_N} z^i.$$

Let Y be a centered Gaussian m-vector, whose covariance matrix is $\Sigma_D = (\Sigma_{s,t})_{s,t\in D}$. We bound the Prohorov distance between s_n and Y. For a real ε such that $0 \leq \varepsilon \leq 1$, we define the class of functions $\mathcal{F}_\varepsilon$ by

$$\mathcal{F}_\varepsilon = \{f \in \mathcal{C}_b^3(\mathbb{R}^m)/\ 0 \leq f \leq 1, ||f^{(i)}||_\infty \leq 2m^{-1/2}\varepsilon^{-i} \text{ for } i = 1,2 \text{ or } 3\}, \tag{10.6.8}$$

where $f^{(i)}$ is the differential of i-th order of f and the norm $||\cdot||_\infty$ is the operator sup-norm w.r.t to the norm $\|\cdot\|_1$. A bound of the Prohorov distance between s_n and Y is given by:

$$\pi(s_n, Y) \leq 4m^{1/2}\varepsilon(1+\log^{1/2}(\varepsilon)) + 2\sup_{F_\varepsilon}|\mathbb{E}(f(s_n) - f(Y)|). \tag{10.6.9}$$

Proposition 10.6. *Define m, p and q as functions of n which converge to infinity with n, negligible with respect to n, q negligible with respect to p, and ε as a function of n which converges to zero as n tends to infinity. For a sufficiently large n, in the case of η-dependence:*

$$\begin{aligned}
\pi(s_n, Y) \leq\ & 4m^{1/2}\varepsilon(1+\log^{1/2}(1/\varepsilon)) && (10.6.10)\\
+\ & K_1\varepsilon^{-1}m^{1/2}q^{1/2}p^{-1/2} && (10.6.11)\\
+\ & 2K_5 m^{1/2}\varepsilon^{-2}n^{3/2}\epsilon(q) && (10.6.12)\\
+\ & 2K_6 m^{3/2}\varepsilon^{-3}n\epsilon(q) && (10.6.13)\\
+\ & 2K_7 m\varepsilon^{-3}p^{d/2}n^{-1/2} && (10.6.14)\\
+\ & 2K_8 m^{3/2}\varepsilon^{-2}p^{-1}q. && (10.6.15)
\end{aligned}$$

In the case of κ-dependence, the terms involving K_5 and K_6 are replaced by:

$$\begin{aligned}
+\ & 2K_5 m^{1/3}p^{d/3}\varepsilon^{-7/3}n^{4/3}\epsilon(q)\\
+\ & 2K_6 m^{4/3}p^{d}\varepsilon^{-11/3}n\epsilon(q).
\end{aligned}$$

The values of the constants K_i are given in the proof.

Proof of proposition 10.6. We use the Bernstein blocking technique (Bernstein, 1939, [13]. Assume that the Euclidean division of n by $(p+q)$ gives a quotient a and a remainder r. Denote $\overline{j} = (j, \ldots, j)$. Define $K = \{-a-1, \ldots, a+1\}^d$; for $i \in \{-a, \ldots, a\}^d$, we define the blocks $P_i = [(p+q)(i-\overline{1}), \ldots, (p+q)i - q\overline{1}]$. These blocks are separated with bands of width q. We complete the construction with at most $4a+4$ incomplete blocks on the boundary, also separated with bands of width q, and associate each of them with a corresponding index in the boundary of K. Denote Q the set of indices that are in the separating bands. Note that the cardinality of Q is less than $d(2a+1)qp^{d-1}$. We order the set of blocks P by the lexicographic order of their index in K. We define the variables $(u_i)_{i=1,\ldots,(2a+1)^d}$ and v exactly as in section 8.2.2. Consider an independent sequence of centered Gaussian vectors $(y^i)_{i=1,\ldots,k}$, such that each y^i has the same covariance matrix as u^i. Let $\varepsilon > 0$ and $f \in \mathcal{F}_\varepsilon$ defined by (10.6.8). We decompose

$$\begin{aligned}
f(s_n) - f(Y) =\ & f(s_n) - f\left(\sum_{i=1}^k u^i\right) && (10.6.16)\\
+\ & f\left(\sum_{i=1}^k u^i\right) - f\left(\sum_{i=1}^k y^i\right) && (10.6.17)\\
+\ & f\left(\sum_{i=1}^k y^i\right) - f(Y). && (10.6.18)
\end{aligned}$$

Consider the left hand side of (10.6.16).

$$\begin{aligned}\left|\mathbb{E}\left(f(s_n)-f\left(\sum_{i=1}^k u^i\right)\right)\right| &\leq \mathbb{E}\left(\left|f\left(\sum_{i=1}^k u^i+v\right)-f\left(\sum_{i=1}^k u^i\right)\right|\right)\\ &\leq 2m^{-1/2}\varepsilon^{-1}\sum_{s=1}^m \mathbb{E}(|v_s|)\\ &\leq 2m^{1/2}\varepsilon^{-1}\mathbb{E}(|v_1|^2)^{1/2}.\end{aligned}$$

Because $\epsilon(r)$ is decreasing,

$$\begin{aligned}\mathbb{E}(|v_1|^2) &= \frac{1}{n}\sum_{i,j\in Q}|\mathrm{Cov}\,(x_i(t_1),x_j(t_1))| \leq \frac{1}{n}\sum_{i,j\in Q}\epsilon(\|i-j\|)\\ &\leq \frac{\#Q}{n}\sum_{r=0}^N (2r+1)^{d-1}\epsilon(r) \leq K_1\frac{\#Q}{n} \leq K_1\frac{q}{p},\end{aligned}$$

where K_1 is a constant depending on the sequence $\epsilon(r)$. We get

$$\left|\mathbb{E}\left(f(s_n)-f\left(\sum_{i=1}^k u^i\right)\right)\right| \leq K_1(mq/p)^{1/2}\varepsilon^{-1}. \tag{10.6.19}$$

Now, we apply Lemma 7.1 to the difference (10.6.17). We first, give a bound for T_1. Assume that ξ is η-dependent. Define $G=\partial f(u^1+\cdots+u^{i-1})/\partial x_s$ and $H=u_s^i$. We apply lemma 10.1 for $d_G\leq p^d i$, $d_H=p^d$, $\mathrm{Lip}\,(G)\leq 2m^{-1/2}\varepsilon^{-2}$, $\mathrm{Lip}\,(H)\leq n^{-1/2}$, $\|G\|_\infty\leq 2m^{-1/2}\varepsilon^{-1}$ and $\|H\|_\infty\leq p^d n^{-1/2}$. Because of the respective order of the parameters, we simplify (10.2.2):

$$\left|\mathrm{Cov}\left(\frac{\partial f(u^1+\cdots+u^{j-1})}{\partial x_s},u_s^j\right)\right| \leq 2\phi(G,H)\epsilon(q) \leq 4m^{-1/2}\varepsilon^{-2}p^{2d}i\,n^{-1/2}\epsilon(q).$$

Summing over j and s:

$$|T_1| \leq \sum_{j=1}^i |\mathbb{E}(f^{(1)}(u^1+\cdots+u^{j-1})\cdot u^j)| \leq 4m^{1/2}\varepsilon^{-2}\epsilon(q)n^{3/2}.$$

Now we give a bound for T_2. Define $G=\frac{\partial^2 f(u^1+\cdots+u^{i-1})}{\partial x_i^s\partial x_i^t}$ and $H=u_s^i u_t^i$. We apply Lemma 10.1 for $d_G=i^2p^{2d}$, $d_H=2p^d$, $\mathrm{Lip}\,(G)\leq 2m^{-1/2}\varepsilon^{-3}$, $\mathrm{Lip}\,(H)\leq p^d n^{-1}$, $\|G\|_\infty\leq 2m^{-1/2}\varepsilon^{-2}$ and $\|H\|_\infty\leq p^{2d}n^{-1}$. Keeping the larger exponents in each term:

$$\mathrm{Cov}\left(\frac{\partial^2 f(u^1+\cdots+u^{j-1})}{\partial x_s\partial x_t},u_s^j u_t^j\right) \leq 4m^{-1/2}\varepsilon^{-3}p^{2d}jn^{-1}\epsilon(q).$$

Summing over j, s and t:

$$|T_2| \leq \sum_{j=1}^{i} \left| \mathbb{E}\left(f^{(2)}(u^1 + \cdots + u^{j-1}) \cdot (u^j, u^j)\right)\right| \leq 2m^{3/2}\varepsilon^{-3}\epsilon(q)n.$$

Assume that ξ is κ-dependent. Using the same bounds for functions G and H and relation (10.2.3):

$$|T_1| \leq \sum_{j=1}^{i} |\mathbb{E}\left(f^{(1)}(u^1 + \cdots + u^{j-1}) \cdot u^j\right)| \leq 4m^{1/3}p^{d/3}\varepsilon^{-7/3}n^{4/3}\epsilon(q),$$

$$|T_2| \leq \sum_{j=1}^{i} \left|\mathbb{E}\left(f^{(2)}(u^1 + \cdots + u^{j-1}) \cdot (u^j, u^j)\right)\right| \leq 4m^{4/3}p^d\varepsilon^{-11/3}n\epsilon(q).$$

The term of third order is bounded using the third order moment. Using Jensen inequality:

$$\mathbb{E}|u^1|_2^3 \leq m^{1/2} \sum_{s=1}^{m} \left(\mathbb{E}|u_s^1|^4\right)^{3/4}.$$

Substituting 1 to the bound $C_\xi|t-s|$ and choosing $R = 1 \vee (4 + b^{-1}\log(C))$, Equation (10.6.5) becomes $V_4(N) \leq p^d \left(3R(1 \vee b^{-2})\right)^{4d} 4!(4d)!$. Then

$$\left(\mathbb{E}|u_s^1|^4\right)^{3/4} \leq K_7 p^{3d/2} n^{-3/2},$$

so that the last term is less than

$$\|f^{(3)}\|A \leq K_7 m\varepsilon^{-3}p^{d/2}n^{-1/2} \tag{10.6.20}$$

We conclude that, if ξ is η-dependent:

$$\begin{aligned} |\mathbb{E}(f(\sum_{j=1}^k u_j) - f(\sum_{j=1}^k y_j))| \quad &\leq \quad K_5 m^{1/2}\varepsilon^{-2}n^{3/2}\epsilon(q) \qquad (10.6.21)\\ &\quad + K_6 m^{3/2}\varepsilon^{-3}n\epsilon(q) + K_7 m\varepsilon^{-3}p^{d/2}n^{-1/2}, \end{aligned}$$

where $K_5 = 4$, $K_6 = 2$ and K_7 is the constant in (10.6.20). If ξ is κ-dependent:

$$\begin{aligned} |\mathbb{E}(f(\sum_{j=1}^k u_j) - f(\sum_{j=1}^k y_j))| \quad &\leq \quad K_5 m^{1/3}p^{d/3}\varepsilon^{-7/3}n^{4/3}\epsilon(q) \qquad (10.6.22)\\ &\quad + K_6 m^{4/3}p^d\varepsilon^{-11/3}n\epsilon(q) + K_7 m\varepsilon^{-3}p^{d/2}n^{-1/2}. \end{aligned}$$

Now we have to bound the difference (10.6.18). We use the following lemma:

Lemma 10.4. *Let X and Y be two centered Gaussian vector of length m, with respective covariance matrix M and N. Let $0 < \varepsilon < 1$, and $f \in \mathcal{F}_\varepsilon$. Then*

$$|\mathbb{E}(f(X) - f(Y))| \leq m^{-1/2}\varepsilon^{-2}\|M - N\|_1,$$

where $\|A\|_1 = \sum_{i,j} |A_{i,j}|$.

Proof of lemma 10.4. Because X and Y are Gaussian

$$\mathbb{E}(f(X) - f(Y)) = \sum_{k=1}^{n} f\left(Z_k + X_{k,n}/\sqrt{n}\right) - f\left(Z_k + Y_{k,n}/\sqrt{n}\right),$$

where the $X_{k,n}$ and $Y_{k,n}$ are independent copies of X and Y, and $Z_k = (X_{1,n}+\cdots+X_{k-1,n}+Y_{k+1,n}+\cdots+Y_{n,n})/\sqrt{n}$. Using the Taylor expansion:

$$\mathbb{E}(f(X) - f(Y)) = \sum_{k=1}^{n} \frac{1}{2n}\mathbb{E}(f^{(2)}(0)\cdot(X_{k,n}, X_{k,n}) - f^{(2)}(0)\cdot(Y_{k,n}, Y_{k,n}))$$
$$+ \frac{1}{6n^{3/2}}\mathbb{E}\Big(f^{(3)}(V_{k,n})\cdot(X_{k,n}, X_{k,n}, X_{k,n}) - f^{(3)}(W_{k,n})\cdot(Y_{k,n}, Y_{k,n}, Y_{k,n})\Big),$$

$V_{k,n}$, $W_{k,n}$ being random vectors. The term involving $f^{(3)}$ tends to 0, and

$$\mathbb{E}\left(f^{(2)}(0)\cdot(X_{k,n}, X_{k,n}) - f^{(2)}(0)\cdot(Y_{k,n}, Y_{k,n})\right) = \sum_{i,j=1}^{m}\left(\frac{\partial^2 f(0)}{\partial x_i \partial x_j}(M_{i,j} - N_{i,j})\right),$$

so that $|\mathbb{E}(f(X) - f(Y))| \leq m^{-1/2}\varepsilon^{-2}\|M - N\|_1$. □

Apply this lemma to the Gaussian vector Y and $\sum y_i$. The covariance matrix of $\sum y_i$ and Y are respectively $k\frac{p^d}{n}\Sigma_{D,p}$ and Σ_D, where

$$\Sigma_{D,p}(s,t) = \sum_{|j|<p} h(j)\mathrm{Cov}(x_0(s), x_j(t)) \quad \text{and } h(j) = \prod_{l=1}^{d}(1 - j_l/p).$$

Define the matrix M by $M(s,t) = \sum_{|j|<p}\mathrm{Cov}(x_0(s), x_j(t))$. We have that

$$\left\|\frac{kp^d}{n}\Sigma_{D,p} - \Sigma_D\right\|_1 \leq \frac{m^2kqp^{d-1}}{n}\|\Sigma_{D,p}\|_\infty + m^2\|\Sigma_{D,p} - M\|_\infty + m^2\|M - \Sigma_D\|_\infty.$$

Because of the convergence of the series defining Σ, $\|\Sigma_{D,p}\|_\infty$ is uniformly bounded in p and gives the main contribution. Because $1 - h(j) \leq d|j|/p$, and $\sum_{|j|<p}|j|\mathrm{Cov}(x_0(s), x_j(t))$ is bounded, the term $\|\Sigma_{D,p} - M\|_\infty = \mathcal{O}(1/p)$.

$\|M - \Sigma_D\|_\infty$ is the remainder of a geometric series and is smaller. We conclude that the difference (10.6.18) satisfies:

$$\mathbb{E}\Big(f\Big(\sum_{i=1}^{k} y^i\Big) - f(Y)\Big) \leq K_8 m^{3/2} \varepsilon^{-2} q p^{-1}. \tag{10.6.23}$$

Thus collecting the bounds (10.6.19), (10.6.21) or (10.6.22) and (10.6.23), we infer that the distance between the finite dimensional distributions of size m satisfies the inequalities of Proposition 10.6. □

Proof of theorem 10.5. Let D be the set of reals $(i/m)_{i=1,\dots,m}$. The corresponding δ is m^{-1}. We collect the results of Proposition 10.4 (oscillation of the empirical process), Proposition 10.5 (oscillation of the Gaussian limit process) and Proposition 10.6 (distance between fidi repartitions). We use lemma 10.2 to conclude.

The $(1/m)$-oscillation of U_n is less than $K_{10} m^{-1/2} \log^{2d+1}(m)$. The $(1/m)$-oscillation of X is negligible with respect to $m^{-1/2} \log^{2d+1}(m)$. We choose $q = \log(n)$. The terms (10.6.11), (10.6.12) and (10.6.13) are negligible. The minimum rate is obtained for parameters such that $m^{-1/2} \log^{2d+1}(n)$, $m^{1/2} \varepsilon \log^{1/2}(n)$, $m^{5/2} \varepsilon^{-3} p^{d/2} n^{-1/2}$ and $m^{3/2} \varepsilon^{-2} p^{-1} \log(n)$ are of same order with respect to n. The solution is

$$\begin{aligned} m &= n^{\frac{1}{4d+12}} (\log(n))^{\frac{6d^2+17d+5}{4d+12}}, \\ p &= n^{\frac{1}{d+3}} (\log(n))^{\frac{-2d+2}{d+3}}, \\ \varepsilon &= n^{\frac{-1}{4d+12}} (\log(n))^{-\frac{2d^2+9d+1}{4d+12}}. \end{aligned}$$

For this choice, the rate of convergence is $n^{\frac{1}{8d+24}} (\log(n))^{\frac{10d^2+39d+29}{8d+24}}$. □

Chapter 11

Functional estimation

In this chapter we are going to consider some methods of functional estimation. We study the estimation of the marginal density of the one weak dependence sequence $(X_t)_{t\in\mathbb{Z}}$ and also the estimation of the regression function in a two dimensional model $Z_t = (X_t, Y_t)_{t\in\mathbb{Z}}$. We will show that the CLT and uniform *a.s.* convergence results hold under general non causal weak dependence. We also establish sharp results about the **MISE** of these estimators under more restrictive causal dependence. Our principal goal consists in extending to less restrictive notions of weak dependence results already known for mixing sequences. We end the chapter with an overview over the different methods of non parametric estimation in order to extend the field of application of the previous results.

11.1 Some non-parametric problems

For a stationary two dimensional process $(Z_t)_{t\in\mathbb{Z}}$ with $Z_t = (X_t, Y_t)$, an important quantity is the regression function

$$r(x) = \mathbb{E}(Y_0|X_0 = x).$$

Various methods to fit such a function have been developed. Nadaraya-Watson kernel estimates are very popular; see Rosenblatt (1991) [169], Prakasa-Rao (1983) [153], or Robinson (1983) [163], for instance.

Volatility. Among other problems, one may wish to estimate the volatility of financial times series, $v(x) = \mathrm{Var}\,(X_t|X_{t-1} = x)$. The question enters into the regression framework with both $Y_i = X_{i-1}$ and $Y_i = X_{i-1}^2$ since $v(x) = v_2(x) - v_1^2(x)$, where $v_j(x) = \mathbb{E}(X_1^j|X_0 = x)$.

Density. Another important problem of econometric interest is to estimate the marginal density f of a stationary sample. Density kernel estimators built by using a kernel function K are usually defined. Derivatives of the density and regression functions can also be estimated by using analogous procedures. Here we simply set $Y_i \equiv 1$.

Quantiles. Conditional quantiles are linked to the conditional distribution by the relation $F(y \mid x) = \mathbb{P}(X_1 \leq y | X_0 = x)$. More precisely, we denote by $q(t|x) = \inf \{y \,/\, F(y \mid x) > t\}$ the generalized (right-continuous, with left-limits) inverse of the monotone function $y \mapsto F(y \mid x)$. Set $Y_t(y) = \mathbf{1}_{\{X_{t+1} \leq y\}}$. Consistent estimators of the conditional regression $\mathbb{E}(Y_t(y)|X_t = x)$ provides information on the previous conditional quantiles.

Derivatives of a density. An estimator of the derivative $f^{(\nu)}$ of f, where ν is a vector $(\nu_1, \ldots, \nu_d)$ in $\mathbb{N}^d$ and $f^{(\nu)} = \dfrac{\partial^{\nu_1 + \cdots + \nu_d} f}{\partial x_1^{\nu_1} \ldots \partial x_d^{\nu_d}}$.

The kernel estimator $\hat{f}$ of f (see below for a precise statement) gives another estimation $\hat{f}^{(\nu)}$ of $f^{(\nu)}$ as:

$$\frac{\partial^{\nu_1 + \cdots + \nu_d} \hat{f}}{\partial x_1^{\nu_1} \ldots \partial x_d^{\nu_d}} = \frac{m^{1+\nu_1+\cdots+\nu_d}}{n} \sum_{i=1}^{n} \frac{\partial^{\nu_1 + \cdots + \nu_d} K}{\partial x_1^{\nu_1} \ldots \partial x_d^{\nu_d}} \left(m^\delta (x - X_i)\right)$$

11.2 Kernel regression estimates

We now consider a stationary process $(Z_t)_{t \in \mathbb{Z}}$ with $Z_t = (X_t, Y_t)$ where $X_t,\ Y_t \in \mathbb{R}$. The quantity of interest is the regression function $r(x) = \mathbb{E}(Y_0 | X_0 = x)$. Let K be some kernel function integrated to 1, Lipschitz and with a compact support. The kernel estimators are defined by

$$\begin{aligned}
\widehat{f}(x) &= \widehat{f}_{n,h}(x) = \frac{1}{nh} \sum_{t=1}^{n} K\left(\frac{x - X_t}{h}\right), \\
\widehat{g}(x) &= \widehat{g}_{n,h}(x) = \frac{1}{nh} \sum_{t=1}^{n} Y_t K\left(\frac{x - X_t}{h}\right), \\
\widehat{r}(x) &= \widehat{r}_{n,h}(x) = \frac{\widehat{g}_{n,h}(x)}{\widehat{f}_{n,h}(x)}, \text{ if } \widehat{f}_{n,h}(x) \neq 0; \widehat{r}(x) = 0, \quad \text{otherwise.}
\end{aligned}$$

Here $h = (h_n)_{n \in \mathbb{N}}$ is a sequence of positive real numbers. We always assume that $h_n \to 0$, $nh_n \to \infty$ as $n \to \infty$.

Definition 11.1. *Let $\rho = a + b$ with $(a, b) \in \mathbb{N} \times]0, 1]$. Denote by $\mathcal{C}_a$ the set of a-times continuously differentiable functions. The set of ρ-regular functions $\mathcal{C}_\rho$*

is defined as

$$\mathcal{C}_\rho = \Big\{ u : \mathbb{R} \to \mathbb{R} \Big/ u \in \mathcal{C}_a \text{ and } \forall K, \exists A \geq 0, \\ |x|, |y| \leq K \Rightarrow |u^{(a)}(x) - u^{(a)}(y)| \leq A|x-y|^b \Big\}.$$

Assuming $g \in \mathcal{C}_\rho$, one can choose a kernel function K of order ρ (which is not necessarily nonnegative integer) such that the bias b_h satisfies

$$b_h(x) = \mathbb{E}(\widehat{g}(x)) - g(x) = \mathcal{O}(h^\rho), \quad \text{uniformly on any compact subset of } \mathbb{R},$$

see *e.g.* Rosenblatt (1991) [169]*. If, moreover, ρ is an integer with $b = 1$, $\rho = a - 1$, then with an appropriately chosen kernel K of order ρ, $b_h(x) \sim h^\rho g^{(\rho)}(x) \int s^\rho K(s)ds/\rho!$, uniformly on any compact interval. In view of the asymptotic analysis we assume that the marginal density $f(\cdot)$ of X_0 exists and is continuous. Moreover, $f(x) > 0$ for any point x of interest and the regression function $r(\cdot) = \mathbb{E}(Y_0|X_0 = \cdot)$ exists and is continuous. Finally, for some $p \geq 1$, $x \mapsto g_p(x) = f(x)\mathbb{E}(|Y_0|^p|X_0 = x)$ exists and is continuous. We set $g = fr$ with obvious shorthand notation. Moreover, we impose one of the following moment conditions:

$$\mathbb{E}|Y_0|^S < \infty, \qquad \text{for some} \qquad S \geq p, \tag{11.2.1}$$

$$\mathbb{E}e^{a|Y_0|} < \infty, \qquad \text{for some} \qquad a > 0. \tag{11.2.2}$$

11.2.1 Second order and CLT results

We consider first the properties of $\widehat{g}(x)$. The following conditionally centered equivalent of g_2 appears in the asymptotic variance of the estimator $\widehat{r}$,

$$G_2(x) = f(x)\text{Var}\,(Y_0|X_0 = x) = g_2(x) - f(x)r^2(x).$$

Assume that the densities of the pairs (X_0, X_k), $k \in \mathbb{Z}^+$, exist, and are uniformly bounded: $\sup_{k>0} \|f_{(k)}\|_\infty < \infty$. Moreover, uniformly over all $k \in \mathbb{Z}^+$, the functions

$$r_{(k)}(x, x') = \mathbb{E}\left(|Y_0 Y_k| \Big| X_0 = x, X_k = x'\right) \tag{11.2.3}$$

are continuous. Under these assumptions, the functions $g_{(k)} = f_{(k)} r_{(k)}$ are locally bounded.

*Such kernels can be constructed as $K = pd$ for some known compactly supported (continuous) density $d(\cdot)$ and a polynomial p with $d^\circ p < \rho$. If $d(x) \neq 0$ on some infinite set then the system of equations $\int t^j K(t)dt = a_j$ $(0 \leq j \leq p)$ admits a unique solution for all choices of a_j's because the quadratic form $p \mapsto \int p^2(x)d(x)\,dx$ is positive definite.

Theorem 11.1. *Suppose that the stationary sequence $(Z_t)_{t\in\mathbb{Z}}$ satisfies the conditions (11.2.1) with $p = 2$ and (11.2.3). Suppose that $n^\delta h \to \infty$ for some $\delta \in]0,1[$. Then*

$$\sqrt{nh}\ (\widehat{g}(x) - \mathbb{E}\widehat{g}(x)) \xrightarrow[n\to\infty]{\mathcal{D}} \mathcal{N}\left(0, g_2(x)\int K^2(u)\mathrm{d}u\right)$$

and setting $\tilde{g}(x) = \widehat{g}(x) - r(x)\widehat{f}(x)$

$$\sqrt{nh}\,(\tilde{g}(x) - \mathbb{E}\tilde{g}(x)) \xrightarrow[n\to\infty]{D} \mathcal{N}\left(0, G_2(x)\ \int K^2(u)\mathrm{d}u\right)$$

under any of the weak dependence condition formulated below.

For clarity sake we do not precise the assumptions here but the result holds if the decay of the sequence $\theta(n) = \mathcal{O}(n^{-a})$ as $n \to \infty$ for each $a > 0$ is faster than any Riemanian decay. The same holds too for $\eta(n), \kappa(n)$ or $\lambda(n)$.
In order to consider asymptotics for the ratio estimator $\widehat{r}$ we use a method, already used by Collomb (1984) [38], which consists of studying higher order asymptotics. It is the topic of the next subsection.

Theorem 11.2. *Suppose that the stationary sequence $(Z_t)_{t\in\mathbb{Z}}$ satisfies the conditions (11.2.1) with $p = 2$ and (11.2.3). Consider a positive kernel K. Let $f, g \in \mathcal{C}_\rho$ for some $\rho \in]0,2]$, and $nh^{1+2\rho} \to 0$. Then, if $f(x) \neq 0$,*

$$\sqrt{nh}\ \left(\widehat{r}(x) - r(x)\right) \xrightarrow[n\to\infty]{\mathcal{D}} \mathcal{N}\left(0,\ \frac{G_2(x)}{f^2(x)}\int K^2(u)\mathrm{d}u\right)$$

under any of the weak dependence condition formulated below.

Assuming that the sequence $(Z_t)_{t\in\mathbb{Z}}$ is θ-weakly dependent with rate $\mathcal{O}(r^{-a})$ and $a > 3$, Ango Nze, Bühlmann and Doukhan (2002) [6] prove that, uniformly in x belonging to any compact subset of $\mathbb{R}$,

$$\mathrm{Var}\,(\widehat{g}(x)) = \frac{1}{nh}\ g_2(x)\int K^2(u)\ \mathrm{d}u\ +\ \mathrm{o}\left(\frac{1}{nh}\right),$$

and

$$\mathrm{Var}\left(\widehat{g}(x) - r(x)\widehat{f}(x)\right) = \frac{1}{nh}G_2(x)\int K^2(u)\ \mathrm{d}u + \mathrm{o}\left(\frac{1}{nh}\right).$$

The exponential moment assumption can be relaxed. Suppose that the stationary sequence $(Z_t)_{t\in\mathbb{Z}}$ satisfies conditions (11.2.1) and (11.2.3) with $p = 2$, $S > 2$. Former results then hold if the sequence $(Z_t)_{t\in\mathbb{Z}}$ is weak dependent with $\eta(r) = \mathcal{O}(r^{-a})$ and $a > 3S - 4/(S-2) + \frac{2}{\delta}/(S-2)$, or $\lambda(r) = \mathcal{O}(r^{-a})$ and $a > 4S - 4/(S-2) + \frac{2}{\delta}/(S-2)$.

The CLT convergence Theorem 11.1 holds, under the conditions (11.2.1) with $p = 2$ and (11.2.3), if the stationary sequence $(Z_t)_{t\in\mathbb{Z}}$ is η or κ-weak dependent with rate $\mathcal{O}(r^{-a})$ and

$$a > \alpha_j(\delta) = \min\left(\max\left(2+j, 3(2+j)\delta\right), \max\left(2+j+\frac{1}{\delta}, \frac{2+2(2+j)\delta}{1+\delta}\right)\right),$$

where $j = 1$ or $j = 2$ according respectively to η or λ dependence assumption. These results extend Doukhan and Louhichi (2001) [68], valid for the case of the density function $\widehat{f}$, to the estimate $\widehat{g}$ under weak dependence. The first right hand side term is obtained by Bernstein's blocking technique. The second right hand side term results from the application of the Lindeberg method (see Rio (2000) [161]). The CLT convergence Theorem 11.2 relies on the expansion

$$u^{-1} = \sum_{i=0}^{p}(-1)^i \frac{(u-u_0)^i}{u_0^{i+1}} + (-1)^{p+1}\frac{(u-u_0)^{p+1}}{u u_0^{p+1}}, \tag{11.2.4}$$

where $p = 2$, $u = b_n$, $u_0 = \mathbb{E}b_n = 1$, and $\widehat{r}(x) = a_n/b_n$ (if $b_n \neq 0$) with

$$\begin{aligned} a_n &= \sum_{i=1}^{n} Y_i K\left(\frac{x-X_i}{h_n}\right) \Big/ \left(n\mathbb{E}K\left(\frac{x-X_0}{h_n}\right)\right), \\ b_n &= \sum_{i=1}^{n} K\left(\frac{x-X_i}{h_n}\right) \Big/ \left(n\mathbb{E}K\left(\frac{x-X_0}{h_n}\right)\right). \end{aligned}$$

Using the Rosenthal inequalities described in § 4.3 and the aforementioned CLT, we obtain the CLT convergence Theorem 11.2 for the regression function, under conditions (11.2.1) for $p = 2$ and (11.2.3), if the stationary sequence $(Z_t)_{t\in\mathbb{Z}}$ is either η or κ-weakly dependent with rate $\mathcal{O}(r^{-a})$, with $a > \alpha_j(\delta)$ and $a > 3 \vee \frac{9(2+j)}{7-4\delta}$ ($j = 1$ under η and $j = 2$ under κ dependence).
The results stated in Theorem 11.1 and Theorem 11.2 also hold for finite dimensional convergence. The components are asymptotically jointly independent, much in the same way that for i.i.d. sequences.

A rapid sketch of the proof for theorem 11.1. We proceed as in Rio (2000) [161] and more specifically as in Coulon-Prieur and Doukhan (2000) [40] for density estimation in a causal case. The case of non causal coefficients is considered with Bernstein blocks as in Doukhan and Louhichi (2001) [68].
Here for $0 \le t \le n$ we consider Lipschitz truncations at level $M = M(n)$:

$$T_t(x) = \left(Y_t\, \mathbf{1}_{|Y_t|\le M} + M(n)\mathbf{1}_{Y_t > M} - M\mathbf{1}_{Y_t<-M}\right) K\left(\frac{x-X_t}{h}\right).$$

Then for a suitable constant σ_n, $\sigma_n \sum_{i=1}^{n} T_t(x)$ approximates the expression $\sqrt{nh}\left(\widehat{g}(x) - r(x)\widehat{f}(x)\right)$ (after centering).

Now a CLT may be proved for $\xi_t = \sigma_n \sqrt{nh}\,(T_t(x) - \mathbb{E}T_t(x))$. The second assertion is a consequence of the first one, where Y_t is replaced by $Y_t - r(x)$. See Ango Nze *et al.*, 2002, [6], Ango Nze & Doukhan, 2004, [7].

Remarks • A more easy and efficient way to get such results may obtained by using lemma 7.1, see Bardet *et al.* (2006) [10] for density estimation and a paper by Nicolas Ragache will clarify the regression estimation case.
• *Finite repartitions distributions.* Let $\hat{H}$ be an estimate of a function H (among the previously cited). Then a central limit result for

$$Z_n(x) = \sqrt{nh}(\hat{H}(x) - \mathbb{E}\hat{H}(x)) \to \mathcal{N}(0, s^2(x))$$

extends to a multivariate central limit theorem $(Z_n(x_1), \ldots, Z_n(x_k)) \to \mathcal{N}_k(0, \Sigma)$ where Σ denotes the diagonal matrix with entries $(s^2(x_1), \ldots, s^2(x_k))$. The previous process is not tight in $\mathcal{C}[0,1]$ at points for which $s^2(x) \neq 0$.

11.2.2 Almost sure convergence properties

Theorem 11.3. *Let $(Z_t)_{t\in\mathbb{Z}}$ be a stationary sequence satisfying the conditions (11.2.1) with $p = 2$ and (11.2.3). Then under the forthcoming conditions,*

(i) *There exists a sequence $(\epsilon_n)_{n\in\mathbb{N}}$ with $nh/(\epsilon_n \log(n)) \to \infty$ as $n \to \infty$ such that for any $M > 0$,*

$$\sup_{|x|\le M} |\widehat{g}(x) - \mathbb{E}\widehat{g}(x)| = \mathcal{O}\left(\sqrt{\frac{\epsilon_n \log(n)}{nh}}\right), \qquad \textit{almost surely.}$$

(ii) *Assume now that $\inf_{|x|\le M} f(x) > 0$. If $f, g \in \mathcal{C}_\rho$ for some $\rho \in]0,\infty[$, $h \sim (\epsilon_n \log(n)/n)^{1/(1+2\rho)}$, then*

$$\sup_{|x|\le M} |\widehat{r}(x) - r(x)| = \mathcal{O}\left\{\left(\frac{\epsilon_n \log(n)}{n}\right)^{\rho/(1+2\rho)}\right\}, \qquad \textit{almost surely.}$$

Remark. Under conditions **(ii)** of Theorem 11.3, but assuming only the weaker condition about the bandwidth sequence

$$n^\delta h \to \infty, \quad \text{for some } \delta \in]0,1[,$$

we obtain

$$\sup_{|x|\le M} |\widehat{r}(x) - r(x)| = \mathrm{o}(1), \qquad \text{almost surely.}$$

For the sake of simplicity, we shall only consider the geometrically dependent case and we defer a reader to Ango Nze, Bühlmann and Doukhan (2002) [6] for Riemanian decays.

Theorem 11.4. *Let $(Z_t)_{t\in\mathbb{Z}}$ be a stationary sequence satisfying the conditions (11.2.1) with $p = 2$ and (11.2.3), and either η or κ-weak dependent with geometric decay rate.*

(i) *If $nh/\log^4(n) \to \infty$, then for any $M > 0$, almost surely,*

$$\sup_{|x|\le M} |\widehat{g}(x) - \mathbb{E}\widehat{g}(x)| = \mathcal{O}\left(\frac{\log^2(n)}{\sqrt{nh}}\right).$$

(ii) *For any $M > 0$, if $f, g \in \mathcal{C}_\rho$ for some $\rho \in]0,\infty[$, $h \sim \left(\frac{\log^4(n)}{n}\right)^{1/(1+2\rho)}$ and $\inf_{|x|\le M} f(x) > 0$, then, almost surely,*

$$\sup_{|x|\le M} |\widehat{r}(x) - r(x)| = \mathcal{O}\left\{\left(\frac{\log^4(n)}{n}\right)^{\rho/(1+2\rho)}\right\}.$$

Proof of Theorem 11.3. We keep usual notations and denote by C a universal constant (whose value can change from one place to another). Assume that $\mathbb{E}(\exp(a|Y_0|)) < \infty$. Then

$$\mathbb{P}\left(\sup_{|x|\le M} |\widehat{g}(x) - \tilde{g}(x)| > 0\right) \le n\,\mathbb{P}\left(|Y_0| \ge M_0 \log(n)\right) \le Cn^{1-M_0},$$

and, by the Cauchy-Schwarz inequality,

$$\sup_{|x|\le M} \mathbb{E}|\widehat{g}(x) - \tilde{g}(x)| \le \frac{1}{h}\mathbb{E}\left[|Y_0|\mathbf{1}_{\{|Y_0|\ge M_0\log(n)\}}\left|K\left(\frac{x - X_0}{h}\right)\right|\right] \le \frac{h^{1/3}}{h} n^{-M_0}.$$

We can now reduce computations to the case of a density estimator, as in Doukhan and Louhichi (2001) [68]. Assume that the interval $[-M, M]$ is covered by L_ν intervals with diameter $1/\nu$ (here $\nu = \nu(n)$ depends on n and we denote by I_j the j-th interval and x_j the center of the interval). Assume that the relation $h\nu \to \infty$ holds (for $n \to \infty$). Assume that the compactly supported kernel K vanishes if $t > R_0$. Liebscher (1996) [121] exhibits another kernel-type density estimate $\tilde{g}'$ based on an even, continuous, kernel, decreasing on $[0,\infty[$, constant on $[0, 2R_0]$, taking the value 0 at $t = 3R_0$ (with compact support). Then, he proves that

$$\sup_{x\in I_j} |\tilde{g}(x) - \mathbb{E}\tilde{g}(x)| \le |\tilde{g}(x_j) - \mathbb{E}\tilde{g}(x_j)| + \frac{C}{h\nu}\left(|\tilde{g}'(x_j) - \mathbb{E}\tilde{g}'(x_j)| + 2|\mathbb{E}\tilde{g}'(x_j)|\right).$$

Therefore, for any $\lambda > 0$,

$$\mathbb{P}\left(\sup_{x\in[-M,M]} |\widehat{g}(x) - \mathbb{E}\widehat{g}(x)| \geq \frac{2\lambda}{\sqrt{nh}} + \frac{1}{h}h^{1/3}n^{-M_0} + \mathrm{C}\,\frac{\log(n)}{h\nu}\right)$$
$$\leq C'n^{1-M_0} + L_\nu \cdot \sup_{x_1}\mathbb{P}\left(|\tilde{g}(x_1) - \mathbb{E}\tilde{g}(x_1)| \geq \frac{\lambda}{\sqrt{nh}}\right)$$
$$+ L_\nu \cdot \sup_{x_1}\mathbb{P}\left(|\tilde{g}'(x_1) - \mathbb{E}\tilde{g}'(x_1)| \geq \frac{\lambda}{\sqrt{nh}}\right).$$

The exponential inequality (4.3.31) completes the proof of assertion **(i)**. The proof of assertion **(ii)** is standard and we defer to [6] for a proof . □

11.3 MISE for $\tilde{\beta}$-dependent sequences

We now consider the problem of estimating the unknown marginal density f from the observations $(X_1, \ldots, X_n)$ of a stationary sequence $(X_i)_{i\geq 0}$. In this context, Viennet (1997) [185] obtained optimal results for the MISE under the condition $\sum_{k>0} \beta(\sigma(X_0), \sigma(X_k)) < \infty$ for a β-mixing sequence X_n. We wish to extend Viennet's results to sequences satisfying only

$$\sum_{k>0} \tilde{\beta}(\sigma(X_0), X_k) < \infty. \tag{11.3.1}$$

For kernel density estimators , this can be done by assuming only that the kernel K is BV and Lebesgue integrable. For projection estimators, it works only if the basis is well localized, because our variance inequality is less precise than that of Viennet. Note that Condition (11.3.1) is much less restrictive than Viennet's, for it contains many non mixing examples. In particular, since f is supposed to be square integrable with respect to the Lebesgue measure, the distribution function F of X_0 is 1/2-Hölder. Hence, we infer from lemma 5.1 point 3.i) that (11.3.1) holds as soon as $\sum_{k>0}(\tau(\sigma(X_0), X_k))^{1/3} < \infty$. If f is bounded (11.3.1) holds as soon as $\sum_{k>0}(\tau(\sigma(X_0), X_k))^{1/2} < \infty$.

Variance inequalities

We set $\tilde{\beta}(i) = \tilde{\beta}(\sigma(X_0), X_i)$. The main results of this section are the following upper bounds (compare to Theorems 1.2 and 1.3(a) in Rio (2000) [161] for the mixing coefficients $\alpha(\sigma(X_0), \sigma(X_i))$).

Proposition 11.1. *Let K be any BV function such that $\int |K(x)|dx$ is finite. Let $(X_i)_{i\geq 0}$ be a stationary sequence, and define*

$$Y_{k,n} = h^{-1}K(h^{-1}(x - X_k)) \quad \textit{and} \quad f_n(x) = \frac{1}{n}\sum_{k=1}^{n} Y_{k,n}\,. \tag{11.3.2}$$

The following inequality holds

$$nh \int \operatorname{Var}(f_n(x))dx \le \int K^2(x)\, dx + 2\sum_{k=1}^{n-1} \tilde{\beta}(k)\ \|dK\| \int |K(x)|dx\,.$$

We now come to the case of projection estimators defined with more details in the next section:

Proposition 11.2. *Let $(\varphi_i)_{1\le i\le n}$ be an orthonormal system of $\mathbb{L}^2(\mathbb{R},\lambda)$ (λ is the Lebesgue measure) and assume that each φ_i is BV. Let $(X_i)_{i\ge 0}$ be a stationary sequence, and define*

$$Y_{j,n} = \frac{1}{n}\sum_{k=1}^{n} \varphi_j(X_k) \quad \textit{and} \quad f_n = \sum_{j=1}^{m} Y_{j,n}\varphi_j\,. \tag{11.3.3}$$

The following inequality holds

$$n\int \operatorname{Var}(f_n(x))dx \le \sup_{x\in\mathbb{R}}\Big\{\sum_{j=1}^{m}\varphi_j^2(x)\Big\} + 2\sum_{k=1}^{n-1}\tilde{\beta}(k)\ \sup_{x\in\mathbb{R}}\Big\{\sum_{j=1}^{m}\|d\varphi_j\|\,|\varphi_j(x)|\Big\}\,.$$

Remark 4. Since $\tilde{\beta}(\mathcal{M},X) \le \tilde{\phi}(\mathcal{M},X)$, Propositions 11.1 and 11.2 apply to dynamical systems satisfying (3.3.1) with $2\sum_{i=1}^{n-1} a_k$ instead of $\sum_{i=1}^{n-1}\tilde{\beta}(k)$. For kernel estimators this can be also deduced from a variance estimate given in Prieur (2001) [154].

Proof of Proposition 11.1. We start from the elementary inequality

$$\operatorname{Var}(f_n(x)) \le \frac{1}{n}\|Y_{0,n}\|_2^2 + \frac{2}{n}\sum_{i=1}^{n-1}|\operatorname{Cov}(Y_{0,n},Y_{i,n})|\,.$$

Now $h\int \|Y_{0,n}\|_2^2(x)dx = \int (K(x))^2dx$. To complete the proof, we apply Proposition 5.3:

$$\begin{aligned} h&\int |\operatorname{Cov}(Y_{0,n},Y_{i,n})|(x)dx \\ &\le \|dK\|\mathbb{E}\Big(b(\sigma(X_0),X_i)\int |Y_{0,n}(x)|dx\Big) \le \tilde{\beta}(i)\|dK\|\int |K(x)|dx\,. \ \square \end{aligned}$$

Proof of Proposition 11.2. Since $(\varphi_i)_{1\le i\le n}$ is an orthonormal system of $\mathbb{L}^2(\mathbb{R},\lambda)$ we have that

$$\int \operatorname{Var}(f_n(x))dx = \sum_{j=1}^{m}\operatorname{Var}(Y_{j,n})\,.$$

Applying Proposition 5.3, we obtain that

$$\begin{aligned}\operatorname{Var}(Y_{j,n}) &\leq \frac{1}{n}\|\varphi_j(X_0)\|_2^2 + \frac{2}{n}\sum_{k=1}^{n-1}|\operatorname{Cov}(\varphi_j(X_0),\varphi_j(X_k))| \\ &\leq \frac{1}{n}\|\varphi_j(X_0)\|_2^2 + \frac{2}{n}\sum_{k=1}^{n-1}\|d\varphi_j\|\mathbb{E}(|\varphi_j(X_0)|b(\sigma(X_0),X_k))\,.\end{aligned}$$

To complete the proof we sum in j:

$$n\int \operatorname{Var}(f_n(x))dx \leq \sum_{j=1}^{m}\mathbb{E}\varphi_j^2(X_0) + 2\sum_{k=1}^{n-1}\mathbb{E}\Big(b(\sigma(X_0),X_k)\sum_{j=1}^{m}\|d\varphi_j\|\,|\varphi_j(X_0)|\Big)\,.$$

Some function spaces

In this section we recall the definition of the spaces $\operatorname{Lip}^*(s,2,I)$, where I is either $\mathbb{R}$ or some compact interval $[a,b]$ (see DeVore and Lorentz (1993) [59], Chapter 2). Let $I_{rh}=\mathbb{R}$ if $I=\mathbb{R}$ and $I_{rh}=[a,b-rh]$ otherwise. For any $h\geq 0$, let T_h be the translation operator $T_h(f,x)=f(x+h)$ and $\Delta_h=T_h-T_0$ be the difference operator. By induction, define the operators $\Delta_h^r=\Delta_h\circ\Delta_h^{r-1}$. Let λ be the Lebesgue measure on I and $\|.\|_{2,\lambda}$ the usual norm on $\mathbb{L}^2(I,\lambda)$. The modulus of smoothness of order r of a function f in $\mathbb{L}^2(I,\lambda)$ is defined by

$$\omega_r(f,t)_2 = \sup_{0\leq h\leq t}\|\Delta_h^r(f,\cdot)\mathbf{1}_{I_{rh}}\|_{2,\lambda},$$

For $s>0$, $\operatorname{Lip}^*(s,2,I)$ is the space of functions f in $\mathbb{L}^2(I,\lambda)$ such that

$$\|f\|_{s,2,I} = \|f\|_{2,\lambda} + \sup_{t>0}\frac{\omega_{[s]+1}(f,t)_2}{t^s} < \infty.$$

These spaces are Banach spaces with respect to the norm $\|.\|_{s,2,I}$. Recall that $\operatorname{Lip}^*(s,2,I)$ is a particular case of Besov spaces (precisely $\operatorname{Lip}^*(s,2,I)=B_{s,2,\infty}(I)$) and that it contains Sobolev spaces $W_s(I)=B_{s,2,2}(I)$.

Application to Kernel estimators

If f_n is defined by (11.3.2), set $f_h=\mathbb{E}(f_n)$. Let r be some positive integer, and assume that the kernel K is such that: for any f belonging to the Sobolev space $W_r(\mathbb{R})$ we have

$$\int (f(x)-f_h(x))^2dx \leq M_1 h^{2r}\|f^{(r)}\|_2^2\,, \tag{11.3.4}$$

for some constant M_1 depending only on r. From (11.3.4) and Theorem 5.2 page 217 in DeVore and Lorentz (1993) [59], we infer that, for any f in $\mathbb{L}^2(\mathbb{R}, \lambda)$,

$$\int (f(x) - f_h(x))^2 dx \le M_2 (w_r(f,h)_2)^2 ,$$

for some constant M_2 depending only on r. This last inequality implies that, if f belongs to $\mathrm{Lip}^*(s, 2, \mathbb{R})$ for $r - 1 \le s < r$, then

$$\int (f(x) - f_h(x))^2 dx \le M_2 h^{2s} \|f\|^2_{s,2,\mathbb{R}} .$$

This evaluation of the bias together with Proposition 11.1 leads to the following Corollary.

Corollary 11.1. *Let r be some positive integer. Let $(X_i)_{i \ge 1}$ be a stationary sequence with common marginal density f belonging to* $\mathrm{Lip}^*(s, 2, \mathbb{R})$ *with $r-1 \le s < r$, or to $W_s(\mathbb{R})$ with $s = r$. Let K be a BV function satisfying (11.3.4) and such that $\int |K(x)|dx$ is finite. Let f_n be defined by (11.3.2) with $h = n^{-1/(2s+1)}$. If (11.3.1) holds, then there exists a constant C such that*

$$\mathbb{E} \int (f_n(x) - f(x))^2 \, dx \le C n^{-2s/(2s+1)} .$$

Here are two well known classes of kernel satisfying (11.3.4).

Example 1. One says that K is a kernel of order k, if

1. $\displaystyle\int K(x)dx = 1, \quad \int K^2(x)dx < \infty \quad \text{and} \quad \int |x|^{k+1}|K(x)|dx < \infty .$

2. $\displaystyle\int x^j K(x)dx = 0 \;$ for $1 \le j \le k$.

If K is a Kernel of order k, then it satisfies (11.3.4) for any $r \le k+1$. For instance, the naive kernel $K = (1/2)\mathbf{1}_{]-1,1]}$ is BV and of order 1. Consequently Corollary 11.1 applies to functions belonging to $\mathrm{Lip}^*(s, 2, \mathbb{R})$ for $s < 2$, or to $W_2(\mathbb{R})$. A footnote on page 249 proves the existence of such kernels.

Example 2. Assume that the fourier transform K^* of K satisfies $|1 - K^*(x)| \le M|x|^r$ for some positive constant M. Then K satisfies (11.3.4) for this r. For instance, $K(x) = \sin(x)/(\pi x)$ satisfies (11.3.4) for any positive integer r. Unfortunately, it is neither BV nor integrable. Another function satisfying (11.3.4) for any positive integer r is the analogue of the de la Vallée-Poussin kernel $V(x) = (\cos(x) - \cos(2x))/\pi x^2$. This function is BV and integrable, so that Corollary 11.1 applies to any function belonging to $\mathrm{Lip}^*(s, 2, I)$ for $s > 0$.

Application to unconditional systems.

Proposition 11.2 is of special interest for orthonormal systems $(\varphi_i)_{i\geq 1}$ satisfying the two conditions:

P1. There exists C_1 independent of m such that $\max\limits_{1\leq i\leq m} \|d\varphi_i\| \leq C_1\sqrt{m}$.

P2. There exists C_2 independent of m such that $\sup\limits_{x\in\mathbb{R}} \sum\limits_{j=1}^{m} |\varphi_j(x)| \leq C_2\sqrt{m}$.

An orthonormal system satisfying P2 is called *unconditional*. For such systems, we obtain from Proposition 11.2 that

$$n\int \operatorname{Var}(f_n(x))dx \leq m\Big(C_2^2 + 2C_1C_2\sum_{k=1}^{n-1}\tilde{\beta}(k)\Big). \tag{11.3.5}$$

Example 1 (piecewise polynomials). Let $(Q_i)_{1\leq i\leq r+1}$ be an orthonormal basis of the space of polynomials of order r on $[0,1]$ and define the function R_i on $\mathbb{R}$ by: $R_i(x) = Q_i(x)$ if x belongs to $]0,1]$ and 0 otherwise. We consider the regular partition of $]0,1]$ into k intervals $\big(](j-1)/k, j/k]\big)_{1\leq j\leq k}$. Define the functions $R_{i,j}(x) = \sqrt{k}R_i(kx-(j-1))$. Clearly the family $(R_{i,k})_{1\leq i\leq r+1}$ is an orthonormal basis of the space of polynomials of order r on the interval $[(j-1)/k, j/k]$. Let $m = k(r+1)$ and $(\varphi_i)_{i\geq 1}$ be any family such that

$$\{\varphi_i \,/\, 1\leq i\leq m\} = \{R_{i,j} \,/\, 1\leq j\leq k, 1\leq i\leq r+1\}. \tag{11.3.6}$$

The orthonormal system $(\varphi_i)_{i\geq 1}$ satisfies P1 and P2 with

$$C_1 = (r+1)^{-1/2} \max_{1\leq i\leq r+1} \|dR_i\| \quad \text{and} \quad C_2 = (r+1)^{-1/2} \sup_{x\in[0,1]} \sum_{i=1}^{r+1} |R_i(x)|.$$

The case of histograms corresponds to $r = 0$. In that case $\varphi_j = \sqrt{k}\mathbf{1}_{](j-1)/k, j/k]}$. Clearly $C_2 = 1$ and $\|d\varphi_j\| = 2\sqrt{k}$, so that $C_1 = 2$.
Assume now that X_0 has a density f such that $f\mathbf{1}_{[0,1]}$ belongs to $\text{Lip}^*(s, 2, [0,1])$. Suppose that $r > s-1$, and denote by $\bar{f}$ the orthogonal projection of f on the subspace generated by $(\varphi_i)_{1\leq 1\leq m}$. From Lemma 12 in Barron *et al.* (1999) [12] we know that there exists a constant K depending only on s such that

$$\int_0^1 (f(x) - \bar{f}(x))^2 dx \leq Km^{-2s}. \tag{11.3.7}$$

Since $\bar{f} = \mathbb{E}(f_n)$, we obtain from (11.3.5) and (11.3.7) the following corollary.

Corollary 11.2. *Let $(X_i)_{i\geq 1}$ be a stationary sequence with common marginal density f such that $f\mathbf{1}_{[0,1]}$ belongs to* $\text{Lip}^*(s, 2, [0, 1])$. *Let r be any nonnegative integer such that $r > s - 1$ and $k = [n^{1/(2s+1)}]$. Let $(\varphi_i)_{1\leq i\leq m}$ be defined by (11.3.6) and f_n be defined by (11.3.3). If (11.3.1) holds, then there exists a constant C such that*

$$\mathbb{E}\int_0^1 (f_n(x) - f(x))^2 dx \leq Cn^{-2s/(2s+1)}\,.$$

Example 2 (wavelet basis). Let $\{e_{j,k}, j \geq 0, k \in \mathbb{Z}\}$ be an orthonormal wavelet basis with the following convention: $e_{0,k}$ are translate of the father wavelet and for $j \geq 1$, $e_{j,k} = 2^{j/2}\psi(2^j x - k)$, where ψ is the mother wavelet. Assume that these wavelets are compactly supported and have continuous derivatives up to order r (if $r = 0$, the wavelets are supposed to be BV). Let g be some function with support in $[-A, A]$. Changing the indexation of the basis if necessary, we can write $g = \sum_{j\geq 0}\sum_{k=1}^{2^j M} a_{j,k}e_{j,k}$, where $M \geq 1$ is some finite integer depending on A and on the size of the wavelets supports. Let $m = \sum_{j=0}^{J} 2^j M$ and $(\varphi_i)_{i\geq 1}$ be any family such that

$$\{\varphi_i \,/\, 1 \leq i \leq m\} = \{e_{j,k} \,/\, 0 \leq j \leq J, 1 \leq k \leq 2^j M\}. \tag{11.3.8}$$

The orthonormal system $(\varphi_i)_{i\geq 1}$ satisfies P1 and P2.
Assume now that X_0 has a density f belonging to $\text{Lip}^*(s, 2, \mathbb{R})$ with compact support in $[-A, A]$. Denote by $\bar{f}$ the orthogonal projection of f on the subspace generated by $(\varphi_i)_{1\leq i\leq m}$. From Lemma 12 in Barron *et al.*(1999) [12] we know that there exist a constant K depending only on s such that

$$\int_0^1 (f(x) - \bar{f}(x))^2 dx \leq K 2^{-2Js}. \tag{11.3.9}$$

Since $\bar{f} = \mathbb{E}(f_n)$, we obtain the following corollary from (11.3.5) and (11.3.9).

Corollary 11.3. *Let $(X_i)_{i\geq 1}$ be a stationary sequence with common marginal density f belonging to* $\text{Lip}^*(s, 2, \mathbb{R})$ *and with compact support in $[-A, A]$. Let r be any nonnegative integer such that $r > s - 1$ and J be such that $J = [\log_2(n^{1/(2s+1)})]$. Let $(\varphi_i)_{1\leq i\leq m}$ be defined by (11.3.8) and f_n be defined by (11.3.3). If (11.3.1) holds, then there exists a constant C such that*

$$\mathbb{E}\int (f_n(x) - f(x))^2 dx \leq Cn^{-2s/(2s+1)}.$$

Remark 5. More generally, if $\sum_{i=1}^{n} \tilde{\beta}(\sigma(X_0), X_i) = \mathcal{O}(n^a)$ for some a in $[0, 1[$, we obtain the rate $n^{-2s(1-a)/(2s+1)}$ for the MISE in Corollaries 11.1, 11.2 and 11.3. Note that if (11.3.1) holds the rate $n^{-2s/(2s+1)}$ is known to be optimal for i.i.d. observations.

11.4 General Kernels

We are aimed to describe other (linear) estimation procedures obtained by replacement of the kernel $K_n(x,y) = h_n^{-1}K(y - x/h_n)$ by another one.
As a *meta-theorem* we claim that theorems 11.1 and 11.3 remain valid if the convolution kernels are replaced by any of the forthcoming one; moreover those results extend to the estimation questions formulated in § 11.1; the needed assumptions are still a fast enough Riemanian decay of the weak dependence coefficient sequence. The more precise theorem 11.4 requires geometric decay rates.
Even if we do not write definitive convergence results, we provide below all the bounds needed to extend results in § 11.2 to other types of estimators.
Projections. Our discussion will be made of three steps: we consider successively finite order polynomial functions (Tchebichev), infinite order kernels (Dirichlet) and summation methods (Cesaro, developed in the case of De la Vallée-Poussin and Fejer on the one hand and Abel-Poisson on the other hand with Melher), and finally generating functions. For the sake of completeness several classical results are here reminded and we are most of the time working on $\mathbb{R}$ rather than $\mathbb{R}^d$, the generalization being straightforward.
Orthogonal polynomials. Consider an interval I of $\mathbb{R}$ and a real Hilbert space $\mathbb{L}^2(I,\mu)$ with μ dominated by the Lebesgue measure. The Schmidt orthogonalization of the family of monomial functions $\{1, x, x^2, \dots\}$ according to the scalar product of $\mathbb{L}^2(I,\mu)$ leads to the Tchebichev polynomials P_j. Such polynomials verify for all j, $\deg(P_j) = j$. Denoting κ_n the highest coefficient of $P_n(x)$, orthogonality of the families implies, for constant B_{n+2}, a three terms recurrence relation

$$P_{n+2}(x) = \left(\frac{\kappa_{n+2}}{\kappa_{n+1}}x + B_{n+2}\right) P_{n+1}(x) - \frac{\kappa_{n+2}\kappa_n}{\kappa_{n+1}^2} P_n(x).$$

Examples. If $I = (-1,1)$ and $\mu(dt) = dt$ then this gives the Legendre polynomials, if $I = (-\infty,\infty)$ and $\mu(dt) = e^{-t^2}dt$ we obtain the Hermite polynomials and if $I = (0,\infty)$ and $\mu(dt) = e^{-t}dt$ this leads to Laguerre polynomials.
If f is in $\mathbb{L}^2(I,\mu)$, a natural way to estimate f is by mean of projection on an orthonormal polynomial basis through the estimation $\hat{f}_n$ that naturally arises :

$$\hat{f}_n(x) = \frac{1}{n}\sum_{i=1}^{n}\sum_{j=1}^{m(n)} P_j(X_i)P_j(x)$$

The kernel we have to consider is thus given by

$$K_{m(n)}(X_i, x) = \sum_{j=1}^{m(n)} P_j(X_i)P_j(x).$$

Application to functional results with $u_m(X_i, x) = \sqrt{m}K_m(X_i, x)$ as usual. Indeed, the Christoffel-Darboux formula (three terms recurrence relation: see *e.g.* Szegö [182], p. 43), implies that the polynomial functions yield the kernel:

$$K_m(x,y) = \frac{\kappa_m}{\kappa_{m+1}} \frac{P_{m+1}(x)P_m(y) - P_m(x)P_{m+1}(y)}{x-y}$$

where κ_m is the highest coefficient of $P_m(x)$. One remarks that the division by $x-y$ does not change the polynomial character of K_m because $x = y$ is a root of $K_m(x,y)$. Then $K_m(x,y) \sim_\infty \kappa_m t^m$. Thereby we can straightforwardly check that $\sqrt{m}\|K_m(X_i,\cdot)\|_\infty \preccurlyeq m^{1/2}$ and that $\int \sqrt{m}|K_m(X_i,t)|dt \preccurlyeq m^{-1/2}$ on every compact subdomain of I.

An example on $\mathbb{L}^2([-1,1], dt)$ is given by Legendre polynomials also known by the formula:

$$L_k(x) = \frac{(-1)^k}{2^n n!}\frac{d^n}{dx^n}(1-x^2)^n.$$

Dirichlet kernel. Consider now the family of trigonometric polynomials functions $\{\cos(nx), \sin(nx)\}_{n\in\mathbb{N}}$. This family is well-known to be a dense subset of the set of 2π-periodical continuous functions, denoted $\mathcal{C}^\star([0,2\pi], dt)$ by WEIERSTRASS density theorem. If we consider the projection S_n on the trigonometric subspace

$$\text{span}\{\cos kx, \sin nx / \, k \le n\} - T_n$$

then the natural associated kernel is Dirichlet kernel:

$$D_m(x,y) = \sum_{k=-m}^{m} e^{ik(x-y)} = 1 + 2\sum_{k=1}^{m} \cos k(x-y) = \frac{\sin \frac{(2m+1)}{2}(x-y)}{\sin \frac{1}{2}(x-y)}$$

The Dirichlet kernel is also of infinite order but is not positive. Thereby one cannot use the previous monotone convergence argument.

Consider a function f in $\mathcal{C}^{\alpha,\star}([0,2\pi])$, the space of r times differentiable, 2π periodical functions on $[0,2\pi]$. A theorem of Jackson (cf. Doukhan and Sifre [74] p. 217) asserts that

$$\|R_m(f)\| = \sup_x \left| \int D_m(y-x)(f(y)-f(x))dy \right| = \mathcal{O}(m^{-\alpha}\log m)$$

This implies that the optimal bandwidth condition for Fourier approximation in our case is not defined as a power of n. This furthermore implies that whenever f is in $\mathcal{C}^{\alpha,\star}([0,2\pi], R)$ we cannot use an optimal window to control the bias and the covariance of the estimate. Then in that case we need to erase the bias by taking a suboptimal window, *i.e.* a window whose order is such that $R_m(f) = o\left(n^{-\frac{1}{\alpha}}\right)$.

Summation methods. The purpose of summation methods is to transform natural kernel into non-negatives, more regular kernels, thereby turning over the above problems. We first define summation methods by the way of a weight sequence $\{a_{m,j}/\, m \in \mathbb{N},\, 0 \leq j \leq m\}$. Then for all $j \leq m$: $a_{m,j} \to_{m\to\infty} 0$, and $\sum_{j=1}^{m} a_{m,j} = 1$. The weighed kernel is defined as a generalized CESARO's mean

$$K_m^a(x,y) = \sum_{j=0}^{m} a_{m,j} K_j(x,y) = \sum_{j=0}^{m} a_{m,j} \sum_{i=0}^{j} P_i(x)P_i(y)$$

In the sequel we will omit the superscript a. The summations method when well chosen lead to improving result on the bias. Consider now the main classical examples. The DE LA VALLÉE-POUSIN kernel is obtain by considering the means of the FOURIER truncated series:

$$S_{m,n} = \frac{1}{m-n}\left(S_n + \cdots + S_{m+1}\right)$$

The kernel associated to this summation of projection is given by:

$$D_{m,n}(t) = \frac{1}{m-n} \frac{\sin\frac{(m+n)t}{2}\sin\frac{(m-n)t}{2}}{\sin^2\frac{t}{2}}$$

When m takes the value 0, the kernel specializes into the FEJER's kernel F_m of order m, wich is also the result of a CESARO mean of the DIRICHLET Kernel D_m, setting in that case $a_{m,j} = m^{-1}\mathbf{1}_{\{1\leq j\leq m\}}$,

$$\tilde{F}_m(X_k, x) = mD_m^2(X_k - x) = F_m(X_k - x) \sim m^{-1}$$

The Fejer kernel is nonnegative then for all $f \in \mathcal{C}^{1,\star}([0,2\pi])$ an equivalent of the bias is easily found with $a > 0$

$$\begin{aligned}
R_{n,m}(f)(x) &= \frac{1}{n}\int_{[0,2\pi]} \sum_{i=1}^{n} m^{1/2} F_m(y-x) f(y)dy - f(x) \\
&= m^{1/2}\int_{[0,2\pi]} F_m(u) f(x+u)dy - \int_{[0,2\pi]} F_m(u) f(x) du \\
&= m^{1/2}\int_{[0,2\pi]} F_m(u)\left(f(x+u) - f(x)\right) du
\end{aligned}$$

Now let $t = mu$ and expand f to order one according to Taylor's formula; the positivity of the kernel implies $R_{n,m}(f)(x) = \mathcal{O}\left(m^{-3/2}\right)$, thus the optimal bandwidth condition is $m = \mathcal{O}(n^{\frac{1}{6}})$ and we may again adapt the CLT.
For $f \in \mathcal{C}^{\beta,\star}([0,2\pi])$ with $\beta > 1$ use Jackson kernel $J_{p,m} = q_{p,m}^{-1} D_{m-1}^{2p}$ where $q_{p,m}$ normalizes $J_{p,m}$.

Generating function and Abel summation. Another way of adding regularity to kernels is with analytic extension of the generating functions. Note that this is close to the principles of Abel summations. *e.g.* the example of Hermite polynomial functions leads to a kernel known as Melher kernel, see [182].
The Hermite polynomial functions $H_n(x)$ are defined by:

$$H_k(x) = \frac{1}{\sqrt{2^k k! \pi^{1/2}}}(-1)^k \frac{d^k}{d\nu^k}(e^{-\nu^2})(x)$$

The double generating function of this family is of the form:

$$K_t(X_i, x) - \sum_{k=0}^{\infty} t^k H_k(X_i) H_k(x) = \frac{1}{\sqrt{\pi(1-t^2)}} e^{-\frac{1}{2(1-t^2)}\left((X_k^2+x^2)(t^2+1)-4tX_k x\right)}$$

And admits an analytic continuation in $t = 1$. The Mehler kernel $M_m(x)$ is defined by setting $t = 1 - 1/m(n)$ and studying the behavior when $m(n) \to \infty$. Then we once more have a result.
Wavelets basis. Another important class of projection estimates is given by wavelets. Consider a wavelet basis derived from a scaling function $\phi(x)$. Then the projection $\tilde{f}$ of order m of a density f over span$\{\phi_{m,k} = m^{d/2}\phi(m^d x - k) \,/\, k \in \mathbb{Z}\}$ is

$$\tilde{f}(x) = \sum_{k=0}^{\infty} \alpha_{m,k}\phi_{m,k}(x) \qquad \text{where} \qquad \alpha_{m,k} = \int f(t)\phi_{m,k}(t)dt$$

Empirical coefficients are: $\hat{\alpha}_{n,m,k} = n^{-1}\sum_{i=1}^{n} \phi_{m,k}(X_i)$. Thus the projection estimate $\hat{f}$ writes as:

$$\hat{f}_{n,m}(x) = \sum_{k=-\infty}^{\infty} \hat{\alpha}_{n,m,k}\phi_{m,k}(x) = n^{-1} \sum_{k=-\infty}^{\infty} \sum_{i=1}^{n} \phi_{m,k}(X_i)\phi_{m,k}(x)$$

We define the wavelet kernel K_M as

$$K_M(X_i, x) = \sum_{k=-\infty}^{\infty} \phi_{M,k}(X_i)\phi_{M,k}(x)$$

The number of terms in the summation is finite for finite M whenever the function ϕ is assumed to have a compact support. This hypothesis is not necessary and we only assume here that wavelets are regular enough (cf. Daubechies (1988) [41]): $\exists C > 0$, $\exists \alpha > 1$, $\forall x \in \mathbb{R}$: $|\phi(x)| \leq C/(1+|x|)^{\alpha}$. This yields a fast decay of the kernel:

$$\begin{aligned}
\int |\phi(M^d x)|dx &\preccurlyeq \frac{M^{-d}}{(1+M^d|x|)^{\alpha-1}} = \mathcal{O}(M^{-d+d(\alpha-1)}) = \mathcal{O}(M^{d(\alpha-2)}), \\
\int |K_M(X_i,x)|dx &\leq \mathcal{O}(M^{-d}M^{-2d\alpha}) \\
&= \mathcal{O}(M^{d(1-2\alpha)}), \ a.s. \text{ if } x \neq 0, \ \mathbb{P}(X_i = 0) = 0.
\end{aligned}$$

We now set $m = M^{2d(2\alpha-2)}$. It is remarkable that for $x = 0$ (and only for this point if we assume regularity conditions) one must set $m = M^{2d(\alpha-2)}$ hence the speed of convergence is lower. $\delta = 1/2d(1-2\alpha)$ holds in our theorem. We now note K_m instead of K_M. Then $K_m(x,y) = m^d K(mx, my)$ for $K(x,y) = \sum_{k=-\infty}^{\infty} \phi(y-k)\phi(x-k)$ and bias writes,

$$\begin{aligned}
E_m(f)(x) &= \left|\mathbb{E}(\hat{f}_{n,m}(x) - f(x))\right| \\
&\leq \left|\int K_m(y,x)f(y)dy - f(x)\right| \\
&= \left|\int m^d K(m^d y, m^d x)(f(y) - f(x))dy\right| \\
&= \left|\int m^d K(m^d x + t, m^d x)\left(f\left(x + \frac{t}{m}\right) - f(x)\right)dt\right|
\end{aligned}$$

Let now the kernel K have regularity r *i.e.* $\int K(u,u-t)t^v dt = 0$ if $0 < v < r$ and $\int K(u,u-t)t^r dt \neq 0$. If now $f \in \mathcal{C}^r([a,b],\mathbb{R})$, then by the Taylor formula,

$$\begin{aligned}
f\left(x + \frac{t}{m}\right) &= \sum_{j=0}^{n-1} \frac{t^j}{m^j j!} f^{(j)}(x) + \int_x^{x+t/m} \frac{u^{r-1}}{m^{r-1}(r-1)!} f^{(r)} du, \text{ we obtain} \\
E_m(f)(x) &= \frac{1}{m^{r-1}(r-1)!} \int K\left(x, x + \frac{t}{m}\right) \int_x^{x+t/m} u^{r-1} f^{(r-1)}(u) du dt
\end{aligned}$$

If $f^{(r-1)}$ is bounded, the second integral is $\mathcal{O}((t/m)^r)$ as $E_m(f)(x)$.

Remarks. • $E_m(f)(x) \sim h(mx)m^{-r}$ where h is a bounded function when m goes to infinity. The precise characteristics of h depends on the wavelet considered. h in general is a pseudo-periodic function.

• Now consider the quadratic deviation. By positivity of the summed functions, the second moment $V_m(f)(x)$ answers

$$\begin{aligned}
V_m(f)(x) &= m^{2d} \int K^2(m^d x + t, m^d x)\left(f\left(x + \frac{t}{m}\right) - f(x)\right)dt \\
&\sim m \int K^2(u, u+t)dt.
\end{aligned}$$

Chapter 12

Spectral estimation

Parametric estimation from a sample of a stationary time series is an important statistic problem both for theoretical research and for its practical applications to real data. Whittle's approximate likelihood estimate is particularly attractive for numerous models like ARMA, linear processes, etc. mainly for two reasons: first, Whittle's contrast does not depend on the marginal law of the time series but only on its spectral density, and second, its computation time is smaller than other parametric estimation methods such as exact likelihood. Numerous papers have been written on this estimation method after Whittle's seminal paper and in particular Hannan (1973) [102], Rosenblatt (1985) [167] and Giraitis and Robinson (2001) [94] established the asymptotic normality respectively for Gaussian, causal linear, strong mixing processes as well as for ARCH(∞) processes.
This chapter is organized as follows. A first section details the expression of the spectral densities of some standard models. A second section addresses the asymptotic behavior of the empirical periodogram. This is a non consistent estimator of the spectral density but it yields a consistent parametric procedure called Whittle estimation. A last section is aimed to derive the properties of a bandwidth estimate of the spectral density. This estimate is obtained by convoluting the periodogram with an approximate Dirac measure. Its second order properties are easily derived through a simple diagram formula but higher order asymptotic needed to conclude to *a.s.* convergence properties need a dependence tool introduced in § 12.3.2.

12.1 Spectral densities

The following models are already described and their weak dependence properties are checked in chapter 3. Nevertheless, this is an important feature to provide precise expressions of their spectral properties.

Non-causal (two-sided) linear processes. Let X be a zero mean stationary non causal (two-sided) linear time series satisfying:

$$X_k = \sum_{j=-\infty}^{\infty} a_j \xi_{k-j} \quad \text{for} \quad k \in \mathbb{Z}, \tag{12.1.1}$$

with $(a_k)_{k\in\mathbb{Z}} \in \mathbb{R}^{\mathbb{Z}}$ and $(\xi_k)_{k\in\mathbb{Z}}$ a sequence of zero mean i.i.d. random variables such that $\mathbb{E}(\xi_0^2) = \sigma^2 < \infty$ and $\mathbb{E}(|\xi_0|^2) < \infty$. We set $\tilde{a}_k = a_{-k}$ and $\tilde{a} = (\tilde{a}_k)_{k\in\mathbb{Z}}$. Therefore the spectral density of X exists and satisfies:

$$\begin{aligned} f(\lambda) &= \frac{\sigma^2}{2\pi}\left|\sum_{k=-\infty}^{\infty} a_k e^{-ik\lambda}\right|^2 \\ &= \frac{\sigma^2}{2\pi}\sum_{k=-\infty}^{\infty}\left(\sum_{j=-\infty}^{\infty} a_{k-j}a_j e^{-ik\lambda}\right) \\ &= \frac{\sigma^2}{2\pi}\sum_{k=-\infty}^{\infty}(a * \tilde{a})_k e^{-ik\lambda}. \end{aligned} \tag{12.1.2}$$

There exist very few explicit results in the case of two-sided linear processes.

Causal GARCH and ARCH(∞) processes. The famous and from now on classical GARCH(q', q) model, introduced by Engle (1982) [84] and Bollerslev (1986) [23], is given by relations

$$X_k = \rho_k \xi_k \quad \text{with} \quad \rho_k^2 = a_0 + \sum_{j=1}^{q} a_j X_{k-j}^2 + \sum_{j=1}^{q'} c_j \rho_{k-j}^2, \tag{12.1.3}$$

where $(q', q) \in \mathbb{N}^2$, $a_0 > 0$, $a_j \geq 0$ and $c_j \geq 0$ for $j \in \mathbb{N}$ and $(\xi_k)_{k\in\mathbb{Z}}$ are i.i.d. random variables with zero mean (for an excellent survey about ARCH modelling, see Giraitis *et al.* (2005) [93]). Under some additional conditions, the GARCH model can be written as a particular case of ARCH(∞) model, introduced in Robinson (1991) [165], that satisfies

$$X_k = \rho_k \xi_k \quad \text{with} \quad \rho_k^2 = b_0 + \sum_{j=1}^{\infty} b_j X_{k-j}^2, \tag{12.1.4}$$

with a sequence $(b_j)_j$ depending on the family (a_j) and (c_j). Different sufficient conditions can be given for obtaining a m-order stationary solution to (12.1.3) or (12.1.4). Notice that for both models (12.1.3) or (12.1.4), the spectral density is a constant. As a consequence, the periodogram is that of the squared processes; in the GARCH case (see Bollerslev (1986) [23]) is based on the ARMA representation satisfied by $(X_k^2)_{k\in\mathbb{Z}}$. Indeed, if (X_k) is a solution of (12.1.3) or (12.1.4), then (X_k^2) can be written as a solution of a particular case of the bilinear equation (see § 3.4.2):

$$X_k^2 = \varepsilon_k\Big(\gamma b_0 + \gamma\sum_{j=1}^{\infty} b_j X_{k-j}^2\Big) + \lambda_1 b_0 + \lambda_1\sum_{j=1}^{\infty} b_j X_{k-j}^2 \quad \text{for} \quad k \in \mathbb{Z},$$

with $\varepsilon_k = (\xi_k^2 - \lambda_1)/\gamma$ for $k \in \mathbb{Z}$, $\lambda_1 = \mathbb{E}\xi_0^2$ and $\gamma^2 = \mathrm{Var}\,(\xi_0^2)$. Moreover, the time series $(Y_k)_{k\in\mathbb{Z}}$ defined by $Y_k = X_k^2 - \lambda_1 b_0\big(1-\lambda_1\sum_{j=1}^{\infty} b_j\big)^{-1}$ for $k \in \mathbb{Z}$, satisfies the bilinear equation with parameter $c_0 = 0$. Hence, a sufficient condition for the stationarity of $(X_k^2)_{k\in\mathbb{Z}}$ with $\|X_0^2\|_m < \infty$ is

$$(\|\varepsilon_0\|_m + 1)\cdot\sum_{j=1}^{\infty}|b_j| < 1 \quad \Longleftrightarrow \quad \Big(\frac{\|\xi_0^2 - \lambda_1\|_m}{\gamma} + 1\Big)\cdot\sum_{j=1}^{\infty}|b_j| < 1.$$

Set $\sigma^2 = \mathbb{E}(X_0^2 - \rho_0^2)$, the spectral density of $(X_k^2)_{k\in\mathbb{Z}}$ is

$$f(\lambda) = \frac{\sigma^2}{2\pi}\Big|1 - \sum_{j=1}^{\infty} b_j\cdot e^{ij\lambda}\Big|^{-2}.$$

The method developed in Giraitis and Robinson (2001) [94] for establishing the central limit theorem satisfied by the periodogram is essentially *ad hoc* and can not be used for non causal or non linear time series. The recent book of Straumann (2005) [181] also provides an up-to-date and complete overview to this question. Chapter 8 of this book is devoted to the results in Mikosch and Straumann (2002) [131] that studied the case of intermediate moment conditions of order > 4 and < 8 for the special case of GARCH(1,1) processes.

Causal Bilinear processes. Now, assume that $X = (X_k)_{k\in\mathbb{Z}}$ is a bilinear process (see the seminal paper of Giraitis and Surgailis (2002) [95]) satisfying the equation

$$X_k = \xi_k\Big(a_0 + \sum_{j=1}^{\infty} a_j X_{k-j}\Big) + c_0 + \sum_{j=1}^{\infty} c_j X_{k-j} \quad \text{for } k \in \mathbb{Z}, \tag{12.1.5}$$

where $(\xi_k)_{k\in\mathbb{Z}}$ are i.i.d. random variables with zero mean and such that $\|\xi_0\|_p < \infty$ with $p \geq 1$, and a_j, c_j, $j \in \mathbb{N}$ are real coefficients. Assume $c_0 = 0$ and define the generating functions

$$A(z) = \sum_{j=1}^{\infty} a_j z^j \qquad C(z) = \sum_{j=1}^{\infty} c_j z^j$$
$$G(z) = (1 - C(z))^{-1} = \sum_{j=0}^{\infty} g_j z^j \qquad H(z) = A(z)G(z) = \sum_{j=1}^{\infty} h_j z^j.$$

If $\|\xi_0\|_p \sum_{j=1}^{\infty} |h_j| < \infty$, for instance when $\|\xi_0\|_p \cdot \left(\sum_{j=1}^{\infty} |a_j| + \sum_{j=1}^{\infty} |c_j| \right) < 1$, there exists a unique zero mean stationary and ergodic solution X in $\mathbb{L}^p(\Omega, \mathcal{A}, \mathbb{P})$ of equation (12.1.5). For $p \geq 2$, the covariogram of X is

$$R(k) = a_0^2 \|\xi_0\|_2 \Big(1 - \sum_{j=1}^{\infty} h_j^2\Big)^{-1} \sum_{j=0}^{\infty} g_j\, g_{j+k}$$

and $\sum_k |R(k)| < \infty$. The spectral density of X exists and satisfies:

$$f(\lambda) = \frac{a_0^2 \sigma^2}{2\pi(1 - \sum_{j=1}^{\infty} h_j^2)} \sum_{k=-\infty}^{\infty} \sum_{j=0}^{\infty} g_j g_{j+k}\, e^{-ik\lambda},$$

with $\sigma^2 = \|\xi_0\|_2^2$.

Non-causal (two-sided) bilinear and LARCH(∞) processes. The bilinear process $X = (X_k)_{k \in \mathbb{Z}}$ satisfies the equation

$$X_k = \xi_k \Big(a_0 + \sum_{j \in \mathbb{Z}^*} a_j X_{k-j} \Big), \quad \text{for } k \in \mathbb{Z}, \tag{12.1.6}$$

where $(\xi_k)_{k\in\mathbb{Z}}$ are i.i.d. random bounded variables and $(a_k)_{k\in\mathbb{Z}}$ is a sequence of real numbers such that $\lambda = \|\xi_0\|_\infty \cdot \sum_{j \neq 0} |a_j| < 1$. Then the spectral density of X exists and is defined by

$$f(\lambda) = \frac{\sigma^2}{2\pi} \left| 1 - \sum_{j=1}^{\infty} b_j e^{ij\lambda} \right|^{-2}.$$

In the previous expression, the coefficients b_j are not written explicitly as functions of the initial parameters $(a_i)_{i\in\mathbb{Z}}$ and a. By the same way as in the causal case, assume now that $Y = (Y_k)_{k\in\mathbb{Z}}$ satisfies the relation

$$Y_k = \xi_k \sqrt{a_0 + \sum_{j \neq 0} a_j Y_{k-j}^2}, \quad \text{for } k \in \mathbb{Z}, \tag{12.1.7}$$

with the same assumptions on $(\xi_k)_{k\in\mathbb{Z}}$ and $(a_k)_{k\in\mathbb{Z}}$. Then, the time series $(Y_k^2)_{k\in\mathbb{Z}}$ satisfies the relation (12.1.6) and is a stationary process. Then, Y is a stationary process, so-called a two-sided LARCH(∞) process.
The condition on the sequence $(\xi_k)_{k\in\mathbb{Z}}$, *i.e.* i.i.d. random bounded variables, is restricting. However, if it is only a sufficient condition for the existence of a non causal LARCH(∞) process; it seems to be very close to be also a necessary condition, see Doukhan and Wintenberger (2005) [77].

Non-causal linear processes with dependent innovations. Let X be a zero mean stationary non causal (two-sided) linear time series satisfying eqn. (12.1.1) with now a dependent and centered stationary innovation process:

$$\mathbb{E}\xi_0 = 0, \qquad \mathbb{E}\xi_0^2 = \sigma^2.$$

Then denoting by f_ξ the spectral of the process (ξ_t) we have $f_\xi(0) = \sigma^2/(2\pi)$, moreover the spectral density of X exists and satisfies:

$$\begin{aligned} f(\lambda) &= \left|\sum_{k=-\infty}^{\infty} a_k e^{-ik\lambda}\right|^2 f_\xi(\lambda) \\ &= f_\xi(\lambda) \sum_{k=-\infty}^{\infty} (a * \tilde{a})_k e^{-ik\lambda}. \end{aligned} \tag{12.1.8}$$

Examples of interest are orthogonal series like mean zero LARCH(∞) processes. In this case no additional information about ξ may be obtained from the expression of this spectral density.
Another important case is thus that of a process ξ with a non constant spectral density; we previously recalled the precise expression of this spectral density for the case of Bilinear processes introduced by Giraitis and Surgailis (2002) [95].

12.2 Periodogram

Let $X = (X_k)_{k\in\mathbb{Z}}$ be a zero mean fourth-order stationary time series with real values. Denote $(R(i))_i$ the covariogram of X, and $(\kappa_4(i,j,k))_{i,j,k}$ the fourth cumulants of X:

$$\begin{aligned} R(i) &= \mathrm{Cov}(X_0, X_i) = \mathbb{E}\,(X_0 X_i), \\ \kappa_4(i,j,k) &= \mathbb{E}X_0X_iX_jX_k - \mathbb{E}X_0X_i\mathbb{E}X_jX_k \\ &\quad -\mathbb{E}X_0X_j\mathbb{E}X_iX_k - \mathbb{E}X_0X_k\mathbb{E}X_iX_j, \end{aligned}$$

for $(i,j,k) \in \mathbb{Z}^3$.
We will use the following assumption on X:
Assumption M. X is such that:

$$\gamma = \sum_{\ell\in\mathbb{Z}} R(\ell)^2 < \infty \quad \text{and} \quad \kappa_4 = \sum_{i,j,k\in\mathbb{Z}} |\kappa_4(i,j,k)| < \infty. \tag{12.2.1}$$

This assumption is closely linked to weak dependence in the comments following definition 4.1 and by using the notation (4.4.9) with lemma 4.11 and propositions

2.1 and 2.2.
Assumption **M** ensures the existence of X's spectral density $f \in \mathbb{L}^2([-\pi,\pi[)$:

$$f(\lambda) = \frac{1}{2\pi}\sum_{k\in\mathbb{Z}} R(k)\, e^{ik\lambda} \quad \text{for } \lambda \in [-\pi,\pi[.$$

The periodogram of X is defined as:

$$I_n(\lambda) = \frac{1}{2\pi n}\left|\sum_{k=1}^{n} X_k e^{-ik\lambda}\right|^2, \quad \text{for } \lambda \in [-\pi,\pi[.$$

We now rewrite

$$\begin{aligned} I_n(\lambda) &= \frac{1}{2\pi}\sum_{|k|<n} \widehat{R}_n(k) e^{-ik\lambda} \\ \widehat{R}_n(k) &= \frac{1}{n}\sum_{j=1\vee(1-k)}^{(n-k)\wedge n} X_j X_{j+k} \end{aligned}$$

Here $\widehat{R}_n(k)$ is a biased estimate of $R(k)$.
Thus, the periodogram $I_n(\lambda)$ is a natural estimator of the spectral density; unfortunately it is not a consistent estimator, as its variance does not tend to zero as n tends to infinity. However, once integrated with respect to some $\mathbb{L}^2$ function, its behavior becomes quite smoother and can provide an estimation of the spectral density. Now, let $g : [-\pi,\pi[\to \mathbb{R}$ a 2π-periodic function such that $g \in \mathbb{L}^2([-\pi,\pi[)$ and define:

$$\begin{aligned} J_n(g) &= \int_{-\pi}^{\pi} g(\lambda) I_n(\lambda)\, d\lambda, \quad \text{the integrated periodogram of } X \\ \text{and } J(g) &= \int_{-\pi}^{\pi} g(\lambda) f(\lambda)\, d\lambda. \end{aligned}$$

Theorem 12.1 (SLLN). *Let $c > 0$. Assume that the function $g(\lambda) = \sum_{\ell\in\mathbb{Z}} g_\ell e^{i\ell\lambda}$ satisfies $\sum_{\ell\in\mathbb{Z}}(1+|\ell|)^s g_\ell^2 \le c$ for some $s > 1$. If X satisfies Assumption M, then uniformly with respect to such functions g,*

$$J_n(g) \to J(g) \qquad a.s.$$

This theorem will be proved after two lemmas of independent interest.

Lemma 12.1. *If X satisfies Assumption M, then:*

$$n \max_{\ell\ge 0}\left(\mathrm{Var}\left(\widehat{R}_n(\ell)\right)\right) \le \kappa_4 + 2\gamma.$$

Proof of Lemma 12.1. To prove this result, we denote $Y_{j,\ell} = X_j X_{j+\ell} - R(\ell)$, use the identity

$$\mathrm{Cov}(Y_{0,\ell}, Y_{j,\ell}) = \kappa_4(\ell, j, j+\ell) + R(j)^2 + R(j+\ell)R(j-\ell)$$

and deduce from the stationarity of $(Y_{j,\ell})_{j\in\mathbb{Z}}$ when ℓ is a fixed integer:

$$\begin{aligned} n\mathrm{Var}\left(\widehat{R}_n(\ell)\right) &\le \frac{1}{n} \sum_{j=1\vee(1-\ell)}^{(n-\ell)\wedge n} \sum_{j'=1\vee(1-\ell)}^{(n-\ell)\wedge n} |\mathrm{Cov}(Y_{j,\ell}, Y_{j',\ell})| \\ &\le \sum_{j\in\mathbb{Z}} |\mathrm{Cov}(Y_{0,\ell}, Y_{j,\ell})| \\ &\le \sum_{j\in\mathbb{Z}} \left(|\kappa_4(\ell, j, j+\ell)| + 2R(j)^2\right) \\ &\le \kappa_4 + 2\gamma, \end{aligned}$$

by using Cauchy-Schwarz inequality for ℓ^2-sequences. □

Lemma 12.2. *Denote $c_s = \sum_{\ell\in\mathbb{Z}}(1+|\ell|)^{-s}$. If X satisfies assumption of Theorem 12.1, then:*

$$\mathbb{E}|J_n(g) - J(g)|^2 \le \frac{3c}{n}\Big(\gamma + c_s(\kappa_4 + 2\gamma)\Big).$$

Proof of Lemma 12.2. As in Doukhan and León (1989) [66], we use the decomposition:

$$J_n(g) - J(g) = \quad T_1(g) \quad T_2(g) + T_3(g) \text{ with } \begin{cases} T_1(g) = \displaystyle\sum_{|\ell|\ge n} R(\ell)\, g_\ell, \\ T_2(g) = \displaystyle\frac{1}{n} \sum_{|\ell|<n} |\ell|\, R(\ell)\, g_\ell, \\ T_3(g) = \displaystyle\sum_{|\ell|<n} (\widehat{R}_n(\ell) - \mathbb{E}\widehat{R}_n(\ell))\, g_\ell \end{cases} \tag{12.2.2}$$

Remark that $T_3(g) = J_n(g) - \mathbb{E}J_n(g)$. Thus, we obtain the inequality:

$$\mathbb{E}|J_n(g) - J(g)|^2 \le 3(|T_1(g)|^2 + |T_2(g)|^2 + \mathbb{E}|T_3(g)|^2).$$

Cauchy-Schwarz inequality yields:

$$\begin{aligned} |T_1(g)|^2 &\le c \sum_{|\ell|\ge n} (1+|\ell|)^{-s} R(\ell)^2 \le \frac{c}{n} \sum_{|\ell|\ge n} R(\ell)^2, \\ |T_2(g)|^2 &\le \frac{c}{n^2} \sum_{|\ell|<n} |\ell|^2 (1+|\ell|)^{-s} R(\ell)^2 \le \frac{c}{n} \sum_{|\ell|<n} R(\ell)^2. \end{aligned}$$

Hence, $|T_1(g)|^2 + |T_2(g)|^2 \leq \frac{\gamma}{n}$. Lemma 12.1 entails

$$\begin{aligned}
|T_3(g)|^2 &\leq c \sum_{|\ell|<n} (1+|\ell|)^{-s} (\widehat{R}_n(\ell) - \mathbb{E}\widehat{R}_n(\ell))^2, \\
\mathbb{E}|T_3(g)|^2 &\leq c \sum_{|\ell|<n} (1+|\ell|)^{-s} \mathrm{Var}\,(\widehat{R}_n(\ell)) \\
&\leq \frac{1}{n} \sum_{|\ell|<n} (1+|\ell|)^{-s} (\kappa_4 + 2\gamma) \\
&\leq c\,c_s \cdot \frac{\kappa_4 + 2\gamma}{n},
\end{aligned}$$

We combine those results to deduce Lemma 12.2. □

Proof of Theorem 12.1. We derive the strong law of large numbers from a weak $\mathbb{L}^2$-LLN and from Lemma 12.2. The proof follows the scheme of the proof of the standard strong LLN. Set $t > 0$. First, we know that for all random variables X and Y, we have $\mathbb{P}(|X+Y| \geq 2t) \leq \mathbb{P}(|X| \geq t) + \mathbb{P}(|Y| \geq t)$. Thus:

$$\begin{aligned}
\mathbb{P}\left(\max_{k^2 \leq n < (k+1)^2} |J_n(g) - J(g)| \geq 2t \right) &\leq \mathbb{P}(|J_{k^2}(g) - J(g)| \geq t) \\
&+ \mathbb{P}\left(\max_{k^2 \leq n < (k+1)^2} |J_n(g) - J_{k^2}(g)| \geq t \right)
\end{aligned}$$

and

$$\begin{aligned}
\mathbb{P}\left(\max_{n \geq N} |J_n(g) - J(g)| \geq 2t \right) &\leq \sum_{k=[\sqrt{N}]}^{\infty} \mathbb{P}(|J_{k^2}(g) - J(g)| \geq t) \\
&+ \sum_{k=[\sqrt{N}]}^{\infty} \mathbb{P}\left(\max_{k^2 \leq n < (k+1)^2} |J_n(g) - J_{k^2}(g)| \geq t \right) \\
&\leq A_N + B_N. \qquad (12.2.3)
\end{aligned}$$

From Bienaymé-Tchebychev (or Markov) inequality, Lemma 12.2 implies:

$$A_N \leq \frac{C_1}{t^2} \cdot \sum_{k \geq \sqrt{N}} \frac{1}{k^2}, \qquad (12.2.4)$$

with $C_1 \in \mathbb{R}_+$. Now set $\widetilde{R}_n(\ell) = \widehat{R}_n(\ell) - \mathbb{E}\widehat{R}_n(\ell)$. The fluctuation term B_N is more involved and its bound is based on the same type of decomposition as (12.2.2), because for $k^2 < n$:

$$J_n(g) - J_{k^2}(g) = -T_1'(g) + T_2'(g) - T_3'(g),$$

with now,

$$\begin{aligned} T_1'(g) &= \sum_{k^2\le|\ell|<n} R(\ell)\,g_\ell, \\ T_2'(g) &= \frac{1}{n}\sum_{k^2\le|\ell|<n} |\ell|\,R(\ell)\,g_\ell, \\ \text{and } T_3'(g) &= \sum_{|\ell|<k^2} \widetilde{R}_{k^2}(\ell)\,g_\ell - \sum_{|\ell|<n} \widetilde{R}_n(\ell)\,g_\ell. \end{aligned}$$

As previously,

$$|T_1'(g)|^2 + |T_2'(g)|^2 \le c\cdot\frac{\gamma}{k^2}.$$

Set $L_k = \max_{k^2\le n<(k+1)^2} |J_n(g) - J_{k^2}(g)|$ and $T_k^*(g) = \max_{k^2\le n<(k+1)^2} |T_3'(g)|$, then,

$$B_N \le \sum_{k\ge\sqrt{N}} b_k, \quad \text{with} \quad b_k = \mathbb{P}\left(L_k \ge t\right) \le \frac{\mathbb{E}(L_k^2)}{t^2}.$$

Now

$$\mathbb{E}(L_k^2) \le 3(|T_1'(g)|^2 + |T_2'(g)|^2 + \mathbb{E}|T_k^*(g)|^2) \le \frac{3c\gamma}{k^2} + 3\mathbb{E}|T_k^*(g)|^2.$$

Then, for $k^2 \le n < (k+1)^2$ and $\ell \in \mathbb{Z}$,

$$\begin{aligned} \widetilde{R}_n(\ell) &= \frac{k^2}{n}\widetilde{R}_{k^2}(\ell) + \Delta_{\ell,n,k} \\ \Delta_{\ell,n,k} &= \frac{1}{n}\sum_{h=(k^2\wedge(k^2-\ell))+1}^{n\wedge(n-\ell)} Y_{h,\ell}. \end{aligned}$$

Remark that $\widetilde{R}_{k^2}(\ell) = 0$ if $k^2 \le |\ell| \le n$ and thus $\widetilde{R}_n(\ell) = \Delta_{\ell,n,k}$ in such a case. Also note that

$$\Delta_{\ell,k}^* = \max_{k^2\le n<(k+1)^2} |\Delta_{\ell,n,k}| \le \frac{1}{k^2}\sum_{h=(k^2\wedge(k^2-\ell))+1}^{(k^2+2k)\wedge((k^2+2k)-\ell)} |Y_{h,\ell}|,$$

and thus,

$$\begin{aligned} \mathbb{E}(\Delta_{\ell,k}^*)^2 &\le \frac{1}{k^4}\,(2k)^2 \max_{(h,\ell)\in\mathbb{Z}^2}\left(\mathbb{E}(|Y_{h,\ell}|^2)\right) \\ &\le \frac{4}{k^2}\,\mathbb{E}(|X_0|^4). \end{aligned}$$

Write

$$T_3'(g) = \sum_{|\ell|<k^2} \widetilde{R}_{k^2}(\ell)\left(1-\frac{k^2}{n}\right) g_\ell - \sum_{|\ell|<n} \Delta_{\ell,n,k}\, g_\ell$$

$$|T_k^*(g)| \leq \frac{2}{k}\sum_{|\ell|<k^2} |\widetilde{R}_{k^2}(\ell)\, g_\ell| + \sum_{|\ell|<(k+1)^2} \Delta_{\ell,k}^*\, |g_\ell|,$$

and we thus deduce for a constant $C_1 > 0$,

$$\mathbb{E}|T_k^*(g)|^2 \leq 2C_1\left(\frac{4}{k^2}\sup_{\ell\in\mathbb{Z}}\left(\mathrm{Var}\,(\widehat{R}_{k^2}(\ell))\right) + \sup_{\ell\in\mathbb{Z}}\left(\mathbb{E}(\Delta_{\ell,k}^*)^2\right)\right) \leq \frac{C_1 cA}{k^2}$$

for a constant $A > 0$ depending on $\mathbb{E}|X_0|^4$, κ_4, and γ only.
Hence $b_k \leq 3(\gamma + AC_1)/(k^2t^2)$ is a summable series and, with $C_2 > 0$,

$$B_N \leq \frac{C_2}{t^2}\cdot\sum_{k\geq\sqrt{N}}\frac{1}{k^2}. \tag{12.2.5}$$

Then, (12.2.3), (12.2.4) and (12.2.5) imply $\sup_{n\geq N}|J_n(g) - J(g)| \to 0$ in probability, so that $J_n(g) \to J(g)$ a.s. $\square$
Two frames of weak dependence are considered here, the θ-weak dependence property and the non causal η-weak dependence.

12.2.1 Whittle estimation

The previous examples are essentially explicit representations of the spectral density of some commonly used times series. In the case when the coefficients are functions of an unknown finite dimensional parameter β, a way to estimate this parameter is to use the contrast $J(g_\beta^{-1})$ where $f(\lambda) = \sigma^2 g_\beta(\lambda)$ denotes the spectral density of the model according to the value β of the parameter and moreover $g_\beta(0) = 1$. We thus exhibit two parameters, σ^2 and β. Let $(X_1, \ldots, X_n)$ be a sample from X. Define the Whittle maximum likelihood estimators of β^* and σ^{*2}, that are

$$\widehat{\beta}_n = \mathrm{Argmin}_{\beta\in\mathcal{K}}\left\{J_n(g_\beta^{-1})\right\} = \mathrm{Argmin}_{\beta\in\mathcal{K}}\left\{\int_{-\pi}^{\pi}\frac{I_n(\lambda)}{g_\beta(\lambda)}\,d\lambda\right\} \quad \text{and}$$

$$\widehat{\sigma}_n^2 = \frac{1}{2\pi}J_n(g_{\widehat{\beta}_n}^{-1}).$$

In Bardet, Doukhan and León (2005) [11] it is shown that strong consistency of the estimators $\widehat{\beta}_n$ and $\widehat{\sigma}_n^2$ may be proved by using extensions of the previous tools. First a SLLN can be deduced of the one of the previous section and

secondly the CLT is derived by using the previous representation of the periodogram and the results in § 7.1. To our knowledge, the known results about asymptotic behavior of Whittle parametric estimation for non-Gaussian linear processes are essentially devoted to one-sided (causal) linear processes (see for instance, Hannan (1973) [102], Hall and Heyde (1980) [100], Rosenblatt (1985) [168], Brockwell and Davis (1988) [28]). There exist very few results in the case of two-sided linear processes. In Rosenblatt ((1985), p. 52 [168]) a condition for strong mixing property for two-sided linear processes was given, but some restrictive conditions on the process were also required for obtaining a central limit theorem for Whittle estimators: the distribution of random variables ξ_k has to be absolutely continuous with respect to the Lebesgue measure with a bounded variation density, $m > 4 + 2\delta$ with $\delta > 0$ and a central limit theorem obtained with a tapered periodogram (under assumption also $\sum_{m=1}^{\infty} \alpha_{4,\infty}(m)^{\delta/(2+\delta)} < \infty$ where $\alpha_{4,\infty}(m) \geq \alpha_m$ denote a strong mixing coefficient defined now with four points in the future instead of 2 for α'_m). The case of strongly dependent two-sided linear processes was also treated by Giraitis and Surgailis (1990) [95] or Horvath and Shao (1999) [108].
In the case of causal linear processes, it is well known that:

$$\sqrt{n}(\widehat{\beta}_n - \beta^*) \to \mathcal{N}_p\big(0\,,\, 2\pi(W^*)^{-1}\big),$$

$\widehat{\sigma}_n^2$ is a consistent estimate of σ^4 and therefore $\sqrt{n}(\widehat{\sigma}_n^2 - \sigma^*) \to \mathcal{N}\big(0\,, \sigma^{*4}\gamma_4\big)$, with γ_4 the fourth cumulant of the $(\xi_k)_{k\in\mathbb{Z}}$, and $\sqrt{n}(\widehat{\beta}_n - \beta^*)$ and $\sqrt{n}(\widehat{\sigma}_n^2 - \sigma^*)$ are asymptotically normal and independent.

12.3 Spectral density estimation

In this section, $X = (X_n)_{n\in\mathbb{Z}}$ denotes a vector valued stationary sequence (with values in $\mathbb{R}^D$). The spectral matrix density of X writes

$$f_X(\lambda) = \sum_{t\in\mathbb{Z}} \mathrm{Cov}(X_0, X_t)e^{it\lambda}$$

here, $R_t = \mathrm{Cov}(X_0, X_t) = \mathbb{E}X_0^T X_t - \mathbb{E}X_0^T \mathbb{E}X_t$ is a $D \times D-$matrix. This $D \times D-$matrix valued function is estimated by

$$\widehat{f}_X(\lambda) = F_m \star I_n(\lambda) \equiv \int_{-\pi}^{\pi} F_m(\mu) I_n(\lambda - \mu) \frac{d\mu}{2\pi} \quad (12.3.1)$$

$$= \frac{1}{n} \sum_{k,l=1}^{n} \left(1 - \frac{|k-l|}{m}\right)^+ X_k^T X_l e^{i(k-l)\lambda} \quad (12.3.2)$$

$$= \frac{1}{2\pi} \sum_{|s| \le m} \left(1 - \frac{|s|}{m}\right) \widehat{R}_n(s) e^{-is\lambda}, \quad (12.3.3)$$

$$\widehat{R}_n(s) = \frac{1}{n} \sum_{j=1 \vee (1-s)}^{(n-s)\wedge n} X_j^T X_{j+s} \quad (12.3.4)$$

where $F_m(\lambda) = \sum_{|s|<m} \left(1 - \frac{|s|}{m}\right) e^{is\lambda} = \frac{1}{n} \left(\frac{\sin(m+\frac{1}{2})\lambda}{\sin\frac{\lambda}{2}}\right)^2$ is the Fejér Kernel and the matrix periodogram is defined as

$$I_n(\lambda) = \frac{1}{n} \sum_{k,l=1}^{n} X_k^T X_l e^{i(l-k)\lambda}$$

if the sequence X_t is centered at expectation.
Note that in equation (12.3.4), the summation contains $n - |s|$ terms hence this estimate of R_s is biased.
The following relation links the spectral density to the limit variance Σ in the CLT for $X = (X_n)_{n\in\mathbb{Z}}$

$$f_X(0) = \sum_{s\in\mathbb{Z}} \mathrm{Cov}(X_0, X_s).$$

Assume $\sum_s |s|^\rho \|R_s\| < \infty$ where $\|\cdot\|$ is any matrix norm, then

$$\begin{aligned}
\mathrm{bias}(\lambda) &= \widehat{f}_X(\lambda) - f_X(\lambda) \\
\|\mathrm{bias}(\lambda)\| &= \left\| \sum_{|s|<m} \left(1 - \frac{|s|}{m}\right) \left(\left(1 - \frac{|s|}{n}\right) - 1 \right) R_s e^{is\lambda} - \sum_{|s|\ge m} R_s e^{is\lambda} \right\| \\
&\le \frac{1}{n} \sum_s |s| \|R_s\| + \frac{1}{m^\rho} \sum_s |s|^\rho \|R_s\| \\
&= \mathcal{O}\left(\frac{1}{n} + \frac{1}{m^\rho}\right)
\end{aligned}$$

12.3.1 Second order estimate

First recall the 4-th order cumulant of a centered random vector $(x, y, z, t) \in \mathbb{R}^{4D}$ writes, for $1 \le a, b, c, d \le D$,

$$\begin{aligned}
\kappa_4^{(a,b,c,d)}(x,y,z,t) &= \mathbb{E}x^{(a)}y^{(b)}z^{(c)}t^{(d)} \\
&\quad - \mathbb{E}x^{(a)}y^{(b)}\mathbb{E}z^{(c)}t^{(d)} \\
&\quad - \mathbb{E}x^{(a)}z^{(c)}\mathbb{E}y^{(b)}t^{(d)} \\
&\quad - \mathbb{E}x^{(a)}t^{(d)}\mathbb{E}y^{(b)}z^{(c)} \\
\mathbb{E}x^{(a)}y^{(b)}z^{(c)}t^{(d)} - \mathbb{E}x^{(a)}y^{(b)}\mathbb{E}z^{(c)}t^{(d)} &= \kappa_4^{(a,b,c,d)}(x,y,z,t) \\
&\quad - \mathbb{E}x^{(a)}z^{(c)}\mathbb{E}y^{(b)}t^{(d)} \\
&\quad - \mathbb{E}x^{(a)}t^{(d)}\mathbb{E}y^{(b)}z^{(c)}.
\end{aligned}$$

Set now

$$\kappa^{(a,b,c,d)}(i,j,k) = \kappa_4(X_0^{(a)}, X_i^{(b)}, X_j^{(c)}, X_k^{(d)}), \quad r^{(a,b)}(i) = \mathbb{E}X_0^{(a)}X_i^{(b)}. \tag{12.3.5}$$

Assume that for $1 \le a, b, c, d \le D$,

$$\sum_{j\in\mathbb{Z}} |j|^\rho \left| r^{(a,b)}(j) \right| = r_\rho^{(a,b)} < \infty, \tag{12.3.6}$$

$$\sum_{i,j,k\in\mathbb{Z}} \left| \kappa^{(a,b,c,d)}(i,j,k) \right| = \kappa^{(a,b,c,d)} < \infty. \tag{12.3.7}$$

Then we may rewrite

$$\widehat{f}(\lambda) - \mathbb{E}\widehat{f}(\lambda) = \sum_{k=1}^{n} Z_{k,n}(\lambda), \tag{12.3.8}$$

$$Z_{k,n}(\lambda) = \sum_{|s|<m} \delta_{n,k,s}(\lambda) Y_{k,s}, \quad \text{with} \tag{12.3.9}$$

$$Y_{k,s} = X_k^T X_{k+s} - R(s) \tag{12.3.10}$$

$$\delta_{n,k,s}(\lambda) = \left(1 - \frac{|s|}{m}\right) e^{-is\lambda}\, \mathbf{1}_{\{1\le k+s\le n\}} \tag{12.3.11}$$

hence setting $\widehat{f}(\lambda) = \left(\widehat{f}^{(a,b)}(\lambda)\right)_{1\le a,b\le D}$, each coordinate writes for $1 \le a,b,c,d \le D$ as:

$$\begin{aligned}
\left(\operatorname{Var}\widehat{f}(\lambda)\right)^{(a,b,c,d)} &= \operatorname{Cov}\left(\widehat{f}^{(a,b)}(\lambda), \widehat{f}^{(c,d)}(\lambda)\right) \\
&= \frac{1}{n}\sum_{|j|<n}^{n}\left(1-\frac{|j|}{n}\right) C_{j,n}^{(a,b,c,d)}(\lambda), \\
C_{j,n}^{(a,b,c,d)}(\lambda) &= \sum_{|s|<m}\sum_{|t|<m} \delta_{n,0,s}(\lambda)\delta_{n,j,t}(\lambda)\left(1-\frac{|s|}{m}\right)\left(1-\frac{|t|}{m}\right) \\
&\quad \times \left(\mathbb{E}X_0^{(a)}X_s^{(b)}X_j^{(c)}X_{j+t}^{(d)} - \mathbb{E}X_0^{(a)}X_s^{(b)}\mathbb{E}X_j^{(c)}X_{j+t}^{(d)}\right)
\end{aligned}$$

Denote by $\otimes$ the tensor product of two $D\times D$ matrices:

$$(u_{a,b})_{1\le c,d\le D}\otimes (u_{a,b})_{1\le a,b\le D} = (u_{a,b}u_{c,d})_{1\le a,b,c,d\le D}.$$

If the set of such tensors, $\alpha = (\alpha_{a,b,c,d})_{1\le i,j,k,l\le D}$, is equipped with a norm derived from the matrix norm

$$\|(u_{a,b})_{1\le a,b\le D}\| = \max_{1\le a,b\le D}|u_{a,b}|, \tag{12.3.12}$$

then

$$|||\alpha||| = \max_{1\le a,b\le D}\sum_{1\le c,d\le D}|\alpha_{a,b,c,d}| \le D^2 \max_{1\le a,b,c,d\le D}|\alpha_{a,b,c,d}|. \tag{12.3.13}$$

The variance of this estimate is a 4−th order tensor which writes (using notation (12.3.13))

$$\begin{aligned}
\left|\left(\operatorname{Var}\widehat{f}(\lambda)\right)^{(a,b,c,d)}\right| &\le \frac{1}{n}\sum_{|j|<n}\sum_{|s|,|t|<m}|\kappa^{(a,b,c,d)}(s,j,j+t)| \\
&+ \frac{1}{n}\sum_{|j|<n}\sum_{|s|,|t|<m}|r^{(a,c)}(j)||r^{(b,d)}(j+t-s)| \\
&+ \frac{1}{n}\sum_{|j|<n}\sum_{|s|,|t|<m}|r^{(a,d)}(j+t)||r^{(b,c)}(j-s)|,
\end{aligned}$$

Assumptions (12.3.7) implies that $\frac{2m+1}{n}\kappa^{(a,b,c,d)}$ bounds the first sum and after an interversion of summations w.r.t. t and s, assumption (12.3.6), both other terms are bounded above by $\frac{2m+1}{n}r_0^{(a,c)}r_1^{(b,d)}$ and $\frac{2m+1}{n}r_0^{(a,d)}r_1^{(b,c)}$, respectively. This entails

$$\left|\left|\left|\operatorname{Var}\widehat{f}(\lambda)\right|\right|\right| \le \frac{2m+1}{n}\left(\max_{1\le a,b\le D}\sum_{c,d=1}^{D}\kappa^{(a,b,c,d)} + 2\max_{1\le a,b\le D}\sum_{c=1}^{D}r_0^{(a,c)}\sum_{d=1}^{D}r_1^{(b,d)}\right),$$

we thus obtain

Lemma 12.3. *Assume that the vector valued stationary sequence $(X_n)_{n\in\mathbb{Z}}$ satisfies conditions (12.3.7) and (12.3.6) for some $\rho \geq 1$, then (using notation (12.3.12))*

$$\mathbb{E}\left\|\widehat{f}(\lambda) - f(\lambda)\right\|^2 \leq C\left(\frac{1}{n^2} + \frac{1}{m^{2\rho}} + \frac{m}{n}\right)$$

This expression is optimized as $\mathcal{O}\left(n^{-2\rho/(2\rho+1)}\right)$ by setting $m = n^{1/(2\rho+1)}$.

12.3.2 Dependence coefficients

This section is aimed to prove that bounds of higher order moments needed for deriving *a.s.* convergence of spectral density estimates are analogue to those computed for densities estimates in chapter 11, see § 11.2.2.
Spectral estimates are written as second order polynomials of the initial process. We thus need a translation table to compute the properties of the initial process in order to derive asymptotics.
In connection with § 4.3 and § 4.4.1 we now make use of the decorrelation coefficients (2.2.1), the following lemma is a first rough bound which relates those coefficients to those built upon the sequence $Z = (Z_k)_{k\in\mathbb{Z}}$ defined in an analogous way to (12.3.9), for some fixed complex valued sequence $\delta_{k,s} \in \mathbb{C}$ such that $\sup_{k,s} |\delta_{k,s}| \leq 1$ (here the dependence with respect to λ is omitted). In order to obtain a suitable bound of this coefficient, it seems unavoidable to make use of the previous diagram formula.

$$c_{Z,q}(r) = \max_{\substack{k_1 < \cdots \leq k_q \\ k_{l+1} - k_l = r \\ 1 \leq a_1, \ldots, a_q \leq D \\ 1 \leq b_1, \ldots, b_q \leq D}} \left|\mathrm{Cov}\left(Z_{k_1}^{(a_1,b_1)} \cdots Z_{k_l}^{(a_l,b_l)}, Z_{k_{l+1}}^{(a_{l+1},b_{l+1})} \cdots Z_{k_q}^{(a_q,b_q)}\right)\right|$$

Those coefficients are already defined for vector valued sequences in (4.4.7). Set,

$$C = \left|\mathrm{Cov}\left(Z_{k_1}^{(a_1,b_1)} \cdots Z_{k_l}^{(a_l,b_l)}, Z_{k_{l+1}}^{(a_{l+1},b_{l+1})} \cdots Z_{k_q}^{(a_q,b_q)}\right)\right|$$

if $k_{l+1} - k_l = r$, then (in a condensed notation)

$$\begin{aligned} C &= \sum_{|s_1|,\ldots,|s_q| \leq m} \left|\mathrm{Cov}\left(Y_{k_1,s_1}^{(a_1,b_1)} \cdots Y_{k_l,s_l}^{(a_l,b_l)}, Y_{k_{l+1},s_{l+1}}^{(a_{l+1},b_{l+1})} \cdots Y_{k_q,s_q}^{(a_q,b_q)}\right)\right| \\ &\leq \sum_{u=1}^{q} \sum_{\mu_1,\ldots\mu_u} \sum_{|s_1|,\ldots,|s_q| \leq m} \prod_{i=1}^{u} |\kappa_{\mu_i(k,s)}(X^{(a,b)})| \end{aligned}$$

where the previous sum extends on such undecomposable diagrams such that some μ_i is not entirely contained neither in the past nor in the future of the table

$$\text{Past}\quad\begin{cases}(1,1), & (1,2)\\ (2,1), & (2,2)\\ \dots & \dots\\ (l,1), & (l,2)\end{cases}$$

$$\text{Future}\quad\begin{cases}(l+1,1), & (l+1,2)\\ \dots & \dots\\ (q,1), & (q,2)\end{cases}$$

The number of such regular diagrams is thus $q!$ (each one is defined by one term in the past and one term in the future). In the general case, set $\lambda_i = \#\mu_i$ for $1 \le i \le u$, then if μ_i contains v_i terms $\big((j_{i,1},2),\dots,(j_{i,v_i},2)\big)$ from the second column of the table

$$\begin{aligned} C_{\mu_1,\dots,\mu_u} &= \sum_{|s_1|,\dots,|s_q|\le m} \prod_{i=1}^{u} \kappa_{\mu_i(k,s)}(X^{(a,b)}) \\ |C_{\mu_1,\dots,\mu_u}| &\le \prod_{i=1}^{u} \sum_{|s_{j_{i,1}}|,\dots,|s_{j_{i,v_i}}|\le m} |\kappa_{\mu_i(k,s)}(X^{(a,b)})| \end{aligned}$$

For the general case we need to consider sums of cumulants indexed by $\nu(s) = (h_1+s_1,\dots,h_i+s_i,h'_1,\dots,h'_j)$

$$\sum_{|s_1|,\dots,|s_i|\le m} |\kappa_{\mu(s)}(X)|$$

Using stationarity, it is easy to check that this sum is bounded by $(2m+1)^w\kappa_{i+j}$ $(i+j=\#\mu)$ where $w=0$ or 1 according to the fact that $j\ne 0$ or $j=0$. Hence $|C_{\mu_1,\dots,\mu_u}| \le (2m+1)^w \prod_{i=1}^{u} \kappa_{\#\mu_i}$ where w is the number of those μ_i entirely in the second column of the table, hence $w \le \frac{q}{2}$. In fact, the non Gaussian partitions (those with $\mu_i > 2$ for some i) have a contribution of a lower order $\mathcal{O}(m^{[q/2]-1})$. Taking in account that in each partition one term has factors both in the Past and in the Future and lemma 4.11 we thus obtain

Theorem 12.2. *If the sums of cumulants of X, $\kappa_p < \infty$ are finite for each $p \le 4q$ (which holds if $\sum_r (r+1)^{p-2} c^*_{X,p}(r) < \infty$ from lemma 4.12) then there is a constant $K_q > 0$ such that*

$$c_{Z,q}(r) \le K_q (2m+1)^{[q/2]} c^*_{X,2q}(r-2m).$$

Using inequality (4.4.12) we thus derive the main result of this section.

Corollary 12.1. *Let $2p \ge 2$ be an even integer such that $\kappa_q < \infty$ for $q \le 4p$, then*

$$\mathbb{E}\|\widehat{f}(\lambda) - \mathbb{E}\widehat{f}(\lambda)\|^{2p} = \mathcal{O}\left(\left(\frac{m}{n}\right)^p\right)$$

• Note that lemma 4.12 prove that the previous conditions may also be expressed in terms of weak dependence coefficients (4.4.8).
• Moreover the condition

$$\sum_n \left(\frac{m(n)}{n}\right)^q < \infty$$

implies a.s. convergence of such spectral density estimates as this was done in § 11.2.2 for regression estimates.

Chapter 13

Econometric applications and resampling

Essentially few rigorous results are stated in this chapter. We are aimed here to check how weak dependence may be applied in standard applications. This last chapter is more aimed at providing reasonable directions for further investigations of times series. The chapter is organized as follows. In order to provide deep econometric motivations, Section 13.1 includes several situations where various weak dependence conditions arise. After some generic examples including bootstrapping, we consider specific problems including unit root problems and parametric or semiparametric problems in § 13.1.1, 13.1.2 and 13.1.3. A following section 13.2 reviews the question of bootstrap; some models based bootstraps are first considered (see also § 13.2.4). We consider the block bootstrap in § 13.2.1 and § 13.2.2 addresses GMM estimation for which weak dependence provides a complete proof of the results in Hall and Horowitz (1996) [101]. We also mention conditional bootstrap in § 13.2.3 and sieve bootstrap in § 13.2.4. Finally in Section 13.3 we study more completely the problem of limiting variance *(in the central limit theorem)* estimation under η-weak dependence.

13.1 Econometrics

Time series analysis is a major part of econometrics. Here we provide several examples of interest in which it is essential to consider dependent structures instead of simple independence. In some situations like bootstrap mixing notions seem useless (see § 13.2.2). An application concerns linearity tests in time series analysis. Rios (1996) [162] considers a stationary functional autoregressive model (13.2.1) where $r = L + C$ is the decomposition of the autoregression function into a sum of linear (L) and nonlinear (C) components. Local linearity of r is then tested via the null hypothesis

$$\mathbf{H}_0 : \int \left(r''(x)\right)^2 w(x)\ \mathrm{d}x = 0$$

where the weight function w has compact support. Rios (1996) [162] proves that a local linearity test can be handled in the strong mixing case. The function r is assumed to be ρ continuous. Then the plug-in estimator $\widehat{T} = \int \widehat{r}^2(x)w(x)\mathrm{d}x$ converges to $T = \int r^2(x)w(x)\mathrm{d}x$ if $\alpha_n = \mathcal{O}(n^{-a})$ and $a > 2 + 3/\rho$ and the bandwidth condition $h_n \in [n^{-\frac{1}{10}}, n^{-\frac{1}{2\rho-4}}]$. This result may be extended to weak dependence as in Ango Nze *et al.* (2002) [6].
Still another problem of interest is to test the independence of the innovations $(\xi_n)_{n\in\mathbb{Z}}$ in a regression model

$$X_n = aY_n + \xi_n.$$

This can be performed using the Durbin-Watson statistic which is a non correlation test. The latter can be written as a continuous functional of the Donsker line of the sequence $(\xi_n)_{n\in\mathbb{Z}}$.
Some other applications are detailed in the forthcoming subsections.

13.1.1 Unit root tests

Consider a stationary autoregressive sequence $(X_n)_{n\in\mathbb{Z}}$ generated by an i.i.d. sequence $(\xi_n)_{n\in\mathbb{Z}}$,

$$X_n = aX_{n-1} + \xi_n.$$

A classical problem is to test whether there is a unit root (that is $a = 1$).
In the specific context of aggregated time series, the assumption of white noise innovations seems to be rather strong. Phillips (1987) [145] develops unit root tests for mixing and heterogeneously distributed innovations. The ordinary least square estimate $\widehat{a}$ is shown to be a continuous functional of the Donsker line of the sequence $(\xi_n)_{n\in\mathbb{Z}}$. As an application of the functional central limit theorem, Phillips shows that a unit root test can be based on the fact that under the null hypothesis $\mathbf{H}_0 : a = 1$,

$$n\left(\widehat{a} - 1\right) \xrightarrow[n\to\infty]{\mathcal{D}} \frac{1}{2\int_0^1 W_t^2 \mathrm{d}t}\left(W_1^2 - \frac{\sigma_\xi^2}{\sigma^2}\right)$$

where W denotes the standard Brownian motion and $\sigma^2 = \sum_{-\infty}^{\infty} \mathrm{Cov}(\xi_0, \xi_k)$; in the initial i.i.d. frame $\sigma_\xi^2/\sigma^2 = 1$ but this is no more the case for dependent innovations. The author works with stationary strong mixing sequences, and conditions under which the functional CLT result holds are given before. This example, as the author suggests, can be generalized to error sequences $(\xi_n)_{n\in\mathbb{Z}}$ that allow for heteroskedasticity.

13.1.2 Parametric problems

Generalized method of moments (GMM) estimation procedures involve an estimate $\widehat{\theta}_n$, which is a solution of the arg-min problem $J_n(\widehat{\theta}_n) = \min_{\theta \in \Theta} J_n(\theta)$, where

$$J_n(\theta) = \left(\frac{1}{n}\sum_{i=1}^{n} g(X_i, \theta)\right)' \Omega \left(\frac{1}{n}\sum_{i=1}^{n} g(X_i, \theta)\right). \tag{13.1.1}$$

Here $\Theta \subset \mathbb{R}^d$ is a finite dimensional parameter set, and $g(\cdot,\cdot)$ is a given function such that $\mathbb{E}_{\theta_0} g(X_1, \theta_0) = 0$, where θ_0 is the true parameter point. In the time series context, the positive semi-definite matrix Ω is often replaced (see Hall and Horowitz, 1996, equation (3.2) [101]) by an asymptotically optimal weight matrix estimate

$$\begin{aligned}
\Omega_n(\theta) &= \frac{1}{n}\sum_{i=1}^{n} g(X_i, \theta) g(X_i, \theta)' + \sum_{j=1}^{\kappa} H(X_i, X_{i+j}, \theta), \\
H(x, y, \theta) &= g(x, \theta) g(y, \theta)' + g(y, \theta) g(x, \theta)',
\end{aligned}$$

and κ is such that $\mathbb{E} g(X_i, \theta) g(X_j, \theta)' = 0$ if $|i-j| > \kappa$. The statistic to test $\mathbf{H}_0 : \theta = \theta_0$ is $J_n(\theta) = K_n(\widehat{\theta}_n)' K_n(\widehat{\theta}_n)$, where $K_n(\theta) = \frac{1}{\sqrt{n}} \Omega_n(\theta)^{\frac{1}{2}} \sum_{i=1}^{n} g(X_i, \theta)$ (the square root of a symmetric positive matrix is uniquely defined). The following CLT holds under standard mixing assumptions:

$$T_n(\theta) = \sqrt{n}\, \Sigma_n^{-1} \left(\widehat{\theta}_n - \theta\right) \xrightarrow[n\to\infty]{\mathcal{D}} \mathcal{N}_d(0, I_d), \tag{13.1.2}$$

where the diagonal matrix Σ_n has d entries. GMM techniques naturally involve an unknown covariance matrix. In order to estimate such limiting distributions it will be natural to use the bootstrap techniques described in § 13.2.1.

13.1.3 A Semi-parametric estimation problem

We follow the presentation in Robinson (1989) [164]. He considers an economic variable observable at time n which is an $R \times 1$ vector of r.v.'s $(W_n)_{n\in\mathbb{Z}}$. We observe W_n at time $n = 1-P, 2-P, \ldots, T$ where P is nonnegative and T large. Hypotheses of economic interest often involve a subset $X_n = B(W_n', \ldots, W_{n-P}')$ of the array $\left(W_n', \ldots, W_{n-P}'\right)'$; for this B is a $J \times (PR)$ matrix formed from the PR–rowed identity matrix I_{PR} by omitting $PR - J$ rows (which means that in B, $PR - J$ elements of $W_n, \ldots, W_{n-P}$ are deleted). Thus, in B, elements of $W_n, W_{n-1}, \ldots, W_{n-P}$ which are not in X_n are deleted, and X_n can have elements in common with $X_{n+P-1}, \ldots, X_{n+1}, X_{n-1}, \ldots, X_{n-P}$. Let $X_n = (Y_n', Z_n')'$, where Y_n and Z_n are $K \times 1$ and $L \times 1$ vectors ($K + L = J$). The problem of interest is to test the hypothesis $\mathbb{E}(Y_n | Z_n) = 0$ against the alternative $\mathbb{E}(Y_n|Z_n) \neq 0$. This null hypothesis is written in the form $\tau = \int_{\mathbb{R}^L} H(z,z) f^2(z)\mathrm{d}z = 0$ for $M = 0$ and

$$\tau = \int_{\mathbb{R}^L} H(z,z)\left(f(z), f^{(1)}(z)', \ldots, f^{(M)}(z)'\right)' f(z)\mathrm{d}z = 0$$

for some $M > 0$ and some function $H(z,z)$ defined as

$$H(z_1, z_2) = \int_{\mathbb{R}^K \times \mathbb{R}^K} G(x_1, x_2)\mathrm{d}F(y_1|z_1)\mathrm{d}F(y_2|z_2)$$

for some convenient function G and where $F(A|z) = \mathbb{P}(Y_n \in A|Z_n = z)$ for any Borel set A of $\mathbb{R}^K$ and $z \in \mathbb{R}^L$ and $x_1 = (y_1', z_1')'$ and $x_2 = (y_2', z_2')'$. Here $f^{(j)}(z)$ denotes the vector of $j-$partial derivatives of f.
An example of this framework is given by $X_n = (Y_n', Z_n')'$, where $Y_n = (t_n, s_n')'$ and $Z_n = v_n$. The regression model

$$t_n = \beta'(s_n - \mathbb{E}_n s_n) + \gamma' \mathbb{E}_n s_n + u_n \tag{13.1.3}$$

is of common use in econometrics. Here s_n, t_n, v_n are respectively scalars, $p \times 1$ and $q \times 1$; they are observable random sequences while the innovation process (u_n) is centered and unobservable, so that $\mathbb{E}(u_n|s_n, t_n) = 0$; we denote $\mathbb{E}_n(\cdot) = \mathbb{E}(\cdot|v_n)$. In the case of a weakly dependent and stationary innovation process, Robinson (1989) [164] considers the hypothesis $\mathbf{H}_0 : \beta = 0$. In this case, the hypothesis can be written as before and Robinson calculates $\beta = \tau$ where $K = p+1$, $L = 1$, $M = 0$ and $G(x_1, x_2) = (t_1 - t_2)s_1\phi(v_1)$ for some function $\phi : \mathbb{R}^q \to \mathbb{R}$ (usually $\phi \equiv 1$). Robinson considers the statistics $\hat{\lambda} = n\hat{\tau}'\widehat{\Omega}^{-1}\hat{\tau}$ constructed from the $n-$sample $(X_1, \ldots, X_n)$. Here, $\hat{\tau} = \dfrac{1}{n^2 h^L}\sum_{i,j=1}^{n} G(X_i, X_j)k(Z_i - Z_j/h)$ is a $U-$statistics and $\widehat{\Omega}$ is the natural estimator of the covariance matrix of $\hat{\tau}$. One such estimate is $\widehat{\Omega} = \dfrac{1}{n}\sum_{i=1}^{n} c_i c_i'$.
Tapered versions might be preferred (see formula (2.21) of Robinson, 1989 [164]), here $c_i = \sum_{j=1}^{n} d_{i,j} + d_{j,i}$ with $d_{i,j} = G(X_i, X_j)\overline{k}(Z_i - Z_j)/h$, where $\overline{k}(z) = h^{-L}\left(k(z), h^{-1}k^{(1)}(z)', \ldots, h^{-M}k^{(M)}(z)'\right)$. Under $\beta-$mixing assumptions, Robinson proves that the above estimates are $\sqrt{n}-$consistent and satisfy a CLT. Under a natural $\beta-$mixing condition, Robinson proves in fact that the statistic $\widehat{\lambda}$ has asymptotically a χ^2-distribution if $\beta_j = \mathcal{O}(j^{-b})$ where $b > \mu/(\mu - 2)$ under the moment assumption $\sup_{i,j} \mathbb{E}|G(X_i, X_j)|^\mu < \infty$.
The $\beta-$mixing assumption allows to compare the joint distribution of the initial sequence with respect to a sequence of r.v.'s with *independent* blocks. This

reconstruction is due to Berbee's coupling lemma, no matter how big the size of the blocks may be. Yoshihara (1976) [193] derives a covariance inequality that fits to $U-$statistics. A way to get rid of $\beta-$mixing conditions is to consider an independent realization $\tilde{X}_1, \ldots, \tilde{X}_n$ of the trajectory $X_1, \ldots, X_n$. Now a simpler estimator of τ is given by

$$\widetilde{\tau} = \frac{1}{n^2 h^L} \sum_{i,j=1}^{n} G(X_i, \tilde{X}_j) k \left(\frac{Z_i - \tilde{Z}_j}{h} \right).$$

The asymptotic behavior of this expression is easy to derive under alternative weak dependence conditions by using our results because $\widetilde{\tau} = \frac{1}{h^L} \sum_{i=1}^{n} W_{n,h}(X_j)$ is the numerator of a Nadaraya-Watson kernel for the regression estimation problem $\mathbb{E}(s_1(t_1 - t)|v_1 = z)$ in the special case of the previous example. In fact this trick avoids the corresponding coupling construction for $U-$statistics.

13.2 Bootstrap

Consider the following example concerning bootstrap: let a stationary autoregressive sequence be generated by an independent and identically distributed (i.i.d.) and centered sequence $(\xi_n)_{n \in \mathbb{Z}}$:

$$X_n = r(X_{n-1}) + \xi_n. \tag{13.2.1}$$

Standard nonparametric estimation techniques provide an estimate of the autoregression function r. Let $\widehat{r}$ be a convenient estimator of r. Given data $(X_1, \ldots, X_n)$ from the sequence 13.2.1, another autoregressive process can be defined by

$$\widehat{X_n} = \widehat{r}(\widehat{X}_{n-1}) + \xi_n^*. \tag{13.2.2}$$

The innovations (ξ_n^*) are i.i.d. drawn according to the centered empirical measure of the estimated centered residuals,

$$\widehat{\xi}_i = \widetilde{\xi}_i - \frac{1}{n} \sum_{j=1}^{n} \widetilde{\xi}_j, \qquad \widetilde{\xi}_i = X_i - \widehat{r}(X_{i-1}), \qquad 1 \le i \le n.$$

From the first example § 1.5 this is clear that no mixing assumption can be expected for the model (13.2.2). However, our concept of fading memory can still be applied. Bickel and Bühlmann (1999) [18] set up such a new weak dependence condition in order to build critical bootstrap values for a linearity test in linear models. Doukhan and Louhichi (1999) [67] have extended it in order to fit models such as positively dependent sequences, Markov chains (with or without topological assumptions), and Bernoulli shifts. The Bernoulli shifts are defined in Assumption 1 of Hall and Horowitz (1996) [101] and are used throughout

that paper. The above mentioned weak dependence conditions yield standard results concerning convergence in distribution with a $\sqrt{n}$-normalization.
If now the process $X_k = H(\xi_k, \xi_{k-1}, \ldots)$ is defined as a Bernoulli shift a suitable form of the resampled version X_k^* of X_k is $\widehat{H}\left(\xi_k^*, \xi_{k-1}^*, \ldots, \xi_{k-l+1}^*, 0, \ldots\right)$. Here $\widehat{H}$ and ξ_k^* are i.i.d. random variables drawn uniformly with the distribution $\widetilde{F}_n$, the centered version of residuals distribution obtained through filtering. In the simple case of a linear process ($H(z_0, z_1, \ldots) = \sum_k a_k z_k$, see Section 13.2.2); in the general setting, one needs to develop additional estimation procedures. In order to describe the asymptotic properties of such processes one needs to know the limiting asymptotic behaviour of Bernoulli shifts.
Unfortunately such representations are not always known and additional bootstrap procedures have been considered.

13.2.1 Block bootstrap

We describe here the block-bootstrap procedure which is adapted to the times series $(X_i)_{i\in\mathbb{N}}$. Let $b = b(n)$ and $l = l(n)$ denote the number and the length of the blocks. Assume $b \cdot l = n$ and consider l blocks $(X_{(j-1)l+1}, \ldots, X_{jl})$ for $1 \le j \le b$. Blocks $(\widetilde{X}_{(j-1)l+1}, \ldots, \widetilde{X}_{jl})$ are now randomly drawn (uniformly) among those l blocks. A trajectory of the resampled process is obtained by concatenation of those block; however a problem clearly appears to connect those (see Künsch, 1989 [113]).

13.2.2 Bootstrapping GMM estimators

Using the notations in § 13.1.2, let $(\tilde{X}_i^*)_{1\le i\le n}$ denote a block-bootstrap sample and let $g^*(x,\theta) = g(x,\theta) - \mathbb{E}^* g\left(x, \widehat{\theta}_n\right)$. The expectation is taken with respect to the bootstrap distribution. The GMM estimate $\widehat{\theta}_n^*$ solves the arg-min problem

$$J_n^*(\theta) = \left(\frac{1}{n}\sum_{i=1}^n g^*(X_i^*, \theta)\right)' \Omega \left(\frac{1}{n}\sum_{i=1}^n g^*(X_i^*, \theta)\right) \tag{13.2.3}$$

when the matrix Ω is known.
In order to prove the consistency of this bootstrap procedure Hall and Horowitz (1996) [101] propose an uncomplete proof. However, weak dependence will allows us to prove rigorously this consistency. More precisely, if $X_n = h(\epsilon_n, \epsilon_{n-1}, \ldots)$ for some i.i.d. sequence $(\epsilon_i)_{i\in\mathbb{Z}}$, their Assumption 1 is*

$$\mathbb{E}\left\|h(\epsilon_n, \epsilon_{n-1}, \ldots) - h(\epsilon_n, \epsilon_{n-1}, \ldots, \epsilon_{n-m}, 0, 0, \ldots)\right\| \le \frac{e^{-dm}}{d}.$$

*The function $h : \mathbb{R}^{\mathbb{N}^*} \to B$ with B a Banach space with norm $\|\cdot\|$.

This condition holds for linear processes and it is claimed to imply geometric strong mixing in [101]. Andrews's simple example (1984) [2] proves that this does not hold in general. It implies however $\theta-$weak dependence with geometric decay yielding the following useful tail inequality for sums of functions of the sequence $\xi_n = f(X_n, \theta)^\dagger$. This is the main tool to prove the validity of the bootstrap in this dependent setting. A rigorous version of Lemma 1 in Hall and Horowitz (1996) [101] thus follows

Lemma 13.1 (Ango Nze & Doukhan, 2004 [7]). *Let (ξ_n) be a stationary $\eta-$weakly dependent sequence with $\mathbb{E}\xi_n = 0$ such that $\eta(r) = \mathcal{O}(e^{-ar})$ as $r \uparrow \infty$ for some $a > 0$, and $\mathbb{P}(|\xi_1| \geq z) = \mathcal{O}\left(|z|^{-33}\right)$, as $|z| \to \infty$. Then $R_n = n^{-1}\sum_{i=1}^n \xi_i$ satisfies*

$$\lim_{n\to\infty} n\,\mathbb{P}\left(|R_n| > n^{-\frac{2+\epsilon}{5}}\right) = 0.$$

Proof. Set $\xi_{i,n} = \xi_i \mathbf{1}_{\{|\xi_i|\leq n^{1/16}\}} - \mathbb{E}\xi_i \mathbf{1}_{\{|\xi_i|\leq n^{1/16}\}}$ and let $\widetilde{R}_n = \frac{1}{n}\sum_{i=1}^n \xi_{i,n}$. Then

$$\begin{aligned}
\mathbb{P}\left(|R_n| > 2n^{-\frac{2+\epsilon}{5}}\right) &\leq \mathbb{P}\left(|\widetilde{R}_n| > n^{-\frac{2+\epsilon}{5}}\right) + 2n^{1+\frac{2+\epsilon}{5}}\mathbb{E}\left|\xi_i \mathbf{1}_{\{|\xi_i|> n^{1/16}\}}\right| \\
&\leq n^{-\frac{32(2+\epsilon)}{5}}\mathbb{E}|R_n|^{32} + 2n^{1+\frac{2+\epsilon}{5}}\|\xi_1\|_4 \mathbb{P}^{\frac{3}{4}}\left(|\xi_1| > n^{\frac{1}{16}}\right) \\
&= n^{-1}\mathcal{O}\left(n^{\frac{-11+32\epsilon}{5}} + n^{\frac{\epsilon}{5}-\frac{47}{320}}\right) \\
&= o(n^{-1}).
\end{aligned}$$

Following precisely the same steps as in Hall and Horowitz (1996) [101], we thus prove, by only replacing their Lemma 1 by our lemma 13.1, that bootstrapping critical values for GMM estimators is an asymptotically valid procedure.

Remark 13.1. *Theorems 1, 2 and 3 in Hall and Horowitz (1996) [101] seem now to be rigorously proved. Analogous comments fit to a more recent paper by Andrews (2002) [4]. The exponent* 33 *in the previous lemma is unnatural and it will be improved in a forthcoming paper.*

The above procedure can be used for testing the null hypothesis $\mathbf{H}_0 : \theta = \theta_0$ against the bilateral alternative. Under $\mathbf{H}_0$ the studentized statistic $T_n(\theta)$ described in (13.1.2) satisfies with the critical value Q_α,

$$\mathbb{P}\left(|T_n(\theta)| > Q_\alpha\right) = \alpha + \mathcal{O}(1/n).$$

---

†In this equation, θ is really the parameter to be estimated and in order too avoid further confusion with the dependence we shall prove below a result under the weaker η weak dependence.

As in Götze & Hipp (1978) [99], Hall and Horowitz seem to prove that $T_n(\theta)$ and the bootstrap studentized statistic

$$T_n^*(\theta) = \sqrt{n}\,{\Sigma_n^*}^{-1}\left(\widehat{\theta}_n^* - \widehat{\theta}_n\right),$$

have close distributions in the sense that

$$\mathbb{P}\left(\sup_{z\in\mathbb{R}} \left|\mathbb{P}^*\left(T_n^*(\theta) \le z\right) - \mathbb{P}\left(T_n(\theta) \le z\right)\right| > n^{-a}\right) = o\left(n^{-a}\right), \qquad (13.2.4)$$

for a relevant integer $2a$, with $a \ge 1+\xi$, and the range of $\xi \in [0,1]$ is formulated according to the dependence assumptions prescribed. This relation comes from an Edgeworth development. It yields an improved acceptation rule for the test of $\mathbf{H}_0$:

$$\mathbb{P}\left(\left|T_n^*(\theta)\right| > Q_\alpha^*\right) = \alpha + \mathcal{O}\left(n^{-1-\xi}\right).$$

13.2.3 Conditional bootstrap

A simple local conditional bootstrap is investigated by Ango Nze *et al.* (2002) [6]. Consider a stationary process $(X_t, Y_t)_{t\in\mathbb{Z}}$. The local bootstrap for nonparametric regression is defined as follows: the empirical distribution for Y_t given $X_t = x$ writes

$$\widehat{F}(y|x) = \frac{1}{nb}\sum_{t=1}^{n} \mathbf{1}_{\{Y_t\le y\}} K\left(\frac{x - X_t}{b}\right) \Big/ \widehat{f}_{n,b}(x),$$

for a kernel density estimator with bandwidth $b = b(n)$. The bootstrap sample is now defined as $(X_t, Y_t^*)_{1\le t\le n}$ where $Y_t^* \sim \widehat{F}(\cdot\,|X_t)$ is independent of $(Y_s^*)_{s\ne t}$ conditionally to the data for $1 \le t \le n$. In this article, it is shown that the asymptotic properties of the local regression bootstrap estimator $\widehat{r}_{n,h}^*$ with bandwidth $h = h(n)$ constructed from this sample are analogue to those of the regression estimator $\widehat{r}_{n,h}$ contructed from the data $(X_t, Y_t)_{1\le t\le n}$:

$$\sup_{u\in\mathbb{R}} \Big|\mathbb{P}^*\left(\sqrt{nh}\left\{\widehat{r}_{n,h}^*(x) - \mathbb{E}^*\widehat{r}_{n,h}^*(x)\right\} \le u\right)$$
$$-\mathbb{P}\left(\left\{\sqrt{nh}\widehat{r}_{n,h}(x) - \mathbb{E}\widehat{r}_{n,h}(x)\right\} \le u\right)\Big| \xrightarrow{\mathbb{P}}_{n\to\infty} 0,$$

under suitable weak dependence assumptions (see theorem 4 in [6]).

13.2.4 Sieve bootstrap

Bickel and Bühlmann (1999) [18] tackle the problem of the 'sieve bootstrap' for a one sided linear process

$$X_n - \mu = \xi_0 + \sum_{t=1}^{\infty} a_t \xi_{n-t} \qquad (13.2.5)$$

where (ξ_n) is a sequence of i.i.d. random variables (r.v.'s) with $\mathbb{E}\xi_0 = 0$ and the density function f_ξ, and where $\sum_{t=1}^{\infty} |a_t| < \infty$ and $\mu = \mathbb{E}X_n$.

Under the assumption that the function $\Psi(z) = 1 + \sum_{t=1}^{\infty} a_t z^t$ has no root in the closed unit circle, the process (13.2.5) admits an AR(∞) representation

$$(X_n - \mu) + \sum_{t=1}^{\infty} b_t (X_{n-t} - \mu) = \xi_n, \qquad \text{with} \qquad \sum_{t=1}^{\infty} |b_t| < \infty. \tag{13.2.6}$$

The latter process (13.2.6) is fitted with an autoregressive process of finite order $p(n)$ ($p(n)/n \to 0$, $p(n) \to \infty$). Using estimated residuals, the resampling (i.i.d.) innovation process $(\xi_n^*)_{n\in\mathbb{Z}}$ is constructed by smoothing the empirical process based on those residuals by a kernel density estimate of the density f_ξ. Finally, the smoothed sieve bootstrap sample $(X_n^*)_{n\in\mathbb{Z}}$ is defined by resampling the AR($p(n)$) process from innovations $(\xi_n^*)_{n\in\mathbb{Z}}$:

$$(X_n^* - \bar{X}) + \sum_{t=1}^{p(n)} \widehat{b}_t (X_{n-t}^* - \bar{X}) = \xi_n^* \tag{13.2.7}$$

The purpose of [18] was to carry over a weak dependence property (here strong mixing) of the initial sequence $(X_n)_{n\in\mathbb{Z}}$ to the sieve processes $(X_n^*)_{n\in\mathbb{Z}}$ (a classic and a smoothed version were examined in the paper). The goal is unrealistic for the classic bootstrap sample, since the distribution of the bootstrapped innovations is discrete. Proving a mixing property for the smoothed sieve bootstrap sample eludes the efforts of the authors. In the latter case, it nevertheless appears that limit theorems can be proven by another method. It consists in using the following property:

$$\begin{aligned} &\left|\mathrm{Cov}\left(g_1\left(X_{-d_1+1}, \ldots, X_0\right), g_2\left(X_k, \ldots, X_{k+d_2-1}\right)\right)\right| \\ &\qquad\qquad \leq 4 \|g_1\|_\infty \|g_2\|_\infty \, \nu\left(k; \mathcal{C}^{d_1}, \mathcal{C}^{d_2}\right), \end{aligned} \tag{13.2.8}$$

with $d_1, d_2 \in \mathbb{N}$ and for smooth functions g_1, g_2 belonging to the classes $\mathcal{C}^{d_1}$ and $\mathcal{C}^{d_2}$ (see equation (3.1) for the definition of the class $\mathcal{C}^d$ and some examples). The new dependence coefficient ν is less than the strongly mixing coefficient. Bickel and Bühlmann (1999) [18] cannot prove that the sieve sequence (X_n^*) is strongly mixing. A weak dependence condition is now defined by the ν coefficient. The authors prove that it is satisfied by both this sequence and a smooth version of the resampled innovations. For instance, Bickel and Bühlmann prove that if the sequence $(X_n)_{n\in\mathbb{Z}}$ satisfies some regularity conditions ensuring that $\alpha_k = \mathcal{O}\left(k^{-\gamma}\right)$ (recall that $\nu_k \leq \alpha_k$), then the sieve bootstrap process $(X_n^*)_{n\in\mathbb{Z}}$ satisfies a ν mixing condition with a polynomial rate $\nu^*(k; \mathcal{C}^{d_1}, \mathcal{D}^{d_2}) = \mathcal{O}\left(k^{-L\gamma}\right)$ for relevant classes $\mathcal{C}^{d_1}, \mathcal{D}^{d_2}$ and a positive constant L. See Theorem 3.2 on page 422 in Bickel and Bühlmann (1999) [18] for more details.

13.3 Limit variance estimates

The end of this chapter is aimed at providing a very important resampling tool for weakly dependent time series. In this monograph (and elsewhere) extensions of the central limit theorem have been proved for times series. The limiting variance takes a complicated form; the results write as

$$\frac{1}{\sqrt{n}}\sum_{k=1}^{n} X_k \xrightarrow{d}_{n\to\infty} \mathcal{N}(0,\sigma^2), \quad \text{where} \quad \sigma^2 = \sum_{k=-\infty}^{\infty} \mathbb{E}X_0X_k.$$

Functional vector valued versions of such results also arise,

$$\frac{1}{\sqrt{n}}\sum_{ns<j\le nt} X_j \to Z(t) - Z(s), \tag{13.3.1}$$

with $Z(t) \in \mathbb{R}^D$, the D-dimensional centered Gaussian random process such that

$$\mathrm{Cov}(Z(s), Z(t)) = (t \wedge s)\cdot \Sigma, \qquad \Sigma = \sum_{k=-\infty}^{\infty} \mathrm{Cov}(X_0, X_k).$$

For statistical use, one needs to provide self-normalized versions of such results. The expression reduces to $\sigma^2 = \mathbb{E}X_0^2$ for iid sequences and it may be directly estimated by $\frac{1}{n}\sum_{k=1}^{n} X_k^2$ using the method of moments. The first and natural way to achieve this for dependent sequences is to replace σ^2 by some convergent estimator. We shall assume $\eta-$weak dependence; several ways of estimating this quantity are reasonable.

• Recalling that σ^2 is only the value at origin of X_t's spectral density gives a first approach; spectral density estimates from chapter 12 yield as in § 12.3 estimators of σ^2, we defer a reader to Bardet *et al.* (2005) [11] for this approach.

• Another way to estimate this quantity is considered here: we mimick an argument by Carlstein (1986) [34], see also Peligrad and Shao (1994) [141]. This estimation is based on the Donsker invariance principle and a subsampling argument described in section 13.3.2.

We provide below *a.s.* convergence properties of those estimates as well as a CLT; modifications of the previous CLT will make it suitable for applications. To this aim, subsection 13.3.1 relates moments of sums with the cumulants of stationary sequences; this is a tool of an independent interest for several applications, like extensions to multispectra of the results in chapter 12.

We are involved here in a vector valued version of the estimation of σ^2. The main motivation for this is to derive a dependent version of Kolmogorov-Smirnov test. Empirical CLT (for the empirical cdf) are derived in chapter 10:

$$\frac{1}{\sqrt{n}}\sum_{k=1}^{n}\left(\mathbf{1}_{(X_k\le x)} - F(x)\right) \xrightarrow{d}_{n\to\infty} \overline{B}(x),$$

where $\overline{B}(x)$ denotes the centered Gaussian process such that

$$\mathbb{E}\overline{B}(x)\overline{B}(y) = \sum_{k=-\infty}^{\infty} \mathrm{Cov}\left(\mathbf{1}_{(X_0 \le x)}, \mathbf{1}_{(X_k \le y)}\right),$$

A direct extension of the Kolmogorov-Smirnov test is not possible for such dependent sequences because limits are not distribution free. After a convenient discretization, Doukhan & Lang (2002) [63] prove that confidence bounds for such statistics, which are not anymore distribution-free, may however be estimated with the present sharp estimates of the multivariate extension of σ^2.

The following subsection is aimed to derive useful tools in order to precise asymptotics for the previous estimation procedure.

13.3.1 Moments, cumulants and weak dependence

We thus consider a stationary vector valued sequence $(X_n)_{n\in\mathbb{Z}}$ with values in $\mathbb{R}^D$; we equip $\mathbb{R}^D$ with the norm defined as $\|x\| = |x^{(1)}| + \cdots + |x^{(D)}|$ for $x = (x^{(1)}, \ldots, x^{(D)}) \in \mathbb{R}^D$. We assume below that there exists some $b > 2$ such that $\|X_0\|_b = \left(\mathbb{E}\|X_0\|^b\right)^{\frac{1}{b}} < \infty$. For each $u \ge 1$ we identify the sets $\left(\mathbb{R}^D\right)^u$ and $\mathbb{R}^{D\cdot u}$. We use the coefficients $c_{X,q}(r)$ from (4.4.7). We define nonincreasing coefficients, for further convenience,

$$c^\star_{X,q}(r) = \max_{1\le l\le q} c_{X,l}(r)\mu_{q-l}, \qquad \text{with} \qquad \mu_t = \max_{1\le d\le D} \mathbb{E}\left|X_0^{(d)}\right|^t. \tag{13.3.2}$$

Those coefficients (4.4.7) are now linked to the weak dependence coefficients:

Proposition 13.1. *Assume that the stationary and vector valued sequence $(X_n)_{n\in\mathbb{Z}}$ is η-dependent and satisfies $\mu = \max_{1\le a\le D} \|X_0^{(u)}\|_b < \infty$ for some $b > q$ then the coefficients defined in eqn. (4.4.7) satisfy*

$$c_{X,q}(r) \le q4^{\frac{b}{b-1}} \mu^{\frac{(q-1)b}{b-1}} \eta(r)^{\frac{b-q}{b-1}}.$$

Proof of proposition 13.1. Consider integers $1 \le \ell < q$, $1 \le a_1, \ldots, a_q \le D$ and $t_1 \le \cdots \le t_q$ such that $t_{\ell+1} - t_\ell \ge r$, we need to bound (uniformly wrt ℓ, $1 \le a_1, \ldots, a_q \le D$ and $t_1 \le \cdots \le t_q$) the expression

$$c = \left|\mathrm{Cov}\left(X_{t_1}^{(a_1)} \cdots X_{t_\ell}^{(a_\ell)}, X_{t_{\ell+1}}^{(a_{\ell+1})} \cdots X_{t_q}^{(a_q)}\right)\right| = |\mathrm{Cov}(A, B)| \tag{13.3.3}$$

In order to bound (13.3.3), for some $M > 0$ depending on r [to be defined later] we now set $\overline{X}_j = (\overline{X}_j^{(1)}, \ldots, \overline{X}_j^{(D)})$ with $\overline{X}_j^{(a)} = X_j^{(a)} \vee (-M) \wedge M$ for each

$1 \le a \le D$ if $X_j = (X_j^{(1)}, \dots, X_j^{(D)})$. Then we also have,

$$\begin{aligned} c &\le |\mathrm{Cov}(\overline{A}, \overline{B})| + |\mathrm{Cov}(A - \overline{A}, B)| + |\mathrm{Cov}(\overline{A}, B - \overline{B})|, \\ \overline{A} &= \overline{X}_{t_1}^{(a_1)} \cdots \overline{X}_{t_\ell}^{(a_\ell)}, \qquad \overline{B} = \overline{X}_{t_{\ell+1}}^{(a_{\ell+1})} \cdots \overline{X}_{t_q}^{(a_q)}. \end{aligned}$$

Now we note that

$$\begin{aligned} |(A - \overline{A})B| &\le \sum_{i=1}^{\ell} Y_i |X_{t_i}^{(a_i)} - \overline{X}_{t_i}^{(a_i)}|, \\ |\overline{A}(B - \overline{B})| &\le \sum_{i=\ell+1}^{q} Y_i |X_{t_i}^{(a_i)} - \overline{X}_{t_i}^{(a_i)}| \end{aligned}$$

where Y_i writes as the product of $q-1$ factors $Z_{i,j} = |X_{t_j}^{(a_j)}|$ or $|\overline{X}_{t_j}^{(a_j)}|$ for $1 \le j \le q$ and $j \ne i$. It is thus clear that for $\frac{q-1}{b} + \frac{1}{p} = 1$ an analogue representation of the centering terms yields

$$|\mathrm{Cov}(A - \overline{A}, B)| + |\mathrm{Cov}(\overline{A}, B - \overline{B})| \le 2 \sum_{i=1}^{q} \max_{1 \le a \le D} \|X_0^{(a)}\|_b^{q-1} \max_{1 \le a \le D} \|X_0^{(a)} - \overline{X}_0^{(a)}\|_p.$$

Set $h(x) = x \vee (-M) \wedge M$ then $\mathrm{Lip}\, h = 1$ and $\|h\|_\infty = M$ thus $f_\ell(x_1, \dots, x_\ell) = h(x_1) \cdots h(x_\ell)$ is such that $\overline{A} = f_\ell\left(X_{t_1}^{(a_1)}, \dots, X_{t_\ell}^{(a_\ell)}\right), \overline{B} = f_{q-\ell}\left(X_{t_{\ell+1}}^{(a_{\ell+1})}, \dots, X_{t_q}^{(a_q)}\right)$ and $\|f_\ell\|_\infty \le M^\ell$, $\mathrm{Lip}\, f_\ell \le M^{\ell-1}$; from η-dependence we derive,

$$|\mathrm{Cov}(\overline{A}, \overline{B})| \le (\ell \mathrm{Lip}\, f_\ell \|f_{q-\ell}\|_\infty + (q-\ell)\mathrm{Lip}\, f_{q-\ell} \|f_\ell\|_\infty)\, \eta(r) \le q M^{q-1} \eta(r).$$

Now for each real valued random variable $X_1^{(a)}$ and $1 \le a \le D$,

$$\begin{aligned} \mathbb{E}|X_0^{(a)} - \overline{X}_0^{(a)}|^p &= \mathbb{E}|X_0^{(a)} - \overline{X}_0^{(a)}|^p \mathbf{1}_{|X_0^{(a)}| \ge M} \\ &\le 2^p \mathbb{E}|X_0^{(a)}|^p \mathbf{1}_{|X_0^{(a)}| \ge M} \\ &\le 2^p \mathbb{E}|X_0^{(a)}|^b M^{p-b}. \end{aligned}$$

Hence,

$$\|X_0^{(a)} - \overline{X}_0^{(a)}\|_p \le 2 \|X_0^{(a)}\|_b^{\frac{b}{p}} M^{1-\frac{b}{p}},$$

thus setting $\mu = \max_{1 \le a \le D} \|X_0^{(a)}\|_b$, we obtain with the previous inequalities,

$$\begin{aligned} |\mathrm{Cov}(A - \overline{A}, B)| &+ |\mathrm{Cov}(\overline{A}, B - \overline{B})| \le 4q\mu^{q-1+\frac{b}{p}} M^{1-\frac{b}{p}}, \\ |\mathrm{Cov}(A, B)| &\le q\left(4\mu^{q-1+\frac{b}{p}} M^{1-\frac{b}{p}} + M^{q-1}\eta(r)\right) \\ &\le q\left(4\mu^b M^{q-b} + M^{q-1}\eta(r)\right). \end{aligned}$$

The previous expression has an order optimized with $4\mu^b M^{1-b} = \eta(r)$ yielding $c = |\mathrm{Cov}(A,B)| \le q4^{\frac{b}{b-1}}\mu^{\frac{(q-1)b}{b-1}}\eta(r)^{\frac{b-q}{b-1}}$. □

As a first application of this relation, it seems useful to state the following moment inequality, they also entail laws of large numbers.

Corollary 13.1 ($D=1$). *Let $(X_t)_{t\in\mathbb{Z}}$ be a real valued and stationary η-weakly dependent times series. Assume that $\mathbb{E}|X_0|^b < \infty$ for some $b > q$, $\mathbb{E}X_0 = 0$, and*

$$\sum_{r=0}^{\infty}(r+1)^{q-2}\eta(r)^{\frac{b-q}{b-1}} < \infty,$$

then there exists a constant $C > 0$ only depending on q and on the previous series such that

$$\mathbb{E}\left(\sum_{j=1}^{n} X_j\right)^q \le Cn^{[q/2]}.$$

Proof. We note, that $\sum_{r\ge 0}(r+1)^{q-2}c_{X,q}(r) < \infty$ and we use theorem 4.2 together with proposition 13.1 to conclude. □

13.3.2 Estimation of the limit variance

Now let N be some fixed integer and $n = N\tilde{m}$, we rewrite the previous limit (13.3.1) as $\tilde{\Delta}_{i,\tilde{m}} \to \tilde{\Delta}_i$ (which have the same distribution) where we set

$$\tilde{\Delta}_{i,\tilde{m}} = \frac{1}{\sqrt{\tilde{m}}}\sum_{j=i+1}^{i+\tilde{m}} X_j,\ \Delta_i = \sqrt{N}\left(Z\left(\frac{i}{m}+1\right) - Z\left(\frac{i}{m}\right)\right) \stackrel{\mathcal{D}}{=} \Delta \sim \mathcal{N}_D(0,\Sigma).$$

The heuristic in Peligrad and Shao (1995) [142]'s variance estimator consists to substitute Δ_i by $\tilde{\Delta}_{i,\tilde{m}}$ in this last relation. For different values of i the random variables Δ_i are independent only for values with a difference $\ge \tilde{m}$. On another hand if $i = i(n), j = j(n)$ are such that $\lim_{n\to\infty}|i(n)-j(n)|/\tilde{m}(n) = \alpha$ then the couple $(\Delta_{i(n)}, \Delta_{j(n)})$ is Gaussian with $\mathrm{Cov}(\Delta_{i(n)}, \Delta_{j(n)}) = 1-\alpha$, if $0 \le \alpha \le 1$. This implies

$$\lim_{n\to\infty}\mathrm{Var}\left(\frac{1}{\sqrt{n-\tilde{m}(n)}}\sum_{i=1}^{n-\tilde{m}(n)} F(\tilde{\Delta}_{i,\tilde{m}(n)})\right) = \int_0^1 \mathrm{Cov}(F(Z(1),\ F(Z(1+\alpha)-Z(\alpha)))\,d\alpha.$$

This integral may be explicitly computed with the help of an Hermite expansion but it is however somewhat complicates. We thus subsample this sums setting $\Delta_{i,\tilde{m}} = \tilde{\Delta}_{i\tilde{m},\tilde{m}}$, then,

$$\frac{1}{N}\sum_{i=1}^{N} F(\Delta_i) \quad \to_{N\to\infty} \quad \mathbb{E}F(\Delta) \tag{13.3.4}$$

$$\frac{1}{\sqrt{N}}\sum_{i=1}^{N}(F(\Delta_i) - \mathbb{E}F(\Delta_i)) \quad \stackrel{\mathcal{D}}{\to}_{N\to\infty} \quad \mathcal{N}_D\left(0, \operatorname{Var} F\left(\Delta\right)\right). \tag{13.3.5}$$

Examples

- $F(x) = x'x$ yields the estimation of the covariance matrix Σ.
- Let $F(x) = |x'a|$ for a fixed vector $a \in \mathbb{R}^D$, then $\mathbb{E}F(\Delta) = \sqrt{a'\Sigma a}\cdot \mathbb{E}|\mathcal{N}(0,1)|$.
- Setting more generally $F(x_1,\ldots,x_D) = (|x_i + x_j|^2)_{1\le i\le j\le D} \in \mathbb{R}^{D(D+1)/2}$ provides estimates for all the coefficients of the matrix Σ: for this, use the polarity argument

$$\sigma_{i,j} = \frac{1}{2}\left(\operatorname{Var}\left(\Delta^{(i)} + \Delta^{(j)}\right) - \frac{1}{4}\operatorname{Var} 2\Delta^{(i)} - \frac{1}{4}\operatorname{Var} 2\Delta^{(j)}\right).$$

In order to make the heuristic (13.3.5) work, we better consider sequences $\ell = \ell(n)$, $m = m(n)$, $\tilde{m} = m + \ell$ and $N = N(n)$ converging to infinity and such that $n \ge N(m+\ell) - \ell$. Now we set

$$\Delta_{i,m} = \frac{1}{\sqrt{m}}\sum_{j=(i-1)(m+\ell)+1}^{(i-1)(m+\ell)+m} X_j, \tag{13.3.6}$$

and $\mathbb{E}F(\Delta)$ is estimated by

$$\widehat{F}_n = \frac{1}{N(n)}\sum_{i=1}^{N(n)} F(\Delta_{i,m(n)}). \tag{13.3.7}$$

Peligrad and Shao (1995) [142] consider instead that $\mathbb{E}F(\Delta)$ is estimated by

$$\widetilde{F}_n = \frac{1}{n-\tilde{m}}\sum_{i=1}^{n-\tilde{m}} F(\tilde{\Delta}_{i,\tilde{m}(n)}).$$

We need to quote that in this case the sum runs on $1 \le i \le n - \tilde{m}$ which is a number of much larger order than N. In this case, the Gaussian limiting variables are not independent. This makes the limiting distribution a bit more complicated and we thus avoid this estimation by working analogously to Carlstein (1986) [34]. More precisely, this author proves that if $i = i(n), j = j(n)$ vary in such a way that $\liminf_{n\to\infty} |i(n) - j(n)|/\tilde{m}(n) \geq 1$ then

$$\mathrm{Cov}\left(\tilde{\Delta}_{i(n),\tilde{m}(n)}, \tilde{\Delta}_{j(n),\tilde{m}(n)}\right) \to_{n\to\infty} 0$$

This is exactly our situation and the variables $\Delta_{i,m}$ are here asymptotically independent as $n \uparrow \infty$ for any choice of ℓ. The setting in Carlstein is that of a strong mixing sequence and other type of increment are also investigated.
The forthcoming section is aimed to derive the asymptotic behaviour for this estimation.

13.3.3 Law of the large numbers

Lemma 13.2. *Let now $F : \mathbb{R}^D \to \mathbb{R}$ be a Lipschitz function. Assume that the sequence $(X_n)_{n\in\mathbb{Z}}$ is stationary and $\eta-$weakly dependent then*

$$\frac{1}{N}\sum_{i=1}^{N} F(\Delta_{i,m}) \stackrel{\mathbb{L}^2}{\to}_{N\to\infty} \mathbb{E}F(\Delta), \tag{13.3.8}$$

if $\mu = \max_{1\leq a\leq D} \left(\mathbb{E}|X_0^{(a)}|^b\right)^{\frac{1}{b}} < \infty$, $\sum_{r=0}^{\infty}(r+1)^2\eta(r)^{\frac{b-4}{b-1}} < \infty$ *for some* $b > 4$.

Proof of lemma 13.2. The proof follows from two steps.
Step 1. $\mathrm{Var}\left(\frac{1}{N}\sum_{i=1}^{N} F(\Delta_{i,m})\right) \leq \frac{1}{N}\sum_{i=1}^{N} |\mathrm{Cov}(F(\Delta_{1,m}), F(\Delta_{i,m}))| \to_{N\to\infty} 0.$
For this we first write

$$\begin{aligned} F_{i,m} = F(\Delta_{i,m}) &= f(X_{(i-1)(m+\ell)+1}, \ldots, X_{(i-1)(m+\ell)+m}), \quad \text{with} \\ f(x_1, \ldots, x_m) &= F\left(\frac{x_1 + \cdots + x_m}{\sqrt{m}}\right), \qquad x_i \in \mathbb{R}^D, \end{aligned}$$

thus $\mathrm{Lip}\, f \leq \mathrm{Lip}\, F/\sqrt{m}$; this means that covariances will not be directly controlled with weak dependence. We thus have to control

$$|\mathrm{Cov}(F(\Delta_{1,m}), F(\Delta_{i,m}))|, \qquad 1 \leq i \leq N \tag{13.3.9}$$

We first consider the variance term obtained with $i = 1$. If we replace F by $F - F(0)$ the expression (13.3.9) is unchanged so that we assume that $F(0) = 0$ thus $|F_{1,m}| \leq \mathrm{Lip}\, F \left\|\sum_{j=1}^{m} X_j\right\|/\sqrt{m}$ and

$$\mathbb{E}\left\|\sum_{j=1}^{m} X_j\right\|^2 \leq \mathbb{E}\Big(\sum_{k=1}^{D}\Big|\sum_{j=1}^{m} X_j^{(k)}\Big|\Big)^2 \leq D\sum_{k=1}^{D}\mathbb{E}\Big(\sum_{j=1}^{m} X_j^{(k)}\Big)^2.$$

With this relation and $\left|\mathrm{Cov}(X_0^{(k)}, X_r^{(k)})\right| \le 2^{3+\frac{2}{b-1}}\mu^{\frac{b}{b-1}}\eta(r)^{1-\frac{1}{b-1}}$ obtained from proposition 13.1, we derive:

$$\begin{aligned}
\mathbb{E}|F_{1,m}|^2 &\le 2D(\mathrm{Lip}\, F)^2 \sum_{k=1}^{D}\sum_{r=0}^{m-1} \left|\mathrm{Cov}(X_0^{(k)}, X_r^{(k)})\right| \\
&\le 2^{4+\frac{2}{b-1}}\mu^{\frac{b}{b-1}} D^2 \sum_{r=0}^{m-1} \eta(r)^{1-\frac{1}{b-1}} \\
&< \infty, \quad \text{from our assumption.}
\end{aligned}$$

Now consider $\overline{F}(x) = F(x) \vee (-M) \wedge M$ for some $M > 0$ to be defined later, we set $\overline{F}_{i,m} = \overline{F}(\Delta_{i,m})$, then $\overline{F}_{i,m} = \overline{f}(X_{(i-1)(m+\ell)+1}, \ldots, X_{(i-1)(m+\ell)+m})$, where $\overline{f}(x_1, \ldots, x_m) = \overline{F}(x_1 + \cdots + x_m/\sqrt{m})$; thus $\mathrm{Lip}\,\overline{f} \le \mathrm{Lip}\, F/\sqrt{m}$ and $\|\overline{f}\|_\infty \le M$ and we derive,

$$\begin{aligned}
\left|\mathrm{Cov}\left(F_{1,m}, F_{i,m}\right)\right| &\le \left|\mathrm{Cov}\left(\overline{F}_{1,m}, \overline{F}_{i,m}\right)\right| \\
&\quad + \left|\mathrm{Cov}\left(F_{1,m}, F_{i,m} - \overline{F}_{i,m}\right)\right| \\
&\quad + \left|\mathrm{Cov}\left(F_{1,m} - \overline{F}_{1,m}, \overline{F}_{i,m}\right)\right| \\
&\le \left|\mathrm{Cov}\left(\overline{F}_{1,m}, \overline{F}_{i,m}\right)\right| \\
&\quad + (\|F_{1,m}\|_2 + \|\overline{F}_{i,m}\|_2)\|F_{i,m} - \overline{F}_{i,m}\|_2 \\
&\le \left|\mathrm{Cov}\left(\overline{F}_{1,m}, \overline{F}_{i,m}\right)\right| + 2\|F_{1,m}\|_2 \|F_{i,m} - \overline{F}_{i,m}\|_2.
\end{aligned}$$

Then

$$\left|\mathrm{Cov}\left(\overline{F}_{1,m}, \overline{F}_{i,m}\right)\right| \le \frac{2\mathrm{Lip}\, F M}{\sqrt{m}} \eta((i-2)(\ell+m)+\ell).$$

On the other hand we already obtained $\|F_{1,m}\|_2 \le CD$ for some constant $C > 0$ and the following lemma 13.3 with $p = 2$ and $q = 4$ implies with corollary 13.1, $\|\Delta_{i,m}^{(k)}\|_4^4 \le c\mu^4$ for some constant if

$$\sum_{r=0}^{\infty}(r+1)^2\eta(r)^{\frac{b-4}{b-1}} < \infty, \quad \text{and} \quad \|F_{i,m} - \overline{F}_{i,m}\|_2 \le CD^2M^{-1}.$$

Lemma 13.3. *Let $q > p \ge 1$ and $1 \le i \le N$ then if $\Delta_{i,m} = (\Delta_{i,m}^{(1)}, \ldots, \Delta_{i,m}^{(D)})$,*

$$\|F_{i,m} - \overline{F}_{i,m}\|_p \le 2D^{q/p}\mathrm{Lip}^{\,q/p} F \cdot M^{1-q/p} \max_{1\le k\le D} \|\Delta_{i,m}^{(k)}\|_q^{q/p}.$$

If now q is an even integer, then if $\sum_{r=0}^{\infty}(r+1)^{q-2}\eta(r)^{\frac{b-q}{b-1}} < \infty$ the last moment is bounded uniformly wrt i and m. Useful cases are given for $q = 2$ and $q = 4$.

Proof of lemma 13.3. We first quote that $\|F_{i,m} - \overline{F}_{i,m}\|_p \leq 2\|F_{i,m}\,\mathbf{1}_{|F_{i,m}|\geq M}\|_p$, then

$$\begin{aligned}
\mathbb{E}|F_{i,m}|^p\,\mathbf{1}_{|F_{i,m}|\geq M} &\leq M^{p-q}\mathbb{E}|F_{i,m}|^q \\
&\leq \mathrm{Lip}^{\,q}F\cdot M^{p-q}\mathbb{E}|\Delta_{i,m}|^q \\
&\leq D^q\mathrm{Lip}^{\,q}F\cdot M^{p-q}\max_{1\leq k\leq D}\|\Delta_{i,m}^{(k)}\|_q^q \\
\|F_{i,m} - \overline{F}_{i,m}\|_p &\leq 2D^{q/p}\mathrm{Lip}^{\,q/p}F\cdot M^{1-q/p}\max_{1\leq k\leq D}\|\Delta_{i,m}^{(k)}\|_q^{q/p}.
\end{aligned}$$

Now corollary 13.1 implies that $\max_{1\leq k\leq D}\|\Delta_{i,m}^{(k)}\|_q^q$ is uniformly bounded. □

End of the proof of lemma 13.2. We thus have proved that there is some constant $C > 0$ such that

$$\begin{aligned}
\mathrm{Var}\left(\frac{1}{N}\sum_{i=1}^{N}F(\Delta_{i,m})\right) &\leq C\left(D^3M^{-1} + \frac{M}{\sqrt{m}}\eta(\ell) + \frac{D^2}{N}\right), \\
&\leq C'\left(D^{3/2}m^{-1/4}\sqrt{\eta}(\ell) + \frac{D^2}{N}\right).
\end{aligned}$$

The last inequality holds for some $C' > 0$ with the choice $M = D^{3/2}m^{1/4}\eta(\ell)^{-1/2}$.

Step 2. We now need to prove that $\mathbb{E}F(\Delta_{1,m}) \to_{m\to\infty} \mathbb{E}F(\Delta)$. We first note that the condition $\eta(r) = \mathcal{O}(r^{-\alpha})$ for $\alpha > 2 + 1/(b-2)$ from Bardet, León and Doukhan (2005) [11] for CLT to hold is implied by the assumptions in our result. Consider again the truncated function $\overline{F}$ but for some M which may vary with m, $|\mathbb{E}F(\Delta_{1,m}) - \mathbb{E}F(\Delta)| \leq A + B + C$, where $A = \mathbb{E}|F(\Delta_{1,m}) - \overline{F}(\Delta_{1,m})|$, $B = \mathbb{E}|\overline{F}(\Delta_{1,m}) - F(\Delta)|$, $C = \mathbb{E}|\overline{F}(\Delta) - F(\Delta)|$. Now as $M \to \infty$, $A \leq \|F(\Delta_{1,m}) - \overline{F}(\Delta_{1,m})\|_2 = \mathcal{O}(D^2/M)$ converges to 0 (uniformly wrt m), and $C \to_{M\to\infty} 0$, because $\mathbb{E}|F(\Delta)| < \infty$. Now, from the central limit theorem, $B \to 0$ as $m \to \infty$. Thus we may choose a sequence $M = M(m) \uparrow \infty$ such that $\mathbb{E}F(\Delta_{1,m}) \to \mathbb{E}F(\Delta)$.

13.3.4 Central limit theorem

Theorem 13.1. *Set $Z_n = \sum_{i=1}^N x_{i,n}$ with $x_{i,n} = \frac{1}{\sqrt{N}}\left(F(\Delta_{i,m}) - \mathbb{E}F(\Delta_{i,m})\right)$. Assume that the centered and stationary sequence (X_t) satisfies $\mathbb{E}|X_0|^b < \infty$ for some $b > 4$ and this is $\eta-$weakly dependent with $\eta(r) = \mathcal{O}\left(r^{-\alpha}\right)$ for some $\alpha > 2 + 1/(b-2)$. Then,*

$$Z_n \xrightarrow{\mathcal{D}}_{n\to\infty} \mathcal{N}(0, \mathit{Var}\,F(\Delta)),$$

where $N = N(n)$ is the largest number such that $n \geq N(m+\ell) - \ell$, and where $m = m(n) = [m^\gamma]$, $\ell = \ell(n) = [m^\delta]$ satisfy $\alpha > \frac{1}{2}/\delta + \frac{4}{3}(1-\gamma)/\delta$.

Remark. For $\alpha > 2 + 1/(b-2)$ one may choose such rates with $\gamma > \delta > 0$.

Proof. Consider $Z \sim \mathcal{N}(0, \operatorname{Var} F(\Delta))$. For $f \in C^3(\mathbb{R}, \mathbb{R})$, we want to control

$$|\mathbb{E}f(Z_n) - f(Z)| \le |\mathbb{E}f(Z^{(n)}) - f(Z)| + \sum_{i=1}^{N} |U_i|,$$

with $U_i = \mathbb{E}\left(f_i(w_{i,n} + x_{i,n}) - f_i(w_{i,n} + y_{i,n})\right)$, $w_{i,n} = \sum_{j=1}^{i-1} x_{i,n}$ and $f_i(t) = f\left(t + \sum_{j=i+1}^{n} y_{i,n}\right)$, where, as usual, empty sums are set equal to 0 and where $Z^{(n)} = \sum_{i=1}^{N} y_{i,n}$, $y_{i,n} \sim \mathcal{N}(0, \operatorname{Var} x_{i,n})$, and the Gaussian random variables $y_{i,n}$ are set as jointly independent and independent of the process $(X_t)_{t\in\mathbb{Z}}$.

Step 1. *Lindeberg technique.* Notice that $\|f_i^{(j)}\|_\infty \le \|f^{(j)}\|_\infty$ for $j = 0, 1, 2, 3$ and for each $i \le N$. From a Taylor expansion and from the bound $\mathbb{E}|y_{i,n}|^3 \le \mathbb{E}|\mathcal{N}(0,1)|^3 \left(\mathbb{E}x_{i,n}^2\right)^{3/2} \le \mathbb{E}|\mathcal{N}(0,1)|^3 \mathbb{E}|x_{i,n}|^3$, there exists a constant $c > 0$ such that

$$\sum_{i=1}^{N} |U_i| \le \frac{c}{\sqrt{N}} \mathbb{E}|\Delta_{1,m}|^3 + \sum_{i=1}^{N} |\operatorname{Cov}(f_i'(w_{i,n}, x_{i,n})| + \frac{1}{2} \left|\operatorname{Cov}(f_i''(w_{i,n}, x_{i,n}^2)\right|.$$

As mentioned in the previous section the first term is bounded as $\mathcal{O}(1/\sqrt{N})$ if the moment of order 4 of a sum is suitably controlled. In order to make use of the weak dependence condition, we have to truncate the random variables $x_{i,n}$. Set $\overline{F}(t) = F(t) \vee (-M) \wedge M$ for a truncation level $M > 0$ precisely set later, then $\overline{x}_{i,n} = \frac{1}{\sqrt{N}}\left(\overline{F}(\Delta_{i,m}) - \mathbb{E}\overline{F}(\Delta_{i,m})\right)$ writes as a function $\overline{x}_{i,n} = g(X_{(i-1)(m+\ell)+1}, \ldots, X_{(i-1)(m+\ell)+n})$ where

$$g(t_1, \ldots, t_m) = \frac{1}{\sqrt{N}} \left(\overline{F}\left(\frac{t_1 + \cdots + t_m}{\sqrt{m}}\right) - \mathbb{E}\overline{F}(\Delta_{i,m})\right),$$

thus $\|g\|_\infty \le 2M/\sqrt{N}$ and $\operatorname{Lip} g \le \operatorname{Lip} F/\sqrt{Nm}$. By another hand, we may write $f_i^{(j)}(w_{i,n}) = G\left((X_s)_{s \le (i-1)(m+\ell)-\ell}\right)$ where $G : (\mathbb{R}^D)^{(i-1)(m+\ell)} \to \mathbb{R}$ satisfies $\|G\|_\infty \le \|f^{(j)}\|_\infty$, $\operatorname{Lip} G \le \|f^{(j+1)}\|_\infty \operatorname{Lip} F/\sqrt{mN}$ and is a function of less than n variables in $\mathbb{R}^D$.

$$\begin{aligned}
|\operatorname{Cov}(f_i'(w_{i,n}), x_{i,n})| &\le |\operatorname{Cov}(f_i'(w_{i,n}), \overline{x}_{i,n})| + \frac{2\|f'\|_\infty}{\sqrt{N}} \|F(\Delta_{1,m}) - \overline{F}(\Delta_{1,m})\|_1 \\
|\operatorname{Cov}(f_i'(w_{i,n}), \overline{x}_{i,n})| &\le c\sqrt{m} M \eta_{(i-1)(m+\ell)+\ell} \le c\sqrt{m} M \eta_\ell, \text{ with } \eta \text{ dependence,} \\
\|F_{i,m} - \overline{F}_{i,m}\|_1 &\le cD^2 M^{-3}, \qquad \text{from lemma 13.3.}
\end{aligned}$$

With $M = \eta(\ell)^{-1/4}m^{-1/8}N^{-1/8}D^{1/4}$ we thus arrive with some constant $c > 0$ to $|\mathrm{Cov}(f_i'(w_{i,n}), x_{i,n})| \le cD^{1/4}m^{3/8}N^{3/8}\eta(\ell)^{3/4} \le cD^{1/4}n^{3/8}\eta(\ell)^{3/4}$. Analogously

$$\begin{aligned}
\left|\mathrm{Cov}(f_i''(w_{i,n}), x_{i,n}^2)\right| &\le \left|\mathrm{Cov}(f_i''(w_{i,n}), \overline{x}_{i,n}^2)\right| + 2\|f''\|_\infty \mathbb{E}|\overline{x}_{i,n}^2 - x_{i,n}^2| \\
\left|\mathrm{Cov}(f_i''(w_{i,n}), \overline{x}_{i,n}^2)\right| &\le c\left(n\frac{1}{\sqrt{mN}}\frac{M^2}{N} + m\frac{M}{\sqrt{mN}\sqrt{N}}\right)\eta(\ell), \\
&\le c\frac{M^2\sqrt{n}}{N}\eta(\ell), \qquad \text{with } \eta \text{ dependence.} \\
\mathbb{E}|\overline{x}_{i,n}^2 - x_{i,n}^2| &\le \frac{2}{N}\|F(\Delta_{i,m})\|_2 \|F(\Delta_{i,m}) - \overline{F}(\Delta_{i,m})\|_2 \\
&\le c\frac{D^3}{MN}, \qquad \text{from lemma 13.3.}
\end{aligned}$$

Here we choose $M = \eta(\ell)^{-1/3}n^{-1/6}D$ to obtain, for some constant $c > 0$,

$$\left|\mathrm{Cov}(f_i''(w_{i,n}), x_{i,n}^2)\right| \le c\frac{D^2 n^{1/6}}{N}\eta(\ell)^{1/3}.$$

We thus have proved that

$$\sum_{i=1}^{N} |U_i| \le c\left(ND^{1/4}n^{3/8}\eta(\ell)^{3/4} + D^2 n^{1/6}\eta(\ell)^{1/3} + \frac{1}{\sqrt{N}}\right).$$

Now consider $m(n) = [n^\gamma]$ and $\ell(n) = [n^\delta]$. Then $N(n) \sim cn^\beta$ where $\beta = 1 - \gamma$. In order that the previous expression converges to 0, we only need

$$\beta > 0, \quad \alpha > \frac{1}{2\delta}, \quad \alpha > \frac{1}{2\delta} + \frac{4(1-\gamma)}{3\delta}, \quad \delta < \gamma.$$

To conclude, this is enough to choose $\delta < \gamma$, such that those relations hold. If δ and γ are both close enough to 1, $\alpha > 2 + 1/(b-2)$ implies those inequalities.

Step 2. *Gaussian approximation.* To conclude we still need to bound the expression $|\mathbb{E}(f(Z^{(n)}) - f((Z))|$. For this set $Z = \sigma N$ and $Z^{(n)} = \sigma_m N$ for some standard Gaussian random variable N and $\sigma_m^2 = \mathrm{Var}\, F(\Delta_{1,m})$, $\sigma^2 = \mathrm{Var}\, F(\Delta)$. Then

$$|\mathbb{E}(f(Z^{(n)}) - f((Z))| \le \|f''\|_\infty |\sigma - \sigma_m|^2 \le \frac{\|f''\|_\infty}{\sigma}|\sigma^2 - \sigma_m^2|.$$

We need to prove that $\mathbb{E}F^2(\Delta_{1,m}) \to_{m\to\infty} \mathbb{E}F^2(\Delta)$ and we use step 2 in the proof of lemma 13.2. We first note that the condition $\eta(r) = \mathcal{O}(r^{-\alpha})$ with $\alpha > 2 + 1/(b-2)$ from Bardet, León and Doukhan (2005) [11] implies the CLT. We consider the truncated function $\overline{F}$ for some M which may vary with m,

$$|\mathbb{E}F^2(\Delta_{1,m}) - \mathbb{E}F^2(\Delta)| \le A + B + C,$$

where we set $A = \mathbb{E}|F^2(\Delta_{1,m}) - \overline{F}^2(\Delta_{1,m})|$, $B = \mathbb{E}|\overline{F}^2(\Delta_{1,m}) - \overline{F}^2(\Delta)|$, and $C = \mathbb{E}|\overline{F}^2(\Delta) - F^2(\Delta)|$. Now $A \le 2\|F^2(\Delta_{1,m})\|_2 \|F(\Delta_{1,m}) - \overline{F}(\Delta_{1,m})\|_2 = \mathcal{O}(D^2/M)$ (as $M \to \infty$) converges to 0 (uniformly wrt m), and $C \to_{M\to\infty} 0$ because $\mathbb{E}|F^2(\Delta)| < \infty$. From the central limit theorem, $B \to 0$ as $m \to \infty$. Thus we may choose a sequence $M = M(m) \uparrow \infty$ such that $\mathbb{E}F^2(\Delta_{1,m}) \to \mathbb{E}F^2(\Delta)$.

13.3.5 A non centered variant

An alternative more attractive result involves

$$T_n = \sum_{i=1}^{N} t_{i,n}, \quad \text{where} \quad t_{i,n} = \frac{1}{\sqrt{N}} \left(F(\Delta_{i,m}) - \mathbb{E}F(\Delta) \right).$$

A central limit theorem for this non centered quantity looks much more convenient to consider the estimation of the parameter $\mathbb{E}F(\Delta)$. Such results are not considered by Peligrad & Shao (1995) [142]. In order to derive them one still uses the Lindeberg technique with blocks. Hence, for some bounded and C^3 function f we bound again:

$$\begin{aligned} |\mathbb{E}(f(T_n) - f(Z))| &\le |\mathbb{E}(f(T_n) - f(Z_n))| + |\mathbb{E}(f(Z_n) - f(Z))| \\ &\le \sqrt{N}\|f'\|_\infty \, |\mathbb{E}F(\Delta_{1,m}) - \mathbb{E}F(\Delta)| + |\mathbb{E}(f(Z_n) - f(Z))| \end{aligned}$$

with $Z \sim \mathcal{N}(0, \operatorname{Var} F(\Delta))$. This means that the previous convergence relies on the decay rate of the expression $|\mathbb{E}F(\Delta_{1,m}) - \mathbb{E}F(\Delta)|$; as stressed in the forthcoming remark, this is $\mathcal{O}\left(\frac{1}{m}\right)$ if $F(x) = (x'a)^2$ and the previous quantity tends to zero under the assumption: $\lim_{n\to\infty} N/m^2 = 0$.

Examples of functions F

- *Case* $D = 1$. In this case for F such that $\int |F'(x)|/(1+x^2)\,dx < \infty$, Petrov (1996) lemma 5.4 page 152 [144] implies

$$|\mathbb{E}F(\Delta_{1,m}) - \mathbb{E}F(\Delta)| \le c(v_n + \delta) \int \frac{|F'(x)|}{1+x^2} dx$$

  with $v_m = \sup_{x\in\mathbb{R}} |\mathbb{P}(\Delta_{1,m} \le x) - \mathbb{P}(\Delta \le x)| = \mathcal{O}\left(m^{-\lambda}\right)$ for some $\lambda < \frac{1}{8}$ given in Bardet, Doukhan and León (2005) [11] and $\delta = \left|\mathbb{E}\Delta_{1,m}^2 - \mathbb{E}\Delta^2\right| = \mathcal{O}\left(\frac{1}{m}\right)$. This implies that the cases $F(x) = |x|^\beta$ for some $\beta \le 2$ is also obtained if, now, $\lim_{n\to\infty} N/m^{2\lambda} = 0$; Peligrad and Shao (1995) [142] consider the special case $\beta = 1$. Here $\lambda \to \frac{1}{8}$ as $a, b \to \infty$ if $\eta(r) = \mathcal{O}(r^{-a})$ and $\mathbb{E}|X_0|^b < \infty$.

- *Case* $D = 1$. If now $F(x) = x^p$ then theorem 4.6 provides a bound $\mathcal{O}(1/m)$ for each integer p which resembles the Rosenthal inequality in the independent case (see Hall & Heyde, 1980 [100]).

- *Case* $D > 1$. The special case $F(x) = (x'a)^2$, for some $a \in \mathbb{R}^D$ is the simplest multi-dimensional one. In this case, indeed setting $\xi_t = X_t'a$, we still assume $\mathbb{E}\xi_t = 0$ and one may write

$$\mathbb{E}F(\Delta_{1,m}) = \sum_{|s|<m} \left(1 - \frac{|s|}{m}\right) \mathbb{E}\xi_0\xi_s, \quad \mathbb{E}F(\Delta) = \sum_{s=-\infty}^{\infty} \mathbb{E}\xi_0\xi_s,$$

$$\begin{aligned} \text{so that } \ \mathbb{E}F(\Delta) - \mathbb{E}F(\Delta_{1,m}) &= \frac{1}{m} \sum_{|s|<m} |s| \mathbb{E}\xi_0\xi_s + \sum_{|s|\geq m} \mathbb{E}\xi_0\xi_s \\ &= \frac{1}{m} \sum_{s=-\infty}^{\infty} |s| \mathbb{E}\xi_0\xi_s + \sum_{|s|\geq m} \left(1 - \frac{|s|}{m}\right) \mathbb{E}\xi_0\xi_s, \end{aligned}$$

thus

$$\left| \mathbb{E}F(\Delta) - \mathbb{E}F(\Delta_{1,m}) - \frac{1}{m} \sum_{s=-\infty}^{\infty} |s| \mathbb{E}\xi_0\xi_s \right| \leq \sum_{|s|\geq m} |\mathbb{E}\xi_0\xi_s|. \qquad (13.3.10)$$

Using proposition 13.1 now yields if $a = (a^{(1)}, \ldots, a^{(D)})$,

$$|\mathbb{E}\xi_0\xi_s| \leq D\|a\|^2 c_{X,2}(s) \leq 2^{\frac{3b-1}{b-1}} D\|a\|^2 \mu^{\frac{b}{b-1}} \eta(s)^{1-\frac{1}{b-1}}.$$

This previous bound (13.3.10) is thus $\mathcal{O}\left(D\|a\|^2 \sum_{s\geq m} \eta(s)^{1-\frac{1}{b-1}}\right)$.

For fixed D this bound has order $o(m^{-1})$ if $\sum_{s=1}^{\infty} s\eta(s)^{1 \;\; \frac{1}{b-1}} < \infty$, and thus

$$\mathbb{E}F(\Delta) = \mathbb{E}F(\Delta_{1,m}) + \frac{1}{m} \sum_{s=-\infty}^{\infty} |s| \mathbb{E}\xi_0\xi_s + o\left(\frac{1}{m}\right).$$

Remark. The previous results given under η weak dependence will be extended in forthcoming papers under κ and λ-weak dependence.

Bibliography

[1] Aldous, D. J., Eagleson, G. K. (1978) *On mixing and stability of limit theorems.* Ann. Probab. 6, 325-331.

[2] Andrews, D. W. K. (1984) *Non strong mixing autoregressive processes.* J. Appl. Probab. 21, 930-934.

[3] Andrews, D. W. K. (1988) *Laws of large numbers for independent non identically distributed random variables.* Econometric Theory 4, 458-467.

[4] Andrews, D. W. K. (2002) *Higher-order Improvements of a Computationally Attractive k-step Bootstrap for Extremum Estimators.* Econometrica 70, 119-162.

[5] Andrews, W. K., Pollard, D. (1994) *An introduction to functional central limit theorems for dependent stochastic processes.* Int. Statist. Rev. 62, 119-132.

[6] Ango Nze, P., Bühlman, P., Doukhan, P. (2002) *Weak Dependence beyond Mixing and Asymptotics for Nonparametric Regression.* Annals of Statistics, 30-2, 397-430.

[7] Ango Nze, P., Doukhan, P. (2004) *Weak dependence, models and applications to econometrics.* Econometric Theory 20-6, 995-1045.

[8] Bakhtin, Yu., Bulinski, A. V. (1997) *Moment inequalities for sums of dependent multiindexed random variables.* Fundamentalnaya i prikladnaya matematika 3-4, 1101-1108.

[9] Barbour, A. D., Gerrard, R. M., Reinert, G. (2000) *Iterates of expanding maps.* Probab. Theory Rel. Fields **116** 151-180.

[10] Bardet, J. M., Doukhan, P., Lang, G., Ragache, N. (2006) *A Lindeberg central limit theorem for dependent processes and its statistical applications.* Submitted.

[11] Bardet, J. M., Doukhan, P., León, J. R. (2006) *A uniform central limit theorem for the periodogram and its applications to Whittle parametric estimation for weakly dependent time series.* Submitted.

[12] Barron, A. R., Birgé, L., Massart, P. (1999) Risk bounds for model selection by penalization. Probab. Theory and Rel. Fields 113, 301-413.

[13] Bernstein, S. (1939) *Quelques remarques sur le théorème limite Liapounoff.* C. R. (Dokl.) Acad. Sci. USSR, Ser. 24, 1939, 3-8.

[14] Benaïm, M. (1996) *A dynamical system approach to stochastic approximation.* SIAM J. control and optimization 34-2, 437-472.

[15] Benveniste, A., Métivier, M. Priouret, P. (1987) *Algorithmes adaptatifs et approximations stochastiques.* Masson.

[16] Berbee, H. C. P. (1979) *Random walks with stationary increments and renewal theory,* Math. centre tracts, Amsterdam.

[17] Bickel, P., Bühlmann, P. (1995) *Mixing property and functional central limit theorems for a sieve bootstrap in times series.* Techn. Rep. 440, Dept. Stat., U. C. Berkeley.

[18] Bickel, P., Bühlmann, P. (1999) *A new mixing notion and functional central limit theorems for a sieve bootstrap in time series* Bernoulli 5-3, 413-446.

[19] Bickel, P. J., Wichura, M. J. (1971) *Convergence criteria for multiparameter stochastic processes and some applications.* Ann. Math. Stat. 42, 1656-1670.

[20] Billingsley, P. (1968) *Convergence of Probability Measures.* Wiley, New-York.

[21] Billingsley, P. (1995) *Probability and measure.* Third edition. Wiley Series in Probability and Mathematical Statistics. A Wiley-Interscience Publication. John Wiley & Sons, Inc., New York.

[22] Birkel, T. (1988) *Moment bounds for associated sequences.* Ann. Probab. 16, 1184-1193.

[23] Bollerslev, T. (1986) *Generalized autoregressive conditional heteroskedasticity* J. of Econometrics, 31, 307-327.

[24] Bolthausen, E. (1982) *On the central limit theorem for stationary random fields.* Ann. Probab. 10, 1047-1050.

[25] Borovkova, S., Burton, R., Dehling, H. (2001). *Limit theorems for functionals of mixing processes with application to U-statistics and dimension estimation.* Trans. Amer. Math. Soc. 353, 4261-4318.

[26] Bosq, D. (1993) *Bernstein's type large deviation inequalities for partial sums of strong mixing process.* Statistics 24, 59-70.

[27] Bougerol, P. (1993) *Kalman filtering with random coefficients and contractions.* SIAM J. Control Optim. 31, 942-959.

[28] Brockwell, P., Davis, R. (1987) *Time Series: Theory and Methods.* Springer-Verlag.

[29] Bradley, R. C. (1983) *Approximations theorems for strongly mixing random variables.* Michigan Math. J. 30, 69-81.

[30] Bradley, R. C. (2002) *Introduction to strong mixing conditions, Volume 1.* Technical Report, Department of Mathematics, I.U. Bloomington.

[31] Broise, A. (1996) *Transformations dilatantes de l'intervalle et théorèmes limites.* Études spectrales d'opérateurs de transfert et applications, Astérisque 238, 1-109.

[32] Brandière, O., Doukhan, P. (2004) *Dependent noise for stochastic algorithms.* Probab. Math. Statist. 2-24, 381-399.

[33] Bulinski, A. V., Sashkin, A. P. (2005) *Strong invariance principle for dependent multi-indexed random variables.* Doklady. Mathematics 72-11, 503-506, MAIK Nauka/Interperiodica.

[34] Carlstein, E. (1986) *The Use of Subseries Values for Estimating the Variance of a General Statistic from a Stationary Sequence.* Annals of Statistics 14-3, 1171-1179.

[35] Castaing, C., Raynaud de Fitte, P., Valadier, M. (2004). *Young Measures on Topological Spaces.* Mathematics and its applications, Kluwer.

[36] Chen, H-F. (1985) *Stochastic Approximation and Its Apllications.* Non convex Optimization and Its Application 64, Kluwer Academic Plublishers.

[37] Chow, Y. S. (1966) *Some convergence theorems for independent random variables.* Ann. Math. Stat. 37, 1482-1493 and 36-4, 1293-1314.

[38] Collomb, G. (1984) *Propriétés de convergence presque complète du predicteur de noyau.* Z. Wahrsch. Verw. Gebiete 66, 441-446.

[39] Comtet, L. (1970) *Analyse Combinatoire.* Tome 1. Coll. S.U.P. Presses Universitaires de France.

[40] Coulon-Prieur, C., Doukhan, P. (2000) *A triangular CLT for weakly dependent sequences.* Statist. Probab. Lett. 47, 61-68.

[41] Daubechies, I. (1988) *Orthonormal bases of compactly supported wavelets.* Comm. Pure & Appl. Math. 41, 909-996.

[42] Dedecker, J. (2001) *Exponential inequalities and functional central limit theorems for random fields.* ESAIM Probab. & Stat. `http://www.emath.fr/ps/` 5, 77-104.

[43] Dedecker, J., Doukhan, P. (2003) *A new covariance inequality and applications.* Stoch. Proc. Appl. 106-1, 63-80.

[44] Dedecker, J., Merlevède, F. (2002) *Necessary and sufficient conditions for the conditional central limit theorem.* Ann. Probab. 30, 1044-1081.

[45] Dedecker, J., Prieur, C. (2004a) *Coupling for τ-dependent sequences and applications.* Journal of Theor. Probab. 17-4, 861-885.

[46] Dedecker, J., Prieur, C. (2004b) *Couplage pour la distance minimale.* C. R. Acad. Sci. Paris Série 1 338-10, 805-808.

[47] Dedecker, J., Prieur, C., Raynaud De Fitte, P. (2006) *Parametrized Kantorovich-Rubinstein theorem and application to the coupling of random variables.*

[48] Dedecker, J., Prieur, C. (2005) *New dependence coefficients. Examples and applications to statistics.* Prob. Theor. and Rel. Fields 132, 203-236.

[49] Dedecker, J., Prieur, C.(2006) *An empirical central limit theorem for dependent sequences,* in Dependence in probapity and statistics, LNS 187, 105-121, Stoch. Proc. Appl., 117, 121-142.

[50] Dedecker, J., Rio, E. (2000) *On the functional Central Limit Theorem for Stationary processes.* Ann. I. H. P. ser. B 36-1, 1-34.

[51] Deheuvels, P. (1979) *La fonction de dépendance empirique et ses propriétés.* Acad. Roy. Belg., Bull. C1 Sci. 5ème sér. 65, 274-292.

[52] Deheuvels, P. (1981a) *A Kolmogorov-Smirnov Type Test for independence and multivariate samples.* Rev. Roum. Math. Pures et Appl. 26-2, 213-226.

[53] Deheuvels, P. (1981b) *A nonparametric test for independence.* Pub. Inst. Stat. Univ. Paris. 26-2, 29-50.

[54] Dehling, H., Philipp, W. (1982) *Almost sure invariance principles for weakly dependent vector-valued random variables.* Ann. Probab. 10, 689-701.

[55] Dehling, H., Taqqu, M. (1989) *The empirical process of some long-range dependent sequences with applications to U-statistics.* Ann. Statist. 17, 1767-1783.

[56] De Jong, R. M. (1996) *A strong law of large numbers for triangular mixingale arrays.* Statist. and Probab. Lett. 27, 1-9.

[57] Delyon, B. (1990) *Limit theorem for mixing processes.* Tech. Report IRISA, Rennes 1, **546**.

[58] Delyon, B. (1996) *General convergence result on stochastic approximation.* IEEE Trans. Automatic Control 41, 9.

[59] De Vore, R., Lorentz, G. (1993) *Constructive Approximation.* Springer-Verlag, Berlin.

[60] Diaconis, P., Freedman, D. (1999) *Iterated random functions.* SIAM Rev. 41-1, 45-76 .

[61] Doukhan, P. (1994) *Mixing: Properties and Examples.* Lecture Notes in Statistics 85. Springer Verlag.

[62] Doukhan, P. (2002) *Models inequalities and limit theorems for stationary sequences,* in Theory and applications of long range dependence (Doukhan *et alii* ed.) Birkhäuser, 43-101.

[63] Doukhan, P., Lang, G. (2002) *Rates of convergence in the weak invariance principle for the empirical repartition process of weakly dependent sequences.* Stat. Inf. for Stoch. Proc. 5, 199-228.

[64] Doukhan, P., Lang, G., Louhichi, S., Ycart, B. (2005) *A functional central limit theorem for interacting particle systems on transitive graphs.* Available from `http://arxiv.org/abs/math-ph/0509041`.

[65] Doukhan, P., Latour, A., Oraichi, D. (2006) *Simple integer-valued bilinear time series model.* Adv. Appl. Probab. 38-2, 559-578.

[66] Doukhan, P., León, J. R. (1989) *Cumulants for mixing sequences and applications to empirical spectral density.* Probab. Math. Stat. 10.1, 11-26.

[67] Doukhan, P., Louhichi, S. (1999) *A new weak dependence condition and applications to moment inequalities.* Stoch. Proc. Appl. 84, 313-342.

[68] Doukhan, P., Louhichi, S. (2001) *Functional estimation for weakly dependent stationary time series.* Scand. J. Statist. 28, 325-342.

[69] Doukhan, P., Madre, H., Rosenbaum, M. (2006) *Weak dependence beyond mixing for infinite ARCH-type bilinear models.* Statistics.

[70] Doukhan, P., Massart, P., Rio, E. (1994) *The functional central limit theorem for strongly mixing processes.* Ann. I. H. P. ser. B 30-1, 63-82.

[71] Doukhan, P., Neumann, M. (2006) *A Bernstein type inequality for times series.* Submitted to Stoch. Proc. Appl.

[72] Doukhan, P., Oppenheim, G., Taqqu, M., *editors* (2003) *Theory and Applications of Long-range Dependence.* Birkhaüser, Boston (2003).

[73] Doukhan, P., Portal, F. (1983) *Moments de variables aléatoires mélangeantes.* C. R. Acad. Sci. Paris Série 1 297, 129-132.

[74] Doukhan, P., Sifre, J.C. (2001) *Cours d'analyse - Analyse réelle et intégration.* Dunod.

[75] Doukhan, P., Teyssière, G., Winant, P. (2006) *Vector valued ARCH infinity processes,* in *Dependence in Probability and Statistics. Patrice Bertail, Paul Doukhan and Philippe Soulier Editors,* Springer, New York.

[76] Doukhan, P., Truquet, L. (2006) *Weakly dependent random fields with infinite interactions.*
Preprint in `http://samos.univ-paris1.fr/ppub2006.html`.

[77] Doukhan, P., Wintenberger, O. (2005) *An invariance principle for weakly dependent stationary general models.* Preprint in `http://samos.univ-paris1.fr/ppub2005.html`.

[78] Doukhan, P., Wintenberger, O. (2006) *Weakly dependent chains with infinite memory* Preprint in `http://samos.univ-paris1.fr/ppub2005.html`.

[79] Dudley, R. M. (1968). *Distances of probability measures and random variables.* Ann. Math. Statist. 39, 1563-1572.

[80] Dudley, R. M. (1989) *Real analysis and probability.* Wadworsth Inc., Belmont, California.

[81] Duflo, M. (1990) *Méthodes récursives aléatoires.* Masson, Paris (English Edition, Springer, 1996).

[82] Duflo, M. (1996) *Algorithmes Stochastiques,* Collection Math. et Appl. 23, Springer.

[83] Eberlein, E., Taqqu, M., *editors* (1986) *Dependence in Probability and Statistics.* Birkhaüser, Boston.

[84] Engle, R. F. (1982) *Autoregressive conditional heteroskedasticity with estimate of the variance in the U.K. inflation* Econometrica, 15, 286-301

[85] Esary, J., Proschan, F., Walkup, D. (1967) *Association of random variables with applications.* Ann. Math. Statist. 38, 1466-1476.

[86] Fermanian, J-D., Radulovic, D., Wegkamp, M. (2004) *Weak convergence of empirical copula processes.* Bernoulli 10, 847-860.

[87] Fortuyn, C., Kastelyn, P., Ginibre, J. (1971) *Correlation inequalities in some partially ordered sets.* Comm. Math. Phys. 74, 119-12.

[88] Fréchet, M. (1957) *Sur la distance de deux lois de probabilité.* C. R. Acad. Sci. Paris. 244-6, 689-692.

[89] Fuk, D. Kh., Nagaev, S. V. (1971) *Probability inequalities for sums of independent random variables.* Theory Probab. Appl. 16, 643-660.

[90] Garsia, A. M. (1965) *A simple proof of E. Hopf's maximal ergodic theorem.* J. of Maths and Mech. 14, 381-382.

[91] Ghosal, S., Chandra, T. K. (1998) *Complete convergence of martingale arrays.* J. Theor. Probab. 11-3, 621-631.

[92] Giraitis, L., Kokoszka, P., Leipus, R. (2000) *Stationary ARCH−models: dependence structure and central limit theorem.* Econometric Theory, 16, 3-22.

[93] Giraitis L., Leipus, R., Surgailis, D. (2006) *Recent advances in ARCH modelling* in *Long memory in Economics*, Ed. Gilles Teyssière and Alan Kirman, Springer Verlag.

[94] Giraitis L., Robinson P. M. (2001) *Whittle estimation of ARCH models* Econom. Theory 17-3, 608-631.

[95] Giraitis, L., Surgailis, D. (2002) *ARCH-type bilinear models with double long memory.* Stoch. Proc. and their Applic. 100, 275-300.

[96] Goldstein, S. (1979) *Maximal coupling.* Z. Wahrsch. Verw. Gebiete 46, 193-204.

[97] Gordin, M. I. (1969) *The central limit theorem for stationary processes.* Dokl. Akad. Nauk SSSR 188, 739-741.

[98] Gordin, M. I. (1973) *Abstracts of Communication, T.1: A-K.* International Conference on Probability Theory, Vilnius.

[99] Götze, F., Hipp, C. (1978) *Asymptotic expansions in the central limit theorem under moment conditions.* Z. Wahr. und Verw. Gebiete 42-1, 67-87.

[100] Hall, P., Heyde, C. C. (1980) *Martingale limit theory and its applications.* Academic Press.

[101] Hall, P., Horowitz, J. (1996) *Bootstrap critical values for tests based on generalized method of moment estimators.* Econometrica 64-4, 891-916.

[102] Hannan, E. J. (1973) *The estimation of frequency.* J. appl. Probab. 10, 510-519.

[103] Heyde, C. C. (1974) *On the central limit theorem for stationary processes.* Z. Wahrsch. Verw. Gebiete 30, 315-320.

[104] Heyde, C. C., Scott, D. J. (1973) *Invariance principles for the law of the iterated logarithm for martingales and processes with stationary increments.* Ann. Probab. 1, 428-436.

[105] Heyde, C. C. (1975) *On the central limit theorem and iterated logarithm law for stationary processes.* Bull. Austral. Math. Soc. 12, 1-8.

[106] Hofbauer, F., Keller, G. (1982) *Ergodic properties of invariant measures for piecewise monotonic transformations.* Math. Z. 180, 119-140.

[107] Hoffmann-Jorgensen, J. (1974) *Sums of independent Banach space valued random variables.* Studia Math. 52, 159-186.

[108] Horváth, L., Shao, Q.M. (1999) *Limit theorems for quadratic forms with applications to Whittle's estimate.* Ann. Appl. Probab. 9, No.1, 146-187.

[109] Hsu, P. L., Robbins, H. (1947) *Complete convergence and the strong law of large numbers.* Proc. Nat. Acad. Sci. 33, 25-31.

[110] Ibragimov, I. A. (1962) *Some limit theorems for stationary processes.* Theory Probab. Appl. 7, 349-382.

[111] Kallenberg, O. (1997) *Foundations of Modern Probability.* Springer-Verlag, New York.

[112] Kolmogorov, A. N., Rozanov, Y. A. (1960) *On the strong mixing conditions for stationary Gaussian sequences.* Th. Probab. Appl. 5, 204-207.

[113] Künsch, H. R. (1989) *The jacknife and bootstrap for general stationary observations.* Ann. Statist. 17, 1217-1241.

[114] Kushner, H. J., Clark, D. S. (1978) *Stochastic approximation for constrained and uncontrained systems.* Applied math. science series 26, Springer.

[115] Lasota, A., Yorke, J. A. (1974) *On the existence of invariant measures for piecewise monotonic transformations.* Trans. Amer. Math. Soc. 186, 481-488.

[116] Latour, A. (1997) *The multivariate GINAR(p) process.* Adv. Appl. Prob. 29, 228247.

[117] Latour, A. (1998) *Existence and spectral density of the GINAR(p) process.* J. Times Ser. Anal. 19, 439455.

[118] Lee, M. L. T., Rachev, S. T., Samorodnitsky, G. (1990) *Association of stable random variables.* Ann. Probab. 18-4, 1759-1764.

[119] Leonov, V. P., Shiraev, A. N. (1959) *On a method of calculation of semi-invariants.* Th. Probab. Appl. 5, 460-464.

[120] Li, D., Rao, M. B., Jiang, T., Wang, X. (1995) *Complete convergence and almost sure convergence of weighted sums of random variables.* J. Theor. Prob. 8, 49-76.

[121] Liebscher, E. (1996) *Strong convergence of sums of α- mixing random variables with applications to density estimation.* Stoch. Proc. Appl. 65, 69-98.

[122] Liggett, T. M. (1985) *Interacting particle systems.* Springer-Verlag, New York.

[123] Liverani, C., Saussol B., Vaienti S. (1999) *A probabilistic approach to intermittency.* Ergodic theory and dynamical systems, 19, 671-685.

[124] Louhichi, S. (2000) *Weak convergence for empirical processes of associated sequences.* Ann. I. H. P. ser. B 36-5, 547-567.

[125] Louhichi, S. (2003) *Moment inequalities for sums of certain dependent random variables.* Theor. Prob. Appl. 47-4.

[126] Major, P. (1978) *On the invariance principle for sums of identically distributed random variables.* J. Multivariate Anal. 8, 487-517.

[127] Mc Leish, D. L. (1975) *A generalization of martingales and mixing sequences.* Adv. in Appl. Probab. 7-2, 247-258.

[128] Mc Leish, D. L. (1975a) *A maximal inequality and dependent strong laws*, Ann. Probab. 3, 829-339.

[129] Mc Leish, D. L. (1975b) *Invariance principles and mixing random variables.* Z. Wahrsch. Verw. Gebiete 32, 165-178.

[130] Merlevède, F., Peligrad, M. (2002) *On the coupling of dependent random variables and applications.* in Empirical process techniques for dependent data. 171-193. Birkhäuser.

[131] Mikosch, T., Straumann, D. (2002) *Whittle estimation in a heavy-tailed GARCH(1,1) model.* Stoch. Proc. Appl. 100, 187-222.

[132] Mokkadem, A. (1990) *Propriétés de mélange des modèles autorégressifs polynomiaux.* Ann. I. H. P. ser. B 26-2, 219-260.

[133] Móricz, F. A., Serfling, J. A., Stout, W. F. (1982) *Moment and probability bounds with quasi-superadditive structure for the maximum partial sum.* Ann. Probab. 10, 1032-1040.

[134] Nelsen, R. B. (1999) *An introduction to copulas.* Lecture Notes in Statistics 139, Springer.

[135] Newman, C. M. (1980) *Normal fluctuations and the FKG inequalities.* Comm. Math. Phys. 74-2, 119-128.

[136] Newman, C. M. (1984) *Asymptotic independence and limit theorems for positively and negatively dependent random variables.* In Y. L. Tong, Editor, Inequalities in Statistics and Probability, IMS Lecture Notes-Monograph Series 5, 127-140.

[137] Oodaira, H., Yoshihara, K. I. (1971) *Note on the law of the iterated logarithm for stationary processes satisfying mixing conditions.* Kodai Math. Sem. Rep. 23, 335-342.

[138] Ornstein, D. S. (1973) *An example of a Kolmogorov automorphism wich is not a Bernoulli Shift.* Adv. in Math. 10, 49-62.

[139] Peligrad, M. (1983) *A note on two measures of dependence and mixing sequences.* Adv. in Appl. Probab. 15-2, 461-464.

[140] Peligrad, M. (2002) *Some remarks on coupling of dependent random variables.* Stat. and Prob. Lett. 60, 201-209.

[141] Peligrad, M., Shao, Q. M. (1994) *Self-normalizing central limit theorem for sums of weakly dependent random variables.* J. Theoret. Probab. 7, 309-338.

[142] Peligrad, M., Shao, Q. M. (1995) *Estimation of the variance for partial sums for ρ-mixing random variables.* J. of Multiv. Analysis 52, 140-157.

[143] Peligrad, M., Utev, S. (1997) *Central limit theorem for linear processes.* Ann. Probab. 25-1, 443-456.

[144] Petrov, V. V. (1996) *Limit theorems of probability theory: sequences of independent random variables.* Clarendon Press, Oxford.

[145] Phillips, P. C. B. (1987) *Time Series Regression with a Unit Root.* Econometrica 55-2, 277-301.

[146] Pitt, L. (1982) *Positively Correlated Normal Variables are Associated.* Ann. Probab. 10, 496-499.

[147] Pham, T. D., Tran, L. T. (1985) *Some mixing properties of time series models.* Stoch. Process. Appl. 19, 297-303.

[148] Pham, T. D. (1986) *Bilinear Markovian representation and bilinear models.* Stoch. Proc. Appl. 20, 295-306.

[149] Phillips, P.C.B. (1987) *Time series regression with a unit root.* Econometrica 55, 277-301.

[150] Philipp, W. (1986) *Invariance principles for independent and weakly dependent random variables* in *Dependence in probability and statistics* (Oberwolfach, 1985), 225-268. Prog. Probab. Statist. 11, Birkhäuser, Boston.

[151] Pollard, D. (1984) *Convergence of stochastic processes.* Springer-Verlag, Berlin.

[152] Pötscher, B. M., Prucha, I. R. (1991) *Basic structure of the asymptotic theory in dynamic nonlinear econometric models, part 1: consistency and approximation concepts, part 2: asymptotic normality.* Econometric Reviews, 10, 125-216 and 253-325.

[153] Prakasa, R. (1983) *Nonparametric functional estimation.* New York, Academic Press.

[154] Prieur, C. (2001) *Density Estimation For One-Dimensional Dynamical Systems.* ESAIM, Probab. & Statist. 5 www.edpsciences.com/ps/, 51-76.

[155] Prieur, C. (2002) *An Empirical Functional Central Limit Theorem for Weakly Dependent Sequences.* Probab. Math. Statist. 22-2, 259-287.

[156] Rényi, A. (1963) *On stable sequences on events.* Shankhya Ser. A 25, 293-302.

[157] Rio, E. (1993) *Covariance inequalities for strongly mixing processes.* Ann. I. H. P. ser. B 29-4, 587-597.

[158] Rio, E. (1995a) *About the Lindeberg method for strongly mixing sequences.* ESAIM-PS 1, 35-61.

[159] Rio, E. (1995b) *The functional law of iterated logarithm for strongly mixing processes.* Ann. Probab. 23, 1188-1203.

[160] Rio, E. (1996) *Sur le théorème de Berry-Esseen pour les suites faiblement dépendantes.* Probab. Theory Relat. Fields 104, 255-282.

[161] Rio, E. (2000) *Théorie asymptotique pour des processus aléatoire faiblement dépendants.* SMAI, Mathématiques et Applications 31, Springer.

[162] Rios, R. (1996) *Utilisation des techniques non paramétriques et semi paramétriques en statistique des données dépendantes.* Ph. D. Thesis: Université Paris sud, Paris.

[163] Robinson, P. M. (1983) *Nonparametric estimators for time series.* Journal of Time Series Analysis 4, 185-207.

[164] Robinson, P. M. (1989) *Hypothesis Testing in Semiparametric and Nonparametric Models for Econometric Time Series.* The Review of Economic Studies 56-4, 511-534.

[165] Robinson, P. M. (1991) *Testing for strong serial correlation and dynamic conditional heteroskedasticity in multiple regression.* Journal of Econometrics, 47, 67-84

[166] Rosenblatt, M. (1956) *A central limit Theorem and a strong mixing condition.* Proc. Nat. Ac. Sc. U.S.A. 42, 43-47.

[167] Rosenblatt, M. (1961) *Independence and dependence* in Proc. 4th. Berkeley Symp. Math. Stat. Prob. 411-443. Berkeley University Press.

[168] Rosenblatt, M. (1985) *Stationary sequences and random fields.* Birkhaüser.

[169] Rosenblatt, M. (1991) *Stochastic curve estimation*, NSF-CBMS Regional Conference Series in Probability and Statistics, 3.

[170] Rosenthal (1970) *On the subspaces of L^p $(p > 1)$ spanned by sequences of independent random variables.* Israel Journal of Math, 8, 272-303.

[171] Rüschendorf, L. (1985) *The Wasserstein Distance and Approximation Theorems.* Z. Wahr. verw. Gebiete 70, 117-129.

[172] Samorodnitsky, G., Taqqu, M. S. (1994) *Stable Non-Gaussian Random Processes: Stochastic Models with Infinite Variance.* Chapman and Hall, New York.

[173] Saulis, L., Statulevicius, V. A. (1991) *Limit theorems for large deviations.* Birkhäuser.

[174] Shao, Q. M. (1993) *Almost sure invariance principle for mixing sequences of random variables.* Stoch. Proc. Appl. 48 319-334.

[175] Shao, Q. M., Yu, H. (1996) *Weak convergence for weighted empirical processes of dependent sequences.* Ann. Probab. 24-4, 2098-2127.

[176] Shashkin, A. P. (2005) *A weak dependence property of a spin system.* Preprint.

[177] Shixin, E. G. (1999) *On the convergence of weighted sums of L_q-mixingale arrays.* Acta Math. Hungar. 82(1-2), 113-120.

[178] Sklar, A. (1959) *Fonctions de répartition à n dimensions et leurs marges.* Publ. Inst. Statist. Univ. Paris 8, 229-231.

[179] Skorohod, A. V. (1976). *On a representation of random variables.* Theory Probab. Appl. 21, 628-632.

[180] Strassen, V. (1964) *An invariance principle for the law of the iterated logarithm.* Z. Wahrsch. Verw. Gebiete 3, 211-226.

[181] Straumann, D. (2005) *Estimation in conditionally heteroscedastic time series.* Lecture Notes in Statistics 181, Springer, New York.

[182] Szegö, S. (1967) *Orthogonal Polynomials.* A. M. S. 23, New York.

[183] van der Vaart, A. W., Wellner, J. A. (1996) *Weak convergence and empirical processes.* Springer, Berlin.

[184] Valadier, M. (1973) *Désintégration d'une mesure sur un produit.* C. R. Acad. Sc. Paris 276, 33-35.

[185] Viennet, G. (1997) *Inequalities for absolutely regular sequences: application to density estimation.* Prob. Th. Rel. Fields 107, 467-492.

[186] Withers, C. S (1981) *Central limit theorem for dependent variables.* Z. Wahrsch. Verw. Gebiete 57, 509-534.

[187] Wolkonski, V. A., Rozanov, Y. A. (1959, 1961) *Some limit theorems for random functions, Part I:* Theory Probab. Appl. 4, 178-197; *Part II:* Theory Probab. Appl. 6, 186-198.

[188] Wu, W. B. (2005a) *Nonlinear system theory: Another look at dependence.* Proc. Nat. Acad. Sci. USA 102, 14150-14154.

[189] Wu, W. B. (2005b) *On the Bahadur representation of sample quantiles for dependent sequences.* Ann. Statist. 33, 1934-1963.

[190] Wu, W. B. (2005c) *Fourier transforms of stationary processes* Proc. Amer. Math. Soc. 133, 285-293.

[191] Wu, W. B. (2006) *M-estimation of linear models with dependent errors.* Ann. Statist., To Appear.

[192] Wu, W. B., Zhao, Z. (2005) *Inference of Trends in Time Series.* Under Review, JRSSB.

[193] Yoshihara, K. (1973) *The law of the iterated logarithm for the processes generated by mixing processes.* Yokohama Math. J. 21, 67-72.

[194] Young, L. S. (1973) *Recurrence times and rates of mixing.* Israel Journal of Math., 110, 153-188.

[195] Yu, H. (1993) *A Glivenko-Cantelli Lemma and weak convergence for empirical processes of associated sequences.* Probab. Th. Rel. Fields 95, 357-370.

[196] Yu, K. F. (1990) *Complete convergence of weighted sums of martingale differences.* J. Theor. Prob. 3, 339-347.

Index

Lecture Notes in Statistics

For information about Volumes 1 to 135, please contact Springer

136: Gregory C. Reinsel, Raja P. Velu, Multivariate Reduced-Rank Regression. xiii, 272 pp., 1998.

137: V. Seshadri, The Inverse Gaussian Distribution: Statistical Theory and Applications. xii, 360 pp., 1998.

138: Peter Hellekalek and Gerhard Larcher (Editors), Random and Quasi-Random Point Sets. xi, 352 pp., 1998.

139: Roger B. Nelsen, An Introduction to Copulas. xi, 232 pp., 1999.

140: Constantine Gatsonis, Robert E. Kass, Bradley Carlin, Alicia Carriquiry, Andrew Gelman, Isabella Verdinelli, and Mike West (Editors), Case Studies in Bayesian Statistics, Volume IV. xvi, 456 pp., 1999.

141: Peter Müller and Brani Vidakovic (Editors), Bayesian Inference in Wavelet Based Models. xiii, 394 pp., 1999.

142: György Terdik, Bilinear Stochastic Models and Related Problems of Nonlinear Time Series Analysis: A Frequency Domain Approach. xi, 258 pp., 1999.

143: Russell Barton, Graphical Methods for the Design of Experiments. x, 208 pp., 1999.

144: L. Mark Berliner, Douglas Nychka, and Timothy Hoar (Editors), Case Studies in Statistics and the Atmospheric Sciences. x, 208 pp., 2000.

145: James H. Matis and Thomas R. Kiffe, Stochastic Population Models. viii, 220 pp., 2000.

146: Wim Schoutens, Stochastic Processes and Orthogonal Polynomials. xiv, 163 pp., 2000.

147: Jürgen Franke, Wolfgang Härdle, and Gerhard Stahl, Measuring Risk in Complex Stochastic Systems. xvi, 272 pp., 2000.

148: S.E. Ahmed and Nancy Reid, Empirical Bayes and Likelihood Inference. x, 200 pp., 2000.

149: D. Bosq, Linear Processes in Function Spaces: Theory and Applications. xv, 296 pp., 2000.

150: Tadeusz Caliński and Sanpei Kageyama, Block Designs: A Randomization Approach, Volume I: Analysis. ix, 313 pp., 2000.

151: Håkan Andersson and Tom Britton, Stochastic Epidemic Models and Their Statistical Analysis. ix, 152 pp., 2000.

152: David Ríos Insua and Fabrizio Ruggeri, Robust Bayesian Analysis. xiii, 435 pp., 2000.

153: Parimal Mukhopadhyay, Topics in Survey Sampling. x, 303 pp., 2000.

154: Regina Kaiser and Agustín Maravall, Measuring Business Cycles in Economic Time Series. vi, 190 pp., 2000.

155: Leon Willenborg and Ton de Waal, Elements of Statistical Disclosure Control. xvii, 289 pp., 2000.

156: Gordon Willmot and X. Sheldon Lin, Lundberg Approximations for Compound Distributions with Insurance Applications. xi, 272 pp., 2000.

157: Anne Boomsma, Marijtje A.J. van Duijn, and Tom A.B. Snijders (Editors), Essays on Item Response Theory. xv, 448 pp., 2000.

158: Dominique Ladiray and Benoît Quenneville, Seasonal Adjustment with the X-11 Method. xxii, 220 pp., 2001.

159: Marc Moore (Editor), Spatial Statistics: Methodological Aspects and Some Applications. xvi, 282 pp., 2001.

160: Tomasz Rychlik, Projecting Statistical Functionals. viii, 184 pp., 2001.

161: Maarten Jansen, Noise Reduction by Wavelet Thresholding. xxii, 224 pp., 2001.

162: Constantine Gatsonis, Bradley Carlin, Alicia Carriquiry, Andrew Gelman, Robert E. Kass Isabella Verdinelli, and Mike West (Editors), Case Studies in Bayesian Statistics, Volume V. xiv, 448 pp., 2001.

163: Erkki P. Liski, Nripes K. Mandal, Kirti R. Shah, and Bikas K. Sinha, Topics in Optimal Design. xii, 164 pp., 2002.

164: Peter Goos, The Optimal Design of Blocked and Split-Plot Experiments. xiv, 244 pp., 2002.

165: Karl Mosler, Multivariate Dispersion, Central Regions and Depth: The Lift Zonoid Approach. xii, 280 pp., 2002.

166: Hira L. Koul, Weighted Empirical Processes in Dynamic Nonlinear Models, Second Edition. xiii, 425 pp., 2002.

167: Constantine Gatsonis, Alicia Carriquiry, Andrew Gelman, David Higdon, Robert E. Kass, Donna Pauler, and Isabella Verdinelli (Editors), Case Studies in Bayesian Statistics, Volume VI. xiv, 376 pp., 2002.

168: Susanne Rässler, Statistical Matching: A Frequentist Theory, Practical Applications and Alternative Bayesian Approaches. xviii, 238 pp., 2002.

169: Yu. I. Ingster and Irina A. Suslina, Nonparametric Goodness-of-Fit Testing Under Gaussian Models. xiv, 453 pp., 2003.

170: Tadeusz Caliński and Sanpei Kageyama, Block Designs: A Randomization Approach, Volume II: Design. xii, 351 pp., 2003.

171: D.D. Denison, M.H. Hansen, C.C. Holmes, B. Mallick, B. Yu (Editors), Nonlinear Estimation and Classification. x, 474 pp., 2002.

172: Sneh Gulati, William J. Padgett, Parametric and Nonparametric Inference from Record-Breaking Data. ix, 112 pp., 2002.

173: Jesper Møller (Editor), Spatial Statistics and Computational Methods. xi, 214 pp., 2002.

174: Yasuko Chikuse, Statistics on Special Manifolds. xi, 418 pp., 2002.

175: Jürgen Gross, Linear Regression. xiv, 394 pp., 2003.

176: Zehua Chen, Zhidong Bai, Bimal K. Sinha, Ranked Set Sampling: Theory and Applications. xii, 224 pp., 2003.

177: Caitlin Buck and Andrew Millard (Editors), Tools for Constructing Chronologies: Crossing Disciplinary Boundaries, xvi, 263 pp., 2004.

178: Gauri Sankar Datta and Rahul Mukerjee, Probability Matching Priors: Higher Order Asymptotics, x, 144 pp., 2004.

179: D.Y. Lin and P.J. Heagerty (Editors), Proceedings of the Second Seattle Symposium in Biostatistics: Analysis of Correlated Data, vii, 336 pp., 2004.

180: Yanhong Wu, Inference for Change-Point and Post-Change Means After a CUSUM Test, xiv, 176 pp., 2004.

181: Daniel Straumann, Estimation in Conditionally Heteroscedastic Time Series Models, x, 250 pp., 2004.

182: Lixing Zhu, Nonparametric Monte Carlo Tests and Their Applications, xi, 192 pp., 2005.

183: Michel Bilodeau, Fernand Meyer, and Michel Schmitt (Editors), Space, Structure and Randomness, xiv, 416 pp., 2005.

184: Viatcheslav B. Melas, Functional Approach to Optimal Experimental Design, vii., 352 pp., 2005.

185: Adrian Baddeley, Pablo Gregori, Jorge Mateu, Radu Stoica, and Dietrich Stoyan, (Editors), Case Studies in Spatial Point Process Modeling, xiii., 324 pp., 2005.

186: Estela Bee Dagum and Pierre A. Cholette, Benchmarking, Temporal Distribution, and Reconciliation Methods for Time Series, xiv., 410 pp., 2006.

187: Patrice Bertail, Paul Doukhan and Philippe Soulier, (Editors), Dependence in Probability and Statistics, viii., 504 pp., 2006.

188: Constance van Eeden, Restricted Parameter Space Estimation Problems, vi, 176 pp., 2006.

189: Bill Thompson, The Nature of Statistical Evidence, vi, 152 pp., 2007.

190: Jérôme Dedecker, Paul Doukhan, Gabriel Lang, José Rafael León R., Sana Louhichi Clémentine Prieur, Weak Dependence: With Examples and Applications, xvi, 336 pp., 2007.

Springer
the language of science

springer.com

The Nature of Statistical Evidence

Bill Thompson

The purpose of this book is to discuss whether statistical methods make sense. That is a fair question, at the heart of the statistician-client relationship, but put so boldly it may arouse anger. The many books entitled something like Foundations of Statistics avoid controversy by merely describing the various methods without explaining why certain conclusions may be drawn from certain data. But we statisticians need a better answer then just shouting a little louder. To avoid a duel, the authors prejudge the issue and ask the narrower question: "In what sense do statistical methods provide scientific evidence?"

2007. 150 9p. (Lecture Notes in Statistics, Vol. 189) Softcover
ISBN 978-0-387-40050-1

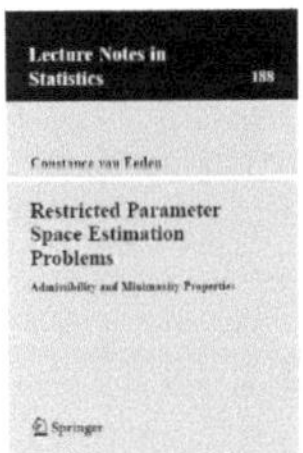

Restricted Parameter Space Estimation Problems

Constance van Eeden

Point process statistics is successfully used in fields such as material science, human epidemiology, social sciences, animal epidemiology, biology, and seismology. Its further application depends greatly on good software and instructive case studies that show the way to successful work. This book satisfies this need by a presentation of the spatstat package and many statistical examples.

2006. 176 pp. (Lecture Notes in Statistics, Vol. 188) Softcover
ISBN 978-0-387-33747-0

Dependence in Probability and Statistics

P. Bertail. P Doukhan, and P. Soulier (Editors)

This book gives a detailed account of some recent developments in the field of probability and statistics for dependent data. The book covers a wide range of topics from Markov chain theory and weak dependence with an emphasis on some recent developments on dynamical systems, to strong dependence in times series and random fields. A special section is devoted to statistical estimation problems and specific applications.

2006. 504 pp. (Lecture Notes in Statistics, Vol. 187) Softcover
ISBN 978-0-387-31741-0

Easy Ways to Order▶ Call: Toll-Free 1-800-SPRINGER • E-mail: orders-ny@springer.com • Write: Springer, Dept. S8113, PO Box 2485, Secaucus, NJ 07096-2485 • Visit: Your local scientific bookstore or urge your librarian to order.

MIX
Papier aus verantwortungsvollen Quellen
Paper from responsible sources
FSC® C105338

If you have any concerns about our products,
you can contact us on
ProductSafety@springernature.com

In case Publisher is established outside the EU,
the EU authorized representative is:
Springer Nature Customer Service Center GmbH
Europaplatz 3, 69115 Heidelberg, Germany

Printed by Libri Plureos GmbH
in Hamburg, Germany